Cast iron technology

Cast Iron Technology

Roy Elliott BSc, PhD
Lecturer, Department of Metallurgy and Materials Science,
University of Manchester, England

Butterworths
London   Boston   Singapore   Sydney   Toronto   Wellington

First published 1988

© **Butterworth & Co. (Publishers) Ltd., 1988**

British Library Cataloguing in Publication Data
Elliott, Roy
    Cast iron technology.
    1. Cast-iron—Metallurgy
    I. Title
    669′.1413   TN710

    ISBN 0-408-01512-8

Elliott, Roy, Ph. D.
    Cast iron technology/Roy Elliott.
         p.   cm.
    Bibliography: p.
    Includes index.
    ISBN 0-408-01512-8:
    1. Iron-founding. 2. Cast iron. I. Title.
    TS230.E63 1988
    672.2–dc19   87-34040
                                     CIP

Photoset by The Alden Press (London & Northampton) Ltd.
Printed and bound in Great Britain by Butler and Tanner, Frome, Somerset

Preface

The foundry industry in general, and iron foundries in particular, have contracted considerably over the past decade. The UK iron founding capacity contracted by 50% between 1975 and 1985. Several factors contributed to this decline. It was precipitated by the world energy and economic crises of the 1970s and was prolonged by the decline of traditional heavy engineering industries, enforced control over emissions and workplace environment, the growth of alternative materials and liability legislation changes.

However, the decade was one of technological innovations. Advances in the understanding of graphite formation led to the use computer analysis of the cooling curve being used to predict structural features in addition to iron composition. Computers became an important feature of new iron foundries. They are an invaluable aid in all aspects of design, solidification simulation, melting operations, production control and statistical quality control. New sand systems and moulding techniques such as the lost foam process added versatility. Compacted irons have been added to the cast iron family and new impetus has been given to the development of spheroidal graphite irons with the promotion of austenitic ductile iron for components such as gears. The new iron industry cannot rely on market growth and the economies of scale it once offered. These are being replaced by sounder practices and more efficient utilization of resources. The way forward is to overcome the long standing prejudice against cast products (the recent certification for the use of spheroidal graphite irons as a substitute for stainless steel in nuclear waste transport containers is an important step in this direction) by demonstrating the versatility of modern metallurgically-advanced, market-led cast iron components manufactured by efficient, attractively priced, near-end shape production methods.

The technological changes of the past decade is the underlying theme of the present book. The importance of the cast iron family as engineering materials is emphasized in the first chapter. Developments in cupola practice, cupola versus electric melting and liquid iron treatments of desulphurization, inoculation and spheroidization are examined in Chapter 2. Chapter 3 uses nucleation and growth concepts to explain the solidification structures of grey and white iron and explains how the foundryman can exercise control over the structure, and hence, properties of an iron. The cooling curve and its use as a control tool are an important feature of this chapter. Chapter 4 describes solid state transformations of importance in cast iron technology. These include malleabilizing and various heat treatment processes as well as the theory and practice of austempering. Chapter 5 is devoted to new foundry technology and explores running and gating design, feeding, moulding materials and a selection of casting processes. Chapter 6 demonstrates typical microstructural features of the cast iron family.

An attempt is made to relate theory and practice throughout the text. It is hoped that this is sufficiently evident to attract practising foundryman to some of the more academic aspects of the subject.

Roy Elliott

Contents

Chapter 1

An introduction to cast irons

The nature of cast irons

Joseph Glanville wrote of cast iron, 'Iron seemeth a simple metal but in its nature are many mysteries.' Many, but not all, of these mysteries have been solved over the past three hundred years using the combined skill of the foundryman and the knowledge of the scientist to provide today's design engineer with a family of casting alloys that offer a virtually unique combination of low cost and engineering versatility. The various combinations of low cost with castability, strength, machinability, hardness, wear resistance, corrosion resistance, thermal conductivity and damping are unequalled among casting alloys.

Cast iron or 'iron' is a Fe–C–Si alloy that always contains minor ($< 0.1\%$) and often alloying ($> 0.1\%$) elements and is used in the as-cast condition or after heat treatment. *Table 1.1* compares the characteristics of the various types of iron with those of cast steel and *Figure 1.1* shows the relationship between members of the family of unalloyed irons. *Figure 1.2* compares the C and Si composition ranges of cast irons and steel and shows that all irons contain C in excess of its solubility limit in austenite. This is defined by the lower line in the figure. The higher C (2–4%) and Si (1–3%) contents confer castability. The phase diagram in *Figure 1.3* shows that irons solidify at approximately 1150 °C compared to about 1500 °C for steel. Liquid iron is more fluid and less reactive with moulding materials. The formation of low density graphite in grey iron is accompanied by an expansion which may be used with a rigid mould to feed secondary shrinkage thus eliminating the need for conventional feeders. These characteristics promote castability and make the casting of complex shapes at relatively low cost possible.

The microstructure and properties of cast iron also differ from those of steel. The excess C precipitates during solidification by a eutectic reaction either as the thermodynamically stable graphite phase (grey iron) and/or the metastable cementite phase (mottled or white iron). Whether the stable or metastable phase forms depends on the nature and treatment given to the liquid, in particular, its graphitization potential, inoculation treatment and the cooling rate. Si increases the graphitization potential strongly and is always present in higher concentrations in grey irons. Al, B ($< 0.15\%$), Cu, Ni and Ti ($< 0.25\%$) also increase the graphitization potential and Bi, B ($> 0.15\%$), Cr, Mn, Mo, Te, V and Ti ($> 0.25\%$) decrease it. The graphitization potential of liquid malleable iron is reduced with trace additions of Bi and Te to allow thicker components to be cast white before graphitization by heat treatment. Graphite is difficult to nucleate in liquid iron so the graphitization

Table 1.1 A comparison of the characteristics of various types of cast iron with those of a 0.3% cast steel

Property	S.I.	M.I.	F.I.	C.S.	W.I.
Castability	1	2	1	4	3
Machinability	2	2	1	3	–
Reliability	1	3	5	2	4
Vibration damping	2	2	1	4	4
Surface hardenability	1	1	1	3	–
Elastic modulus	1	2	3	1	–
Impact resistance	2	3	5	1	–
Wear resistance	2	4	3	5	1
Corrosion resistance	1	2	1	4	2
Strength/Weight ratio	1	4	5	3	–
Cost of manufacture	2	3	1	4	3

S.I. Spheroidal iron
M.I. Malleable iron
F.I. Flake iron
C.S. 0.3% C steel
W.I. White iron
1 = Best; 5 = Worst

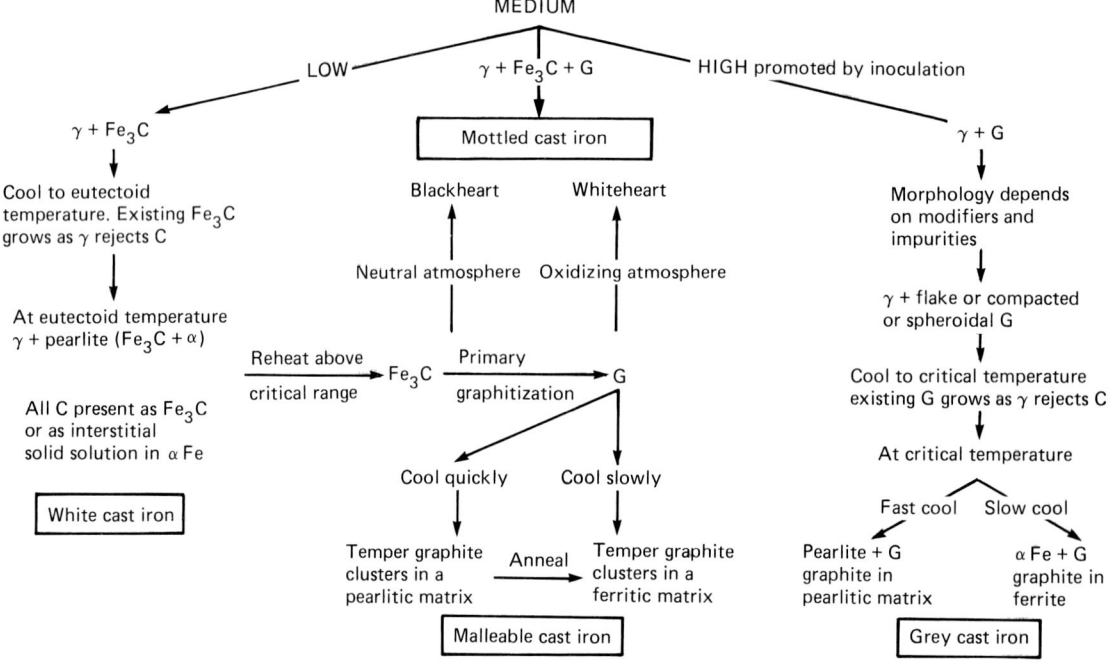

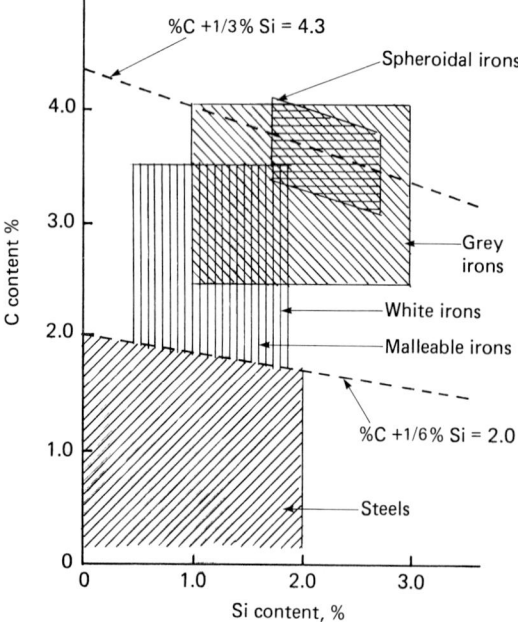

Figure 1.1 The cast iron family

Figure 1.2 C and Si composition ranges of various cast irons and steel

potential, which is the iron's ability to solidify grey, is realized by inoculating the iron to provide graphite nuclei. Ferrosilicon is used frequently for this purpose. However, neither Fe nor Si are inoculants. They act as carriers for other elements such as Ca, Ba, Al, Sr, B and Zr which are present in relatively low concentrations.

Inoculation is one of the most important and widely practised processes in cast iron production. It is used to avoid chill formation in thin sections, helps to counteract the effect of variations in raw materials and melting practice and is usually the final process in the ladle treatment of spheroidal irons. Rapid cooling favours metastable carbide formation (chill). The upper line in *Figure 1.2* defines the eutectic composition. The second phase in the eutectic reaction is austenite, which forms the matrix and transforms during cooling in the solid state, as in steels, to one of several structures depending on the alloying additions and the cooling rate. Whereas C content is of primary importance in selecting the heat treatment temperature of a steel, Si is the significant element in determining the critical temperature range for an iron.

Many of the mystiques referred to by Glanville can be traced to the complicated and considerable influence of minor elements on structure and

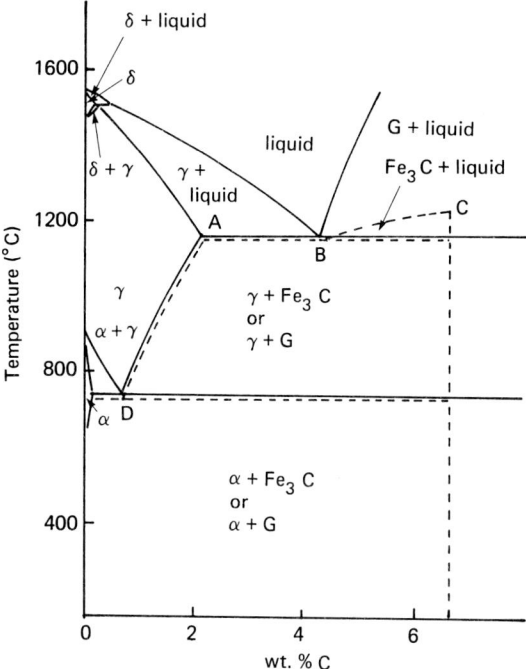

Figure 1.3 The iron-carbon phase diagram

	Fe – G system		Fe – Fe₃C system		
	A	B	C	D	
(———) Fe – G	2.09	4.25		0.68	%C
	1154	1154		739	°C
(– – –) Fe – Fe₃C	2.12	4.31	6.68	0.76	%C
	1148	1148	1226	727	°C

properties. This is illustrated in *Table 1.2* (ref. 1). The influence of individual elements can be multiple and the effects both beneficial and detrimental[2-4]. For example, inoculation is achieved as a result of the action of minor elements as explained later. Trace additions of Mg to a low S iron modify the graphite growth process to promote spheroidal graphite. This

1.2 The influence of trace elements on the hardness and strength of a grey iron (after ref. 1)

Element (%)	Casting 1	Casting 2
C	3.39	3.39
Si	2.01	1.94
Cr	0.14	0.06
Cu	0.20	0.13
Sn	0.01	< 0.01
As	0.02	0.01
Mo	0.045	0.025
N	0.006	0.0055
Ti	< 0.01	0.02
Hardness HB	207	165
Tensile strength N/mm²	195	167

only occurs over a narrow concentration range. Lower concentrations result in compacted/vermicular or even flake graphite; higher concentrations promote spheroid degeneracy or carbide formation. N in malleable iron stabilizes carbides, reduces mottling tendency and retards first and second stage graphitization. N levels above 0.014% can lead to severe production problems[5]. N acts as a pearlite stabilizer in flake irons and exerts a strong carbide stabilizing effect. It promotes compacted/vermicular rather than flake graphite in thick sections at levels > 0.008% (see also *Figure 6.12* in Chapter 6). All these influences increase the tensile strength. A 0.003 to 0.008% N increase raises the ultimate tensile strength (U.T.S.) of a flake iron from 260 to 300 N/mm² (ref. 6). In practice, N with Sn and Sb has been used to reduce the ferrite content in gravity-cast brake drums and to strengthen piston ring irons. However, high N levels (~ 0.02%) are a cause of blowhole and fissure defects[7]. In this connection, it has been shown that a N–V combination can eliminate unsoundness at the same time as increasing the tensile strength[8]. The effect of N in spheroidal iron is less pronounced because the Mg treatment tends to flush it from the melt[9]. Pb as low as 0.0015% in the presence of H leads to Widmanstätten graphite formation and a reduction in mechanical properties[10] (see also *Figures 6.15–6.18* in Chapter 6). These are only a few of the compositional effects that must be considered in producing irons, but they emphasize the need to control minor elements and the importance of their selective use to control structure and properties. Indeed, this is a most important aspect of cast iron technology.

The properties of cast irons depend on the form of C precipitation and the matrix structure. The C precipitated in the eutectic reaction is not a major contributor to mechanical strength. It is responsible for several properties not displayed by steels. Carbides contribute hardness and abrasion resistance. Graphite contributes machinability, wear resistance, damping and thermal conductivity depending on its shape. The mechanical properties of cast iron are derived mainly from the matrix. This is why irons are often described in terms of their matrix structure, for example, as ferritic or pearlitic types. Important matrix structures include;

Ferrite. This is a Fe–C solid solution in which appreciable Si and some Mn, Cu and Ni may dissolve. It is relatively soft (although dissolved Si makes it harder than in plain C steel), ductile, of low strength and with poor wear resistance, good fracture toughness, relatively good thermal conductivity and good machinability. A ferritic matrix can be produced as-cast but is often the result of an annealing treatment.

Pearlite. This is a mixture of ferrite and Fe₃C, which forms from austenite by a eutectoid reaction and derives its name from its mother of pearl appearance.

It is relatively hard, shows moderate toughness, reduced thermal conductivity and good machinability. Several mechanical properties increase as the pearlite spacing decreases. The C content of pearlite in unalloyed steel is 0.8%. It is variable in cast iron depending on the iron composition and cooling rate. It can be as low as 0.5% in the high Si irons.

Ferrite-Pearlite. This mixed structure is often used to obtain properties intermediate between the extremes described above.

Bainite. This structure can be produced, as-cast, in alloyed (Ni and Mo) irons when it is known as acicular iron or, more reliably, by an austemper heat treatment (see *Figures 6.34* and *6.35* in Chapter 6). The properties of austempered spheroidal irons have been reported since the 1950s but the cost effective role that these alloys can play, particularly in automotive engineering, and gearing and transmission components has been appreciated only recently. The benefits of austempered spheroidal irons include;

1. high tensile strength coupled with toughness, ductility and good fatigue resistance;
2. good resistance to wear and scuffing which is retained under poor lubrication;
3. high noise damping capacity giving quiet operation;
4. good casting characteristics;
5. near net shape formability even with highly complex shapes;
6. good machinability as-cast and
7. a 10% weight saving against steel.

Considerable interest is being shown in this modern engineering material as a replacement for forged steel components[11–15].

Austenite. A high alloy content is required to retain this phase during cooling. High alloy flake and spheroidal irons (i.e. Ni-resist) have excellent heat, corrosion and non-magnetic properties[16]. This matrix can show good toughness, creep resistance and stress rupture properties up to 800 °C and a wide range of thermal expansivity depending on the Si content (see also *Figures 6.38, 6.39* and *6.41* in Chapter 6).

Most casting alloys are specified by composition. However, this is not possible for many cast irons because, as *Figure 1.2* shows, the composition of the various types of iron overlap and the properties of a particular type depend on the solidification conditions. Consequently, the majority of cast iron specifications define the mechanical properties for a particular section size. Hence, an appropriate section size must be defined for a casting in order to select a specification.

The foundryman must control the melting and casting sequence with composition as a variable to achieve the required properties in the casting. Composition is specified for special purpose high alloy

irons and concentration limits are sometimes imposed in order to control the effects of specific elements. Sulphur[17] and various minor elements[3], including P, fall into this category.

P has limited solubility in austenite and segregates positively during solidification. This can result in phosphide formation in the last areas to solidify (see also *Figures 6.3–6.7* in Chapter 6). Phosphides are similar to carbides. They cause machining problems but they cannot be eliminated by heat treatment. High P content has been associated with the segregation of Mo, Cr, V and W. This leaves areas of the matrix depleted of these alloying elements and reduces the matrix strength[18]. Embrittlement of ferritic spheroidal cast iron that has undergone cooling or long range annealing between 350 °C and 550 °C has been attributed to P segregation[19]. Phosphide eutectic liquid at cell boundaries creates a mushy state that is difficult to feed. This condition creates high eutectic solidification forces in the outer solidifying shell, causing mould wall movement and a requirement for feed metal in the final stages of solidification. This can lead to shrinkage porosity. An upper limit is usually specified for P for these reasons except when it is used for a specific purpose such as promoting castability in low grade irons. The thermal shock resistance of grey irons depends on the total C content and a minimum is quoted for this reason.

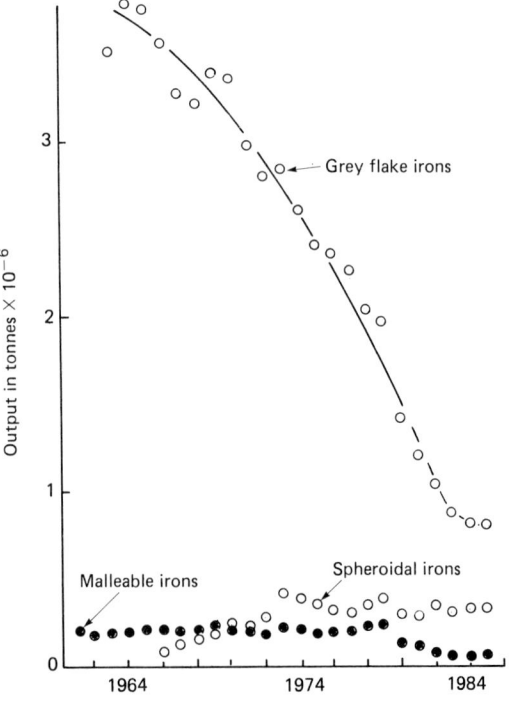

Figure 1.4 Cast iron production in the United Kingdom over the period 1960–1985

Types of cast iron

Cast irons remain the most used casting alloy despite a considerable reduction in their production during the past decade as shown in *Figure 1.4*. Their popularity stems from an ability to cast complex shapes at relatively low cost and the wide range of properties that can be achieved by careful control over composition and cooling rate without radical changes in production methods. Cast irons divide into two main groups, general purpose alloys which are used for the majority of engineering applications and the special purpose, white and alloy cast irons which are used for applications involving extremes of heat, corrosion or abrasion.

General purpose cast irons

These irons form the largest group of casting alloys and can be classified according to the graphite morphology into flake, malleable, spheroidal and compacted/vermicular types. Specifications exist for all types except the relatively new, compacted/vermicular irons. Specifications used in the different countries are indicated in *Table 1.3*. *Table 1.4* lists the British Standard Specifications. The first grade number denotes the tensile strength and the second, the minimum per cent elongation. Several bodies define standards in the USA notably the American Society for Testing and Materials (ASTM) and the Society of Automotive Engineers (SAE). The Iron Castings Society publishes a summary of specifications as shown in *Table 1.5*. These tables include composition specifications and also indicate the major uses of cast irons.

Most mechanical properties can be related to a tensile strength specification as shown in *Table 1.6* although tensile or yield strength may not be the only design criterion. Not all engineering properties increase with grade number and the selection of a particular grade or even type should be made to obtain the best combination of properties for the application.

Material selection for diesel engine components illustrates this point[20]. Acceptable material for cylinder heads must possess high thermal stability, high mechanical strength and hardness. These three requirements are not easily satisfied in a single material. A compromise solution has to be sought for each application and there are over 40 compositions

Table 1.3 A selection of International specifications for flake and spheroidal irons

Country				← Increasing tensile strength					Standard
Britain	Gr 400	Gr 350	Gr 300	Gr 260		Gr 220	Gr 180	Gr 150	BS 1452 1977
France	Ft 40D	Ft 35D	Ft 30D		Ft 25D		Ft 20D	Ft 15D	NFA 32–101 1965
Germany	GG 40	GG 35	GG 30		GG 25		GG 20	GG 15	DIN 1691 1964
ISO	Gr 40	Gr 35	Gr 30		Gr 25		Gr 20	Gr 15	R 185 1961
Italy		G 35	G 30		G 25		G 20	G 15	UN 5007 1969
Japan		FC 35	FC 30		FC 25		FC 20	FC 15	JIS G5501 1976
Sweden	SISO 140	SISO 135	SISO 130		SISO 125		SISO 120	SISO 115	MNC 705 1976
USA	Class 60 B	Class 55 B	Class 50 B 45 B	Class 40 B	Class 35 B	Class 30 B	Class 25 B	Class 20 B	ASTM 48 1974
USSR	SC 40–60	SC 36–56	SC 32–52	SC 28–48	SC 24–44	SC 21–40	SC 18–36	SC 15–32	GOST 1412 1970
				Tensile strength increasing →					
Britain			Gr 350–22	Gr 420–12	Gr 500–7	Gr 600–3	Gr 700–2	Gr 800–2	BS 2789 1985
France			FGS 370–17	FGS 400–12	FGS 500–7	FGS 600–3	FGS 700–2	FGS 800–2	NFA 32–201 1976
Germany	GGG 35.3	GGG 40.3	GGG 40		GGG 50	GGG 60	GGG 70	GGG 80	DIN 1693 1973
ISO			Gr 370–17	Gr 400–12	Gr 500–7	Gr 600–3	Gr 700–2	Gr 800–2	1083 1976
Italy			GS 370–17	GS 400–12	GS 500–7	GS 600–2	GS 700–2	GS 800–2	UN 14544 1974
Japan			FCD 40		FCD 45 50	FCD 60	FCD 70		JIS G5502 1975
Sweden	SISO 717–15	SISO 717–12	SISO 717–02		SISO 727–02	SISO 732–03	SISO 737–01		MNC 706E 1977
USA			60–40–18	65–45–12	80–55–06		100–70–03	120–90–02	ASTM 536 1972
USSR			Vch 38–17	Vch 42–12	Vch 45–5 50–2	Vch 60–2	Vch 70–3	Vch 80–3	GOST 7293 1970

Table 1.4 British specifications for general purpose cast irons

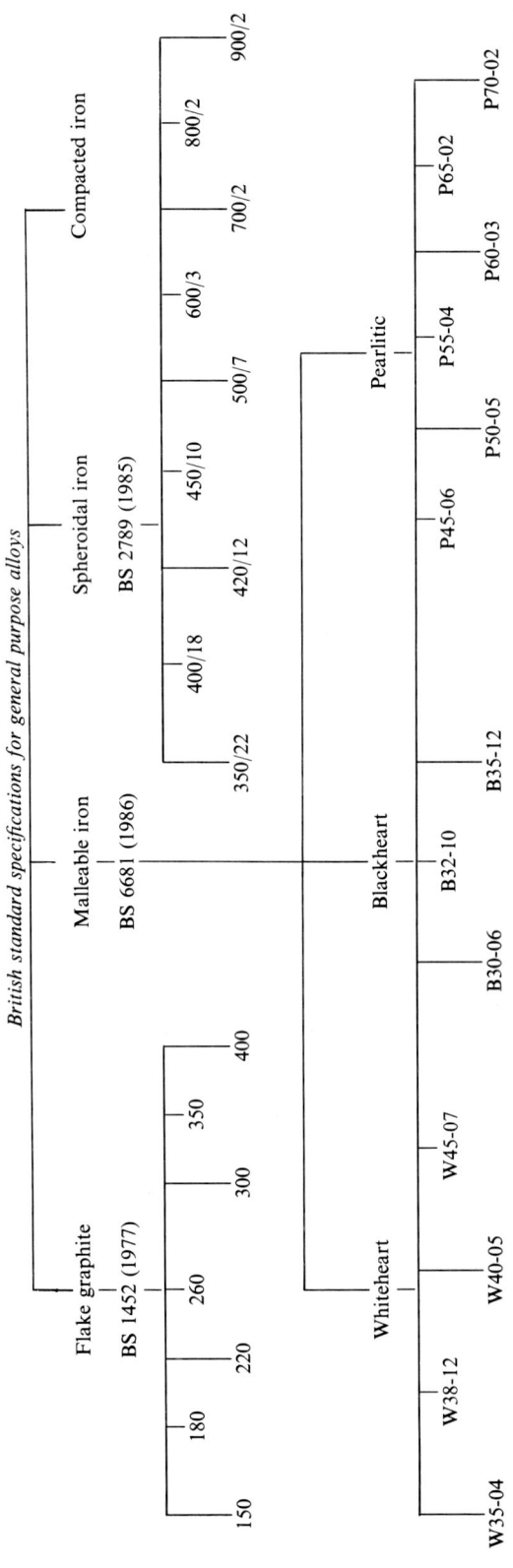

British standard specifications for general purpose alloys

Table 1.5 (part 1) A summary of USA standards for grey irons

	Spec. No.	Use	Grade	Minimum Tensile p.s.i.	Other requirements	Typical applications
ASTM	A126–73 (1979)	Valves, flanges pipe fittings	A B C	21 000 31 000 41 000	0.15% S maximum 0.75% P maximum	Stock fittings and castings not requiring critical tensile evaluation
ASTM	A48–76	Castings intended for use where strength is a major consideration	20 25 30 35 40 45 50 55 60	20 000 25 000 30 000 35 000 40 000 45 000 50 000 55 000 60 000	1) Test bar size must be related in cooling rate to the critical section of the casting. 2) At least two test bars shall be cast and prepared to each casting. Lot size designated. 3) Test bars cast in dry silica sand the same as the casting. 4) Tensile test under true axial loading. 5) Hardness, composition, microstructure, pressure tightness, radiographic soundness requirements to be agreed.	Thin sections requiring good finish machinability and close dimensions. General machinery, light compressors. Machine tools, medium gear blanks, heavy compressors. Dies, crankshafts, high pressure cylinders large gears, heavy duty machine tool parts.
				Brinell H.	*Total C%*	
			G1800 G2500	187 max 170–229	Microstructure ferritic-pearlitic pearlitic-ferritic	Machinability: high strength not required. Small cylinder blocks and heads, pistons, clutch plates, gear boxes.
ASTM	A159–77	Automotive grey iron castings	G2500a G3000	170–229 187–241	3.4 minimum	A graphite size 2–4 15% maximum ferrite pearlitic — Brake drums and clutch plates. Cylinder blocks, heads, liners, pistons flywheels, medium duty brake drums.
SAE	J431	Cast in sand moulds for automobile and truck parts.	G3500 G3500b G3500c G4000	207–255 207–255 207–255 217–269	3.4 minimum 3.5 minimum	A graphite size 3–5; 5% maximum ferrite or carbide A graphite size 3–5; 5% maximum ferrite or carbide pearlite — Truck cylinder blocks and heads heavy flywheels. Brake drums and clutch plates for heavy service. Extra heavy duty service brake drums. Diesel engine parts; heavy castings.
ASTM ASME	A278–75 (1980) SA278	Pressure parts for temperatures up to 345 °C	40 50 60 70 80	40 000 50 000 60 000 70 000 80 000	C.E. = 3.8 maximum C.E. = %C + 0.3 (%Si + %P) %P = 0.25 maximum %S = 0.12 maximum Castings and test bars must be stress relieved	Valve bodies, papermill drier rolls, chemical process equipment, pressure vessel castings
ASTM	A319–71 (1980) Non pressure parts for elevated temperature use	For superior thermal shock Average thermal shock resistance Higher strength at temperature	I II III	Low strength Above 30 000 As high as 40 000 C.E. Minimum C 3.81–4.4 3.5 3.51–4.1 3.2 3.2–3.8 2.8	When Cr is present each class divided A 0.2–0.4% Cr B 0.41–0.65% Cr C 0.66–0.95% Cr D 0.96–1.20% Cr	Stoker and fire box parts, grate bars, process furnace parts, ingot moulds, glass moulds, metal meltings pots.

Table 1.5 (part 2) A summary of USA standards for malleable irons

	Spec. No.	Use	Grade	Minimum tensile p.s.i.	Minimum yield p.s.i.	Elongation p.s.i. %	Microstructure	Applications
ASTM ASME	A47-77 SA47	Ferritic malleable Iron castings	32510 35018	50 000 53 000	32 000 35 000	10 18	Temper carbon in ferrite	General engineering at normal and elevated temperatures good machinability, excellent shock resist.
ASTM ASME	A197-79 SA197	Cupola malleable Iron		40 000	30 000	5	Free of primary graphite	Pipe fittings and valve parts for pressure service.
ASTM	A220-76	Pearlitic Malleable Iron Castings	40010 45008 45006 50005 60004 70003 80002 90001	60 000 65 000 65 000 70 000 80 000 85 000 95 000 105 000	40 000 45 000 45 000 50 000 60 000 70 000 80 000 90 000	10 8 6 5 4 3 2 1	Temper carbon in necessary matrix without primary cementite or graphite	General engineering service at normal and elevated temperatures. Dimensional tolerance range is stipulated.
ASTM	A338-61 (1977)	Flanges, pipe fittings	Property requirements specified					Railroad, marine and heavy duty service up to 345 °C.

	Spec. No.	Use	Grade	Brinell H.	Heat Treatment	Microstructure	Applications
ASTM	A602-70 (1976)	Automotive Malleable Iron	M3210	156 max	Annealed	Ferritic	Good machinability, mounting brackets
SAE	J158a	Castings	M4504	163-217	Air Q. and Temper	Ferrite and Tempered pearlite	Compressor crankshafts, hubs
			M5003	187-241	Air Q and Temper	Ferrite and Tempered pearlite	For selective hardening
			M5503	187-241	Liq Q and Temper	Tempered martensite	For machinability and induction hardening
			M7002	229-269	Liq Q and Temper	Tempered martensite	For strength; connecting rods.
			M8501	269-302	Liq Q and Temper	Tempered martensite	High strength, wear resistance as in gears.

Table 1.5 (part 3) A summary of USA standards for spheroidal irons (after the Iron Castings Soc.)

	Spec No	Uses	Grade	Minimum tensile p.s.i.	Minimum yield p.s.i.	Elongation %	Heat treatment	Other requirements	Uses
ASTM	A536–80	Maximum shock resistance used at sub zero temperature widely used normal service	60–40–18	60 000	40 000	18	Annealed	Composition is subordinate to mechanical properties The content of any element may be specified by mutual agreement	Pressure castings such as valve and pump bodies machinery subjected to shock and fatigue loading crankshafts, gears, rollers
			65–45–12	65 000	45 000	12			
		Suitable for flame and induction hardening	80–55–06	80 000	55 000	6			
		Best combination for strength, wear resistance and surface hardening	100–70–03	100 000	70 000	3	Normalized		High strength gears, automotive and machine parts Pinions, gears, rollers and slides.
		Maximum strength and wear resistance	120–90–02	120 000	90 000	2	Quenched and tempered		
SAE	J434C	For automotive and allied industries	D–4018		*Brinell Hardness of Casting* 170 maximum		Annealed	Microstructure > 80% spheroidal graphite of types I and II and be substantially free of primary cementite	Steering knuckles Disc brake calipers Crankshafts Gears Rocker arms
			D–4512		156–217				
			D–5506		187–255		Normalized Q and tempered		
			D–7003		241–302				
			DQ and T		as specified				

Composition requirements (bottom specifications):

	Spec No	Uses	Grade	Minimum tensile p.s.i.	Minimum yield p.s.i.	Elongation %	Heat treatment		T.C	Si	P	S	C.E	BNH	Uses
ASTM ASME ASTM	A395–80 SA395 A475–70 (1976)	Pressure contg. parts at elevated temperature Spheroidal castings for paper mill dryer rolls up to 230 °C	60–40–18	60 000	40 000	18	ferritized	Minimum Maximum	3.0	2.50	0.08		3.8	143	Valves and fittings for steam and chemical plant Paper mill dryer rolls
			80–60–03	80 000	60 000	3	Used as cast BHN/201	Minimum Maximum	3.0	3.0	0.08	0.05	4.5	187	
U.S. Military	MIL-1-24137 (ships) amended	Navy shipboard and other applications requiring shock resistance	Class A	60000	45000	15	ferritized to BHN 190 maximum	Minimum Maximum	3.0	2.5	0.06		4.3	190	Shipboard electric equipment, engine blocks, pumps, gears valves, clamps etc.

Table 1.6 The general direction of increase of mechanical and physical properties of general purpose cast irons

| Property | Unit | Grey flake irons | | | |
		G150	G220	G300	G400
Tensile minimum 30 mm cast bar	N mm^{-2}	150	220	300	400
Compressive strength	N mm^{-2}	600	768	960	1200
Design direct tensile maximum	N mm^{-2}	38	55	75	100
Design direct compression depends on stiffness	N mm^{-2}	up to 156	229	312	416
Modulus of elasticity	GN m^{-2}	100	120	135	145
Typical hardness range	BHN	136–167	167–204	202–247	251–307
Thermal conductivity 100 °C	W m^{-1}K^{-1}	52.5	50.1	47.4	44
Fatigue limit Wohler; unnotched 8.4 mm	N mm^{-2}	68	99	135	152
Design fatigue maximum	N mm^{-2}	23	33	45	51
Machinability		←—————————————			
Better machined finish		————————————→			
Thermal shock resistance		←—————————————			
Damping capacity		←—————————————			
Producibility–castability		←—————————————			
Wear resistance		————————————→			
Possibility of thinner sections		←—————————————			

| | | Malleable iron | | | |
| | | Blackheart | | Pearlitic | |
		B 30–06	B 35–12	P 45–06	P 70–02
Tensile 15 mm cast test bar	N mm^{-2}	300	350	450	700
Design direct tension	N mm^{-2}	102 maximum	120 maximum	120 maximum	241
Design compression	N mm^{-2}	100 maximum	130 maximum	175	330
Thermal conductivity 100 °C	W m^{-1}K^{-1}	49	49	45	38.9
Ductility		←—————————————			
Wear resistance		————————————→			
Machinability		←—————————————			
Hardenability		————————————→			
Impact strength		←—————————————			

| | | Spheroidal irons | | |
		350/22	500/7	800/2
Tensile minimum	N mm^{-2}	350	500	800
Design direct tension	N mm^{-2}	125	145	198–270
Design direct compression	N mm^{-2}	148	204	271–367
Typical hardness range	BHN	116–140	172–216	> 259
Modulus of elasticity	GN m^{-2}	169	169	175
Fatigue limit Wohler Unnotched bar 10.6 mm diameter	N mm^{-2}	180	224	304
Design fatigue maximum	N mm^{-2}	62	75	101
Thermal conductivity 100 °C	W m^{-1}K^{-1}	36.5	35.5	32.5
Ductility		←—————————————		
Wear resistance		————————————→		
Machinability		←—————————————		
Hardenability		————————————→		
Impact strength		←—————————————		

in use[21]. Thermal loading is a prime consideration in small engines used in cars and Cr–Mo–Ni alloyed flake irons with high C contents of 3.7–3.8% ensure good thermal conductivity with adequate strength. With high mechanical loading and moderate to high thermal loading, as in medium speed or slow running engines, a change of grade to high strength Cr–Mo flake irons with a lower C content of 3.2–3.3% C provides the change in balance of properties. When mechanical stress is the major consideration it can be advantageous to change iron type to a ferritic compacted/vermicular or ferritic spheroidal iron. However, such a change can lower the thermal loading capacity and, possibly, the permissible operating temperature.

Grey flake irons

Grey flake irons are the most used of the general purpose engineering irons. Their name derives from the characteristic grey colour of the fracture surface and the graphite morphology. They are relatively inexpensive and easy to produce because, in contrast to other irons, they have composition tolerances which are easily satisfied and few foundry problems arise from feeding and shrinkage provided the moulds are prepared correctly. They are readily machined and the machined surfaces are resistant to sliding wear. They have high thermal conductivity, low modulus of elasticity and an ability to withstand thermal shock. This makes them suitable for castings subjected to local or repeated thermal stressing. Their main disadvantage is their section sensitivity and low strength in heavy sections. This must be taken into account when designing castings to withstand service stresses.

Irons of the composition given below satisfy a low and a high grade specification in a medium size, uniform section sand casting.

G 150 (C.E.V. 4.5) 3.1–3.4% C; 2.5–2.8% Si; 0.5–0.7% Mn; 0.15% S; 0.9% P.
G 350 (C.E.V. 3.6) 3.1% C max; 1.4–1.6% Si; 0.6–0.75% Mn; 0.12% S.

C.E.V. is the carbon equivalent value, which is equal to the C content of the iron plus the amount of C equivalent to the added elements.

The properties of flake irons depend on the size, amount and distribution of the graphite flakes and the matrix structure. These, in turn, depend on the C and Si content (C.E.V. = %C + 1/3% Si + 1/3% P) as well as minor and alloying additions and processing variables, such as method of melting, inoculation practice and cooling rate. The G 150 iron above is of a higher C.E.V. and is rich in P. These characteristics promote castability, which is the main attribute of the lower grade iron. They usually have a ferritic matrix and find extensive use in non-critical components as indicated in *Table 1.5*. They can be produced consistently by the correct melting charge selection. Often hardness measurements on selected areas of the component are a sufficient quality test. Higher grades have a pearlitic matrix which can be achieved by heat treatment. However it is preferable to achieve a pearlitic matrix, carefully balancing alloying elements. The higher grades are used in components such as engine blocks, cylinder heads, moulds and hydraulic valve bodies. They require more careful control over composition and processing to obtain the desired graphite and matrix properties. Quality control usually involves melt analysis, mechanical testing of cast test bars and radiography.

Flake morphologies are divided into five classes by ASTM specification A247 as shown in *Figure 1.5*. Type A is a random distribution of flakes of uniform size (see also *Figures 6.3* and *6.8* in Chapter 6) and is preferred for mechanical applications. A high degree of nucleation that promotes eutectic solidification close to the equilibrium graphite eutectic is necessary for the formation of A-type graphite. The flake distribution is refined when changing from A- to D-type as the undercooling increases. Inadequate nucleation may be the reason for this and in extreme cases may result in chill formation. Correct inoculation of a liquid with adequate graphitization potential avoids these undesirable structures.

Type B graphite forms a rosette pattern. The eutectic cell size is large because of the low degree of nucleation. Fine flakes form at the centre of the rosette because eutectic solidification begins at a large undercooling. Recalescence raises the eutectic growth temperature resulting in a coarse, radially growing flake structure.

Type C flakes occur in hypereutectic irons and form with coarse primary Kish graphite[22,23]. This may influence the size of the eutectic cell and distribution of eutectic graphite. It may also reduce the tensile properties and cause pitting on machined surfaces but it can be beneficial when thermal conductivity is important.

Type D is fine undercooled graphite which forms when solidification occurs at a large undercooling (see also *Figure 6.13* in Chapter 6). This structure forms in the presence of Ti (ref. 24) and in rapidly cooled irons that contain sufficient Si to ensure a graphitizing potential that is high enough to avoid chill formation at the high cooling rate. Although finer flakes increase the strength of the eutectic, this morphology is not desirable because it interferes with the formation of a fully pearlitic matrix by providing short diffusion paths for C, hence aiding ferrite formation.

Type E graphite forms in strongly hypoeutectic irons of low C.E.V. that form a strong primary austenite dendrite structure before undergoing eutectic solidification. This morphology is classified as interdendritic with preferred orientation. ASTM A247 also provides standards for measuring flake size.

One of the main disadvantages of flake irons is the dependence of their structure and mechanical properties on cooling rate which makes them particularly section sensitive. Cooling rate can be related to section thickness only for the simplest castings. Such a relationship is shown in *Figure 1.6*. Tensile strengths quoted in the specifications for a 30-mm diameter cast test bar are shown in *Table 1.4*. If the casting is of simple shape but varying in thickness, it

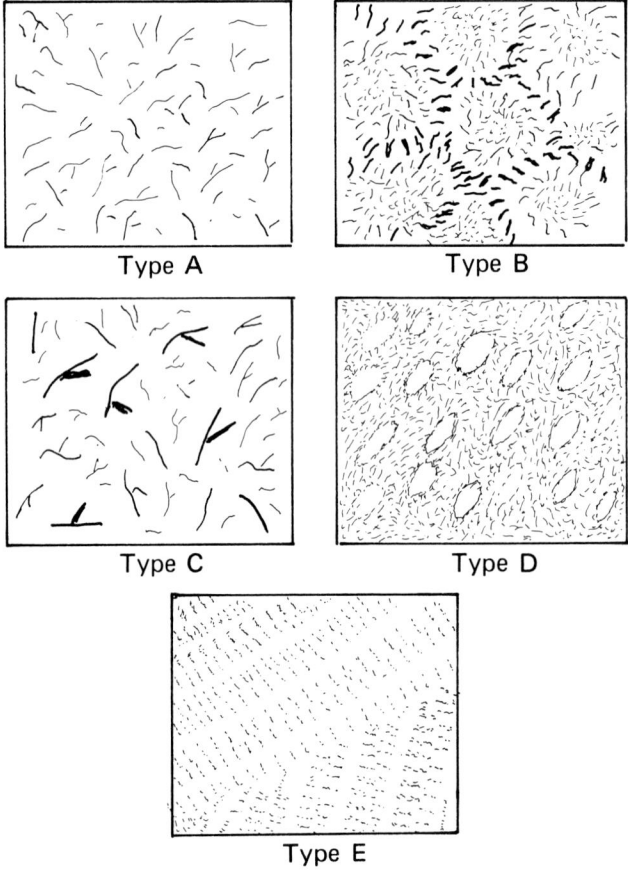

Figure 1.5 Types of flake graphite defined by ASTM A247 (A) uniform distribution, random orientation; (B) rosette grouping, random orientation; (C) superimposed flake size, random orientation; (D) interdendritic, random orientation; (E) interdendritic, preferred orientation

is often sufficient to select the iron grade by considering the thickness of critical sections. However, in complex, heavily cored castings other factors can influence cooling rate including;

1. the location of sections that may act as a heat source (heavy sections) or a heat sink (thin sections);
2. the location of a section with respect to the thermal centre of the casting or heavily cored sections and
3. pouring temperature, gating, feeder and runner design and thermal capacity of the mould.

These influences can affect both graphite and matrix structures. For example, if they cause a thick section to cool through the eutectoid temperature range at the same rate as a thin section, more ferrite will form in the thin section because C diffusion in the solid state is aided by the finer graphite distribution formed on solidification. The increased ferrite content will reduce strength. This type of section sensitivity has been reported in cast iron engine blocks. The thin cylinder wall displays a ferritic matrix and the thick bearing saddle displays a largely pearlitic matrix. This emphasizes that strength predictions of sections in a complex casting require a knowledge of the specific cooling rate in each section.

Malleable cast irons

Malleable cast irons differ from other irons in this group because they are cast white when their structure consists of metastable carbide in a pearlitic matrix. High temperature annealing followed by suitable heat treatment produces a final structure of graphite aggregates, known as temper C clusters, in a matrix which can be ferritic or 'pearliltic' depending on the composition and heat treatment. Specifications in *Tables 1.5* and *1.7* define mechanical properties but composition falls in a narrow range defined by the requirements of the malleabilizing process. The tradi-

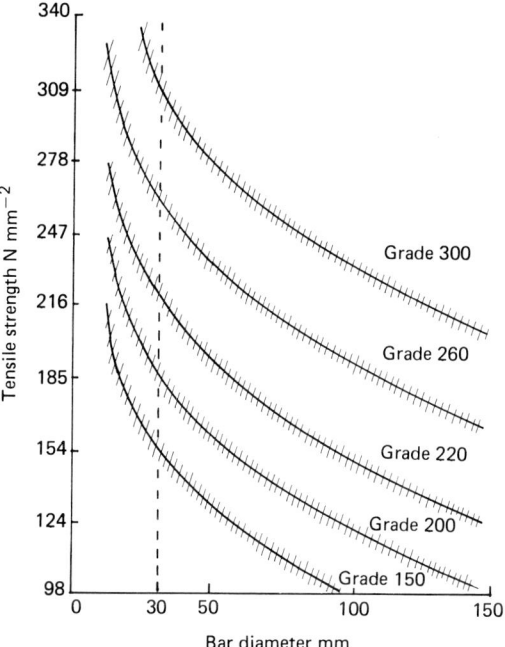

Figure 1.6 Variation of tensile strength with section thickness for grey flake irons

tional European Whiteheart was introduced by Reaumur in 1722 and American Blackheart by Seth Boyden in 1820.

Typical compositions are given in *Table 1.8*. The composition used is selected to satisfy the needs for cost effective and good annealability and foundry production. Increased Si content promotes good and rapid annealability (short cycle process, *Table 1.8*) and improves mechanical strength. However, metallurgical and energy considerations can limit the Si content[25]. C and Si contents are adjusted to ensure freedom from graphite formation on solidification. However, irons of high Si and low C content require more feeding, thus reducing the casting yield, and need high metal temperatures which increase production costs. Consequently, high C contents are used for very small section castings, such as pipe fittings, in order to obtain the required fluidity without using excessive metal temperatures.

Minor elements are of consequence. Bi and Te promote white iron formation on solidification and B and/or Al improve annealability by increasing the number of graphite clusters. Mn content and the Mn/S ratio must be controlled to facilitate annealing. The Mn content may be increased to about 1% to produce a pearlitic iron and Cu, Ni and/or Mo may be added to obtain higher strength or to increase wear and corrosion resistance. The concentration of deleterious elements P, Sb, Sn and Cr must be controlled.

The Whiteheart process is a combined decarburization and graphitization process performed in an oxidizing atmosphere. This was originally done by packing castings into iron ore mixtures but more recently it is carried out in continuous gas ovens which allow higher temperatures of approximately 1070 °C to be used with shorter annealing times. The two reactions produce a C gradient in the casting, the outer layer normally displays a ferritic structure without graphite and the centre temper C clusters in a pearlitic matrix. Small castings may be fully decarburized and are referred to as weldable malleable irons.

The Blackheart process involves annealing in a neutral atmosphere and only graphitization occurs. Slow cooling after the annealing process results in a final uniform structure of temper C clusters in a ferritic matrix. The higher strength 'pearlitic' grades are produced by;

1. increasing the Mn content to $\approx 1\%$;
2. arrested annealing, quenching and tempering and
3. annealing, reheating and quenching with or without subsequent tempering.

The latter process, in particular, is very versatile and the austenite that exists after the graphitizing process can transform into pearlite, bainite or martensite to give the wide range of properties specified for 'pearlitic' malleable irons.

The main reason for selecting malleable irons are low cost and ease of machining. Their uses include car and agricultural components, pipe fittings, mining and electrical fittings, valve components, hardware, tools etc. Malleable irons have limitations in section size, damping capacity, impact resistance and hot and cold cracking and warpage.

Spheroidal cast irons

The invention of this iron, also known as ductile, nodular and spherulitic is usually credited to presentations made at the American Foundryman's Congress in Philadelphia in 1948[26] although there is evidence of spheroidal iron picks having been cast in China over 2000 years ago[27]. Rapid developments occurred in the 1950s followed by increased usage in the 1960s. As *Figure 1.4* shows, production has been maintained in the last decade despite a fall in the overall iron production.

A typical spheroidal base iron composition is 3.7% C, 2.5% Si, 0.3% Mn, 0.01% S, 0.01% P and 0.04% Mg. Note the higher C.E.V., much reduced S content and the presence of Mg compared to a flake iron. Mg or Ce can be used to produce spheroidal graphite but the former is more adaptable and economical. A combined addition results in an improved Mg recovery and allows a reduction in the amount of Mg

Table 1.7 British specification, BS 6681 (1986) for malleable irons

Type	Grade	Diameter of test bar mm	Tensile N mm^{-2} minimum	0.2% PS N mm^{-2} minimum	Elongation % minimum	Hardness maximum
Whiteheart	W 35–04	9	340	–	5	230
		12	350	–	4	
		15	360	–	3	
	W 38–12	9	320	170	15	200
		12	380	200	12	
		15	400	210	8	
	W 40–05	9	360	200	8	220
		12	400	220	5	
		15	420	230	4	
	W 45–07	9	400	230	10	220
		12	450	260	7	
		15	480	280	4	
Blackheart	B 30–06	12	300	–	6	150
		15	300	–	6	
	B 32–10	12	320	190	10	150
		15	320	190	10	
	B 35–12	12	350	200	12	150
		15	350	200	12	
Pearlitic	P 45–06	12	450	270	6	150–200
		15	450	270	6	
	P 50–05	12	500	300	5	160–220
		15	500	300	5	
	P 55–04	12	550	340	4	180–230
		15	550	340	4	
	P 60–03	12	600	390	3	200–250
		15	600	390	3	
	P 65–02	12	650	430	2	210–260
		15	650	430	2	
	P 70–02	12	700	530	2	240–290
		15	700	530	2	

Table 1.8 Typical compositions for traditional whiteheart, blackheart and short cycle blackheart irons

Element (%)	Whiteheart traditional	Blackheart traditional	Blackheart short cycle
Total C	3.0–3.7	2.2–3.0	2.35–2.45
Si	0.4–0.9	0.7–1.2	1.5–1.6
Mn	0.2–0.4	0.2–0.5	0.35
S	0.3 maximum	0.05–0.16	0.12
P	0.1 maximum	0.12 maximum	0.05
Cr	0.1 maximum	0.1 maximum	0.1 maximum

bearing modifier used. Mg can be added in several forms including metal, Ni–Mg, Ni–Si–Mg, Fe–Si–Mg alloy or Mg coke[28]. Methods of addition include ladle transfer, the covered ladle technique, the porous plug stirring method and the in-mould technique[29]. Additions are made shortly before casting in a quantity sufficient to produce a residual Mg content of 0.03 to 0.05%. The molten iron must be inoculated

simultaneously with, or subsequent to, the Mg addition (post inoculation). Mg is a deoxidizer and a desulphurizer. Consequently, it will only modify the graphite morphology when the O and S contents are low. Deoxidizers such as C, Si and Al present in the liquid iron ensure a low O content but a desulphurization process is often necessary to reduce the S content. Too low a S level in the base iron can decrease the spheroid count by removing potential nuclei for graphite formation. Too high a level results in excessive Mg usage and dross formation. Formulae are available for calculating the correct Mg addition, for example;

$$\text{Mg} = \frac{3/4 \ (\text{initial S content}) + \text{residual Mg} \ (0.03\text{–}0.05\%)}{\text{expected Mg recovery}}$$

An iron of the composition given above cast into a medium section sand mould would show a spheroidal graphite structure in a predominantly ferritic matrix and satisfy the lower strength grades of the spheroidal graphite (S.G.) iron specifications. Spheroidal irons are inferior to flake irons with respect to physical properties such as thermal conductivity but exhibit better mechanical properties. They make suitable replacements for steel and flake iron, for example in pipes[30], and are complementary to malleable irons[31]. They are less section sensitive than flake irons but are very sensitive to minor and trace elements, an influence that can be magnified by segregation during solidification. Close control must be exercised over composition during production to ensure cast structures with acceptable graphite shape and desired matrix structure. The latter ranges from fully ferritic to pearlitic or bainitic as-cast (see also *Figures 6.22, 6.34, 6.35* in Chapter 6), but, as *Figure 1.7* shows, a much wider range of properties can be achieved by heat treatment. Although this adds to the cost, some foundries use heat treatment to produce a wide range of properties from an iron of fixed composition.

Compositional control cannot be over emphasized and the complex and interrelated behaviour of minor and alloying elements in both spheroidal and flake irons[32] is, as yet, incompletely understood. However, its importance is illustrated in the description of considerations in controlling spheroidal iron composition given below. The following structural requirements have to be satisfed simultaneously;

1. freedom from carbides (chill and intercellular);
2. correct shape and distribution of graphite and
3. required matrix structure

Compositional effects are influenced by the various steps in the production sequence, for example, pouring temperature, time and sequence used at each stage, cooling rate and section size. Considerations must be modified depending on whether sand or permanent moulding is being used[33] and special considerations are required for large castings[34].

The first consideration which is often written into the specification is freedom from carbides in the as-cast state[35,36]. Failure in this respect results in areas of extreme hardness which impair mechanical properties and can prohibit machining. This necessitates heat treatment which increases costs and, if not performed correctly, can lead to distortion. A high liquid graphitization potential and effective inoculation reduce the risk of chill carbide formation. This requirement is opposite to that required for malleable iron and is achieved using higher C and Si levels as shown in *Figures 1.2* and *1.8*. The need to maintain a minimum C.E.V. of at least 4.3 is widely accepted as a requisite for a carbide-free structure and good spheroid quality. If the C.E.V. is too low, the graphitizing potential is reduced. A low C.E.V. with a high Si content to maintain the potential is associated with excessive shrinkage. A value in excess of 4.65 can lead to spheroid flotation and degeneracy, particularly in

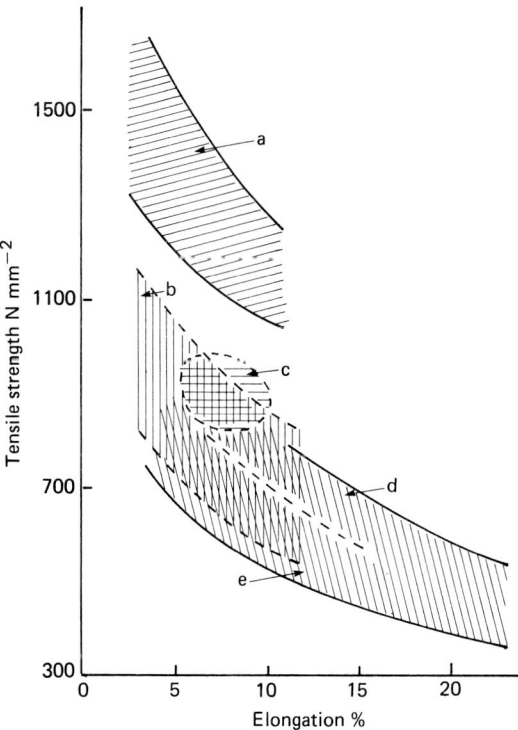

Figure 1.7 Strength and ductility that can be achieved in spheroidal irons in the following conditions: a, austempered; b, quenched and tempered; c, normalized; d, as-cast bullseye ferritic; e, as-cast or annealed

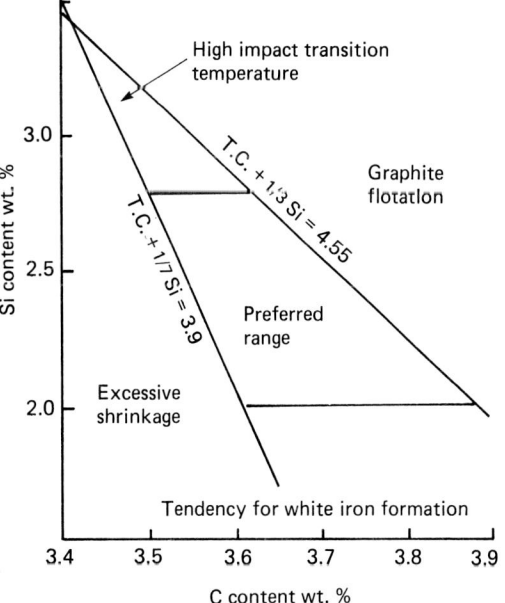

Figure 1.8 Typical C and Si concentration ranges for spheroidal irons and factors that limit these ranges

heavy sections[37]. Recommended C.E.V.'s for irons of various thicknesses for sand and permanent moulds are shown in *Figure 1.9*.

The second consideration is the choice of C and Si levels for the chosen C.E.V. This is made with due regard for the effect of Si on properties. On the credit side, Si increases the graphitization potential and refines the graphite distribution. On the debit side, it promotes ferrite-reducing strength, thus increasing impact transition temperature and decreasing thermal conductivity. It also increases ferrite strength, which is its as-cast ductility and hardness, particularly in the annealed state. General recommendations for C and Si levels are given in *Table 1.9* when there is no restriction on either element as may be imposed by the melting method.

Mn is a moderately strong carbide promoter and must be limited for this reason. The recommended level depends on the Si content and section thickness as shown in *Figure 1.10*. The carbide forming tendency of Mn can be countered in thin sections by increasing the Si content but not in heavy sections because Mn segregates to cell boundaries whereas Si segregates negatively to austenite dendrites.

There is a general tendency for graphite-promoting elements (Si, Co, Ni, Cu) to segregate negatively and for carbide-forming elements (Mo, Ti, V, Cr, Mn, P) to segregate positively. The degree of segregation depends on the element and the solidification

Table 1.9 General recommendations for C and Si contents of unalloyed and low alloyed carbide-free spheroidal irons (after ref. 29, p. 45)

Wall thickness mm		As-cast Pearlitic		Matrix Ferritic	
minimum	maximum	Total C%	Si%	Total C%	Si%
3	3	4.00	3.75	3.95	4.00
	6	3.70	3.75	3.65	4.00
	12	3.50	3.75	3.35	4.00
	25	3.20	3.75	3.15	4.00
	50	3.05	3.75	3.00	4.00
	75	3.00	3.75	2.95	4.00
	> 100	3.00	3.75	2.95	4.00
6	6	3.90	3.25	3.75	3.75
	12	3.70	3.25	3.55	3.75
	25	3.40	3.25	3.25	3.75
	50	3.25	3.25	3.10	3.75
	75	3.20	3.25	3.05	3.75
	> 100	3.20	3.25	3.05	3.75
12	12	3.85	2.75	3.70	3.25
	25	3.55	2.75	3.40	3.25
	50	3.40	2.75	3.25	3.25
	75	3.35	2.75	3.20	3.25
	> 100	3.35	2.75	3.20	3.25
25	25	3.60	2.50	3.50	3.00
	50	3.45	2.50	3.35	3.00
	75	3.40	2.50	3.30	3.00
	> 100	3.40	2.50	3.30	3.00
50	50	3.60	2.10	3.40	2.75
	75	3.55	2.10	3.35	2.75
	> 100	3.55	2.10	3.35	2.75
75	75	3.60	2.00	3.40	2.50
	> 100	3.60	2.00	3.40	2.50
> 100	> 100	3.60	2.00	3.40	2.50

conditions[38]. Generally recommended maximum levels for carbide-forming elements are 0.05% Cr, 0.03% V, 0.003% B, 0.003% Te and 0.01–0.75% Mo. The higher levels of Mo are for alloying purposes and are usually balanced by lower Mn additions. Spheroidizing elements are also carbide forming elements and should not be used in excess of that required to modify the graphite.

Every S.G. iron producer aims for a fine and uniform distribution of perfectly shaped spheroids. This promotes good mechanical properties. Graphite distribution is conveniently measured as the number of spheroids per mm². Distribution and morphology can be measured metallographically[39] by using simple comparator charts, magnetic techniques[40] and ultrasonic velocity measurements[41]. Spheroid count is analogous to cell count in flake iron but is usually much higher. Although increasing C.E.V. increases spheroid count, high counts are the result of effective inoculation of a liquid with a high graphitization potential. Response to nucleation differs from iron to

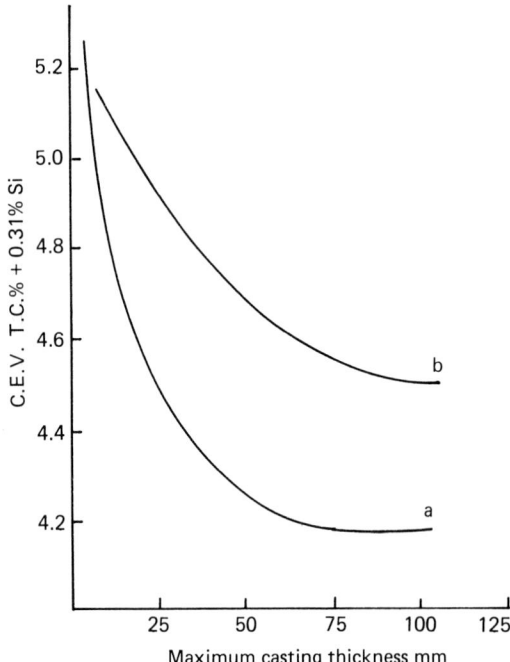

Figure 1.9 Suggested C.E.V. values for spheroidal iron as as a function of thickness for: a, sand castings: b, permanent moulded castings

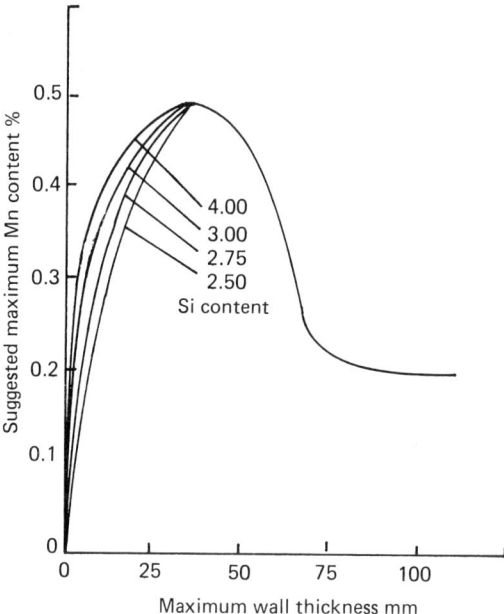

Figure 1.10 Suggested maximum Mn content for different Si contents as a function of maximum casting wall thickness

iron. In general, the higher the superheat, the poorer the response. The higher the Si content prior to treatment, the better the response.

Inoculants also fade, and this places emphasis on processing time and sequence. In-mould inoculation has been introduced to overcome fade[38,42]. Several classifications of graphite shape exist[43]. For example, specification ASTM A247 defines seven basic graphite morphologies as shown in *Figure 1.11*. Types I and II have little effect on iron properties. Alloy specifications often stipulate an excess of 85% types I and II. Type III is cluster C often found in malleable iron after annealing. Type IV is typical of compacted/vermicular irons. Types V and VI are degenerate spheroidal shapes that often impair mechanical properties[45]. Type VII is flake graphite which is divided into various types which were described on page 12.

Spheroid degeneracy is also measured by a shape factor defined as the ratio between the average cross-section area of graphite sections and the smallest circle into which the same average graphite inclusion can be fitted. Mechanical properties show little correlation with factors between 1.0 and 0.75 but deteriorate with decreasing factors below 0.75. Karsay has criticized the ASTM classification[46] and considers a simple microstructural classification, as shown in *Figure 1.12* to be more meaningful. Intercellular flake graphite is particularly damaging to mechanical properties. Several trace elements are known to cause degeneracy. Most of these elements are highly surface

active and tend to concentrate at interfaces where they exert a considerable influence on graphite morphology.

Originally elements were classified into beneficial, neutral and deleterious as shown in *Table 1.10*. Several studies[2,45] have shown that such a classification is inadequate. For example, Mg in excess of 0.1% residual content causes intercellular flakes or flake projections from spheroids (spiky graphite). Intercellular chunk graphite-promoting elements segregate during solidification and reach a concentration sufficient for the formation of flake graphite. This occurs particularly in heavy sections. Intracellular chunk graphite forms within cells and the cell boundaries often contain well formed spheroids. This defect is found in heavy sections as well. Elements promoting these two defects are shown in *Table 1.11*.

The undesirable effect of the flake-promoting elements can be neutralized by the addition of an element from column 1 of *Table 1.11* that promotes chunky graphite (Ce)[46-49]. Conversely, chunky graphite can be eliminated and spheroidal graphite restored by the addition of elements from column 2. The elements in the third column cause degeneracy but do not fit into the same simple scheme.

As our understanding of the nucleation and growth of graphite increases, it becomes possible to select minor element combinations for optimization of structure and properties. For example, a new group of inoculants that combine the beneficial effects of Bi and the alkaline and rare earth metals and permit the use of spheroidizing alloys that do not contain Ce or mischmetal have been proposed[50,51]. With these inoculants, the spheroid count is increased, the chilling tendency reduced and as-cast ferritic structures promoted without spheroid degeneracy.

The third consideration is composition control for matrix structure. The matrix provides strength and this can be promoted using minor alloying additions to increase the pearlite/ferrite ratio, to refine the pearlite or to produce alternative austenite transformation products such as bainite. Irons treated in this manner are usually referred to as alloyed irons and are to be distinguished from alloy irons. The considerations described below apply to flake as well as spheroidal irons. Alloyed flake irons received more attention before the development of spheroidal irons so dramatically increased the strength levels of cast irons. However, there has been a resurgence of interest in alloyed flake irons recently, particularly with respect to efforts being made to reduce the weight of engineering components and as a means of overcoming difficulties in producing sound castings. They may also help avoid chill in low C.E.V. irons by increasing the C.E.V. and maintaining the mechanical properties by alloying. Minor and alloying elements have unique characteristics with respect to solid state transformations in addition to those already discussed for the liquid–solid transformation, and all

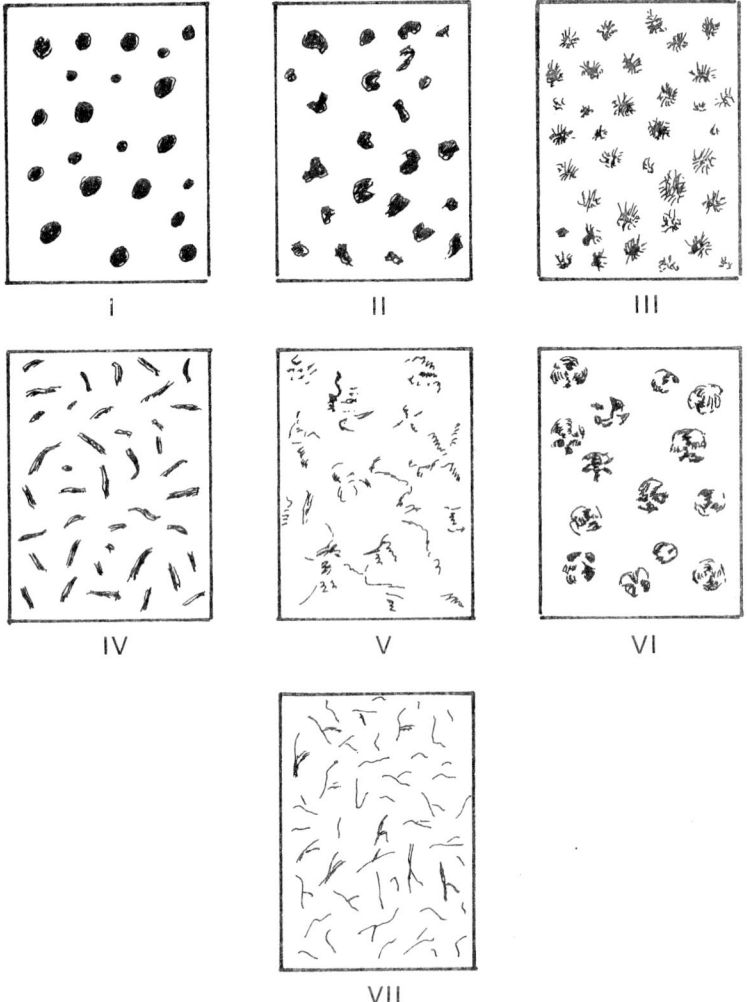

Figure 1.11 Seven types of graphite morphology as defined in ASTM A247

characteristics must be considered in selecting the iron composition. It is not just a question of adding elements to a base composition, but of their selective use, often in combinations, to achieve desired strength with uniform property and machining characteristics or low section sensitivity[32,52].

The benefits of Si in increasing the graphitization potential and reducing the tendency for chill formation are known but, in general, Si has a negative effect on strength. It promotes ferrite in addition to increasing the C.E.V. and reduces pearlite strength by increasing the formation temperature and coarsening the structure. Mn in excess of that required to balance S retards ferrite formation by delaying nucleation and slowing growth. It is a strong pearlite promoter because it stabilizes austenite by increasing the C solubility and it moderately refines pearlite. However, Mn is not used primarily for strengthening because it

can affect nucleation adversely and segregates strongly, causing intercellular carbide formation.

Cr promotes pearlite in a similar way to Mn. Unfortunately, it is also a potent chill and carbide promoter. Whilst Si and inoculation may be used to eliminate chill, carbide is not easily prevented, particularly, the formation of intercellular carbides. Sn (0.04–0.1) and Sb (< 0.03%) have specific characteristics as excellent pearlite promoters. These elements provide a barrier to C diffusion and their effect is so potent that they produce a pearlitic matrix in irons with a fine graphite distribution. However, this is their only role and excess additions can lead to embrittlement. Cu, like Sn, retards C diffusion thus promoting pearlite formation. It only refines pearlite weakly. However, in combination with elements such as Mo it has a much greater hardenability effect. Cu is also a graphitizer and can be used in partial replacement for

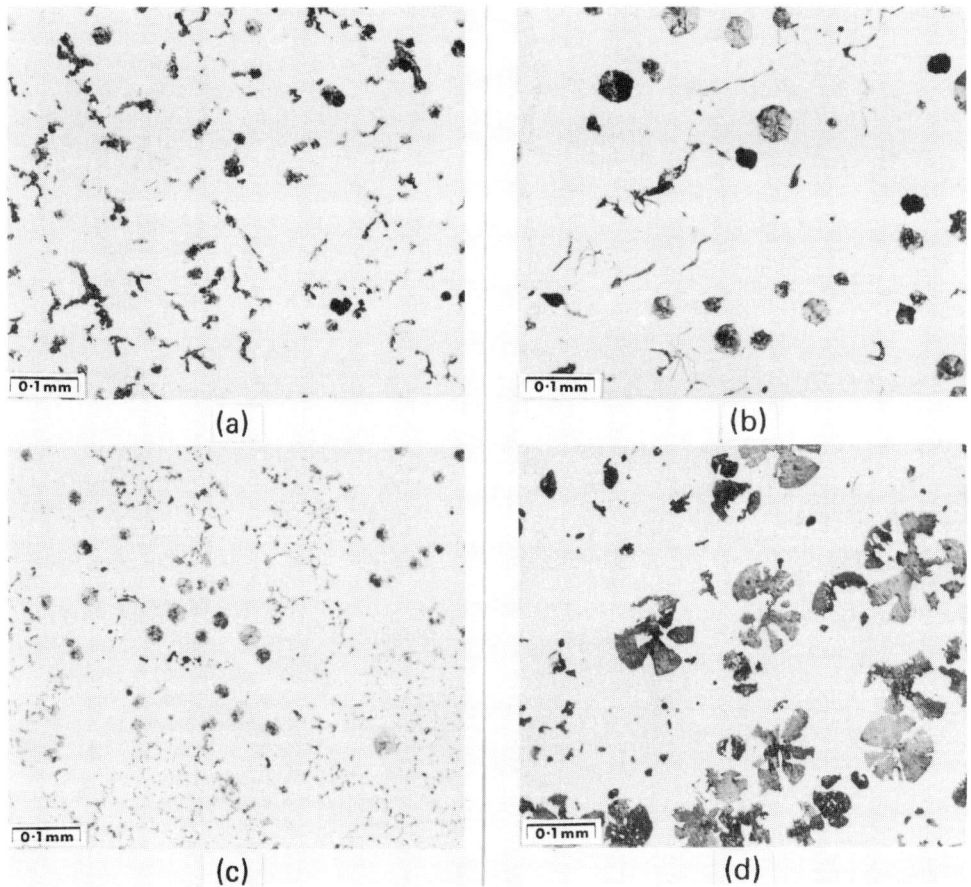

Figure 1.12 Undesirable graphite structures in spheroidal irons; (a), quasi-flake; (b), intercellular flake; (c), chunky graphite; (d), exploded graphite; (after ref. 29, p. 127)

Table 1.10 The influence of minor elements on spheroid structure

Beneficial	Neutral	Deleterious
Mg, Ce, Ca and other spheroidizers	Fe, Si, Ni Mo, C	Al, Sb, As Bi, Pb

Si to minimize chilling tendencies without promoting free ferrite. Ni behaves in a similar way to Cu but is only a weak pearlite promoter. Mo is a potent hardenability element and combines synergistically with Cr, Cu and Ni to provide enhanced hardenability. It retards pearlite formation more than ferrite formation and this can promote ferrite formation in heavy sections. This is why Mo is usually used in combination with pearlite-promoting elements. V is a potent hardenability element and a mild pearlite promoter. However, it is the strongest chill and car-bide-forming element and can only be used in low concentrations (< 0.3%).

Graham[34] has discussed the implementation of the above considerations for large section spheroidal iron castings. Details are given for a 14 500 kg flywheel casting made to ASTM specification 80/60/03 or BS (2789) 600/3 grade. *Table 1.5* (part 3) recommends the following concentrations of main constituents; 3.0% C minimum, 3.0% Si maximum, 0.08% P maximum, 0.05% S maximum.

Table 1.11 Minor elements promoting chunky, intercellular and degenerate spheroidal graphite structures

Chunky	Intercellular flake	Deleterious
Ce, Ca Si, Ni	Bi, Cu, Al Pb, Sb, Sn As, Cd	Zr, Zn, Se Ti, N, S, O

Table 1.12 Composition and mechanical properties of three spheroidal iron flywheel castings (after ref. 34)

Castings	C	Mn	Si	S	P	Ni	Cr	Mg	Cu
1	3.54	0.50	1.75	0.004	0.036	1.2	0.061	0.04	1.0
2	3.57	0.36	1.55	0.004	0.056	1.1	0.06	0.048	0.98
3	3.36	0.56	2.65	0.004	0.03	0.13	0.06	0.025	0.58

Castings	U.T.S. (N/mm^2)	P.S. (N/mm^2)	Elongation %
1	730	448	3
2	796	447	5
3	415	400	1

More detailed considerations resulted in the choice of composition shown for castings 1 and 2 in *Table 1.12*. With reference to *Table 1.9* and for section thicknesses between 10 and 25 cm, the recommended C is 3.4 to 3.6% and the recommended Si is 2.0 to 2.5%. This gives a C.E.V. of approximately 4.3, which will avoid graphite flotation and satisfy *Figure 1.9*. S, Ce, P, Mn, Cr, V and Ti are all deleterious elements in large section spheroidal iron castings and careful selection of raw materials is necessary to avoid introducing these and other trace elements. Limitation of the Mn content to 0.2% as suggested in *Figure 1.10* is difficult if steel scrap is used in the charge.

Differences in composition and properties are evident between castings 1 and 2 in *Table 1.12*. Casting 1 was made from an all steel scrap charge and casting 2 from a charge containing 70% pig iron. Casting 3 is a larger section (approximately 40 cm) casting made with a Ce bearing alloy. The graphite was degenerate and intercellular carbides were present reducing the tensile properties as shown in *Table 1.12*. Hence care was taken with the flywheel casting not to use Ce bearing alloys even in the spheroidizing treatment because insufficient trace elements were present for the Ce to neutralize.

Under these conditions the Ce will produce chunky or exploded graphite. A pearlitic matrix is required to satisfy the mechanical properties and Cu was added for this purpose. Ni is a mild pearlite promoter but is present as a result of having been used as a carrier for Mg. Melting was performed in two induction furnaces over a five hour period. This allows considerable opportunity for adjustment of C and Si levels.

The melt-out composition must be higher in C and lower in Si to allow for the compositional changes that occur during the melt-conditioning treatments of desulphurization, spheroidization and inoculation. Desulphurization is a pre-requisite for the production of high quality, sound castings. Levels of approximately 0.01% were readily achieved using the porous plug method. Good response to inoculation requires a minimum Si content of 1.0%. With this melt-out composition a maximum of 1.8 MgFeSi alloy containing 50% Si could be added for inoculation and spheroidization and still allow a post-inoculation

treatment without exceeding 2.1% Si. However, this addition of Mg was not sufficient to fully spheroidize the structure, and alternative alloys Ni–Mg and Cu–Mg were used for spheroidizing.

As the characteristics of various elements are defined it is becoming apparent that irons are often overalloyed and correct compositional choice often results in a lower alloy content. For example, Si reduction in the base composition while still maintaining C.E.V. can have far reaching effects in reducing alloying additions and cost.

Compacted/vermicular irons

This iron, also referred to as quasi-flake, pseudo nodular, upgraded chunk and semi-ductile, was originally considered as a degenerate form of spheroidal iron and it is only recently that it has been accepted commercially and used to fill the mechanical and physical property void between flake and spheroidal irons. Compacted graphite takes the form of blunt edged stubby flakes that are interconnected within the eutectic cell (see also *Figures 6.10–6.12* in Chapter 6). A compacted iron exhibits superior tensile strength, stiffness and ductility, fatigue life, impact resistance and elevated temperature properties compared to a flake iron with a similar matrix structure[53-57].

Some of these properties are compared in *Figure 1.13* and *Table 1.13*. The interconnected graphite imparts physical properties similar to those of flake irons. The machinability and thermal conductivity are superior to spheroidal irons and, as is evident from *Figure 1.14*, the combination of resistance to crazing, cracking and distortion is superior to either spheroidal or flake irons. A secondary shrinkage of 0.0052 cm/cm compared to 0.0083 cm/cm for spheroidal irons means that the feeding requirements are less exacting than for spheroidal irons. This has led to the use of compacted irons for thick section, heavily-cored castings which require high strength, such as hydraulic valves.

However, it is the combination of physical and mechanical properties that makes compacted irons an

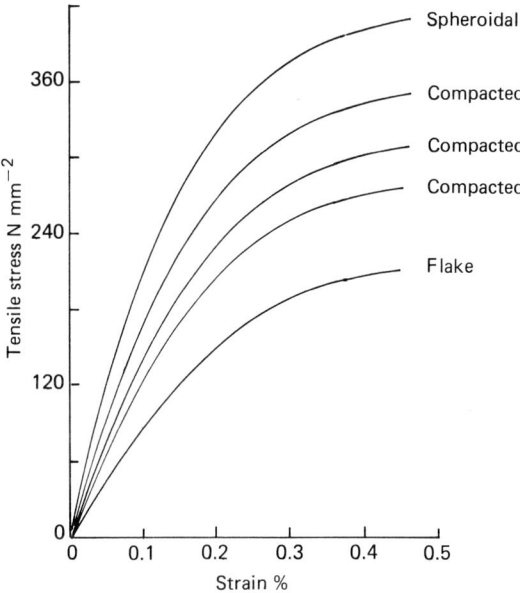

Figure 1.13 The influence of graphite morphology on the stress-strain curve of several cast irons

ideal choice for several applications. Their greater resistance to cracking compared to flake irons and better resistance to distortion than spheroidal irons makes them suitable for ingot moulds. Good elevated temperature properties have resulted in their use for cylinder heads, exhaust manifolds, brake drums, discs and piston rings. For example, a pearlitic compacted

iron was used to reduce a severe thermal fatigue problem encountered with a grade 260 flake iron disc brake during the development of British Rail's 200 km/h high speed train[58].

Specifications for compacted irons have been discussed but, as yet, not defined because of the limited applications for these irons. In general, their inferior mechanical properties and similar production costs to spheroidal irons will limit their use as a replacement for spheroidal irons to cases in which their combined properties are superior. Likewise, the lower production costs of unalloyed flake irons will ensure their continued use. However, where greater strength is required, compacted irons offer a viable alternative to the more expensive alloyed flake irons.

The commercial exploitation of this latest member of the cast iron family depends on successful graphite morphology control in the foundry[59]. Flake graphite must be avoided and spheroidal graphite limited to 10–20%. The degree of graphite compaction is influenced by composition, section size and the process used for graphite structure control. If the C.E.V. increases above 4.4, the amount of spheroidal graphite increases rapidly. Consequently, the C.E.V. should be restricted to about 4.0 before treatment.

Several methods have been used for graphite compaction. N ($\sim 0.015\%$) produces compaction in heavy sections[60] and it increases strength without structural modification in thin sections. Until recently the addition of controlled amounts of N has been difficult. However, this can be achieved with ladle additions of nitrided ferro-manganese (80% Mn; 4% N) with a yield of about 40%. Compaction may occur

Table 1.13 Tensile design stresses at various temperatures and creep and stress to rupture data for various cast irons at 350 °C (after ref. 56)

Temperature °C	Maximum tensile design stress		
	Compacted graphite $N\,mm^{-2}$	Flake graphite Gr 260 $N\,mm^{-2}$	Spheroidal graphite 600/3 $N\,mm^{-2}$
20	103	66	148
100	94	57	133
200	89	60	130
300	92	63	130
Basis for design stress	$0.38 \times 0.1\%$ P.S.	$0.35 \times 0.1\%$ P.S.	$0.45 \times 0.1\%$ P.S.

Creep strain or rupture	Creep and stress to rupture at 350 °C stress ($N\,mm^{-2}$) for creep strain or rupture in 10 000 h				
	Grey iron	Pearlitic malleable	Ferritic spheroidal	Pearlitic spheroidal	Pearlitic compacted
0.1%	100	114	159	178	136
0.5%	165	222	195	270	193
1.0%	–	247	210	297	216
Rupture	182	292	264	370	259

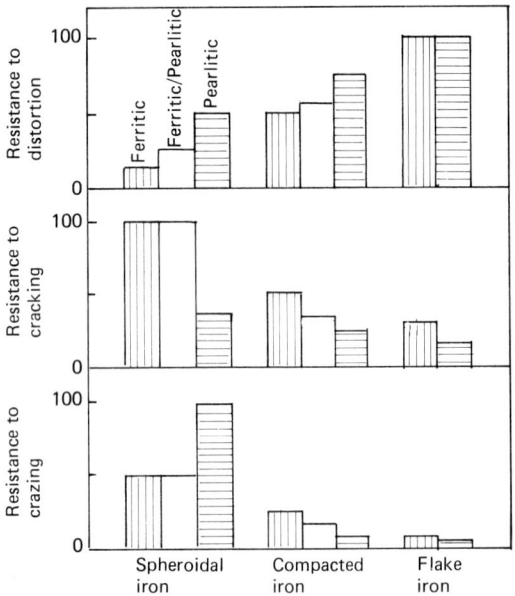

Figure 1.14 Comparison of the performance of heavy section irons of different matrix structure in the BCIRA distortion test (after ref. 56)

unintentionally when large proportions of steel are included in cupola charges and when high N carburizers and steel scrap-based charges are used in electric melting.

Disadvantages of N treatment include difficulty in producing uniform structures in castings with varying section size and the occurrence of unsoundness and fissure defects with excessive N. The effectiveness of N can be controlled by adjusting the Ti and Al levels in the iron. The more commonly practised commercial processes use either Mg or rare earths as an effective spheroidizing agent and rely on either the co-addition of an anti-spheroidizing agent such as Ti or the addition of a controlled amount of spheroidizer[61].

A spheroidizing undertreatment using Mg alone has been used for compacted iron production. However, this presents practical difficulties in controlling fading and overcoming the environmental effects accompanying its addition to molten iron. This method represents a very unstable situation in that slight fluctuations in S content (see *Figure 1.15*) or in residual Mg content (see *Figure 1.16*) promote either flake or spheroidal graphite. Schelleng[62], Dawson and Evans[63], Lalich[64,65] and Loper and co-workers[66,67] have shown that the narrow band of Mg concentration (~ 0.005%) that produces compacted irons can be extended by adding 0.08 to 0.1% Ti, which is an anti-spheroidizing element. Additions of Ce were found to be necessary to produce the compacted structure.

Mg additions are usually made through a Ce bearing Mg ferrosilicon alloy which may or may not

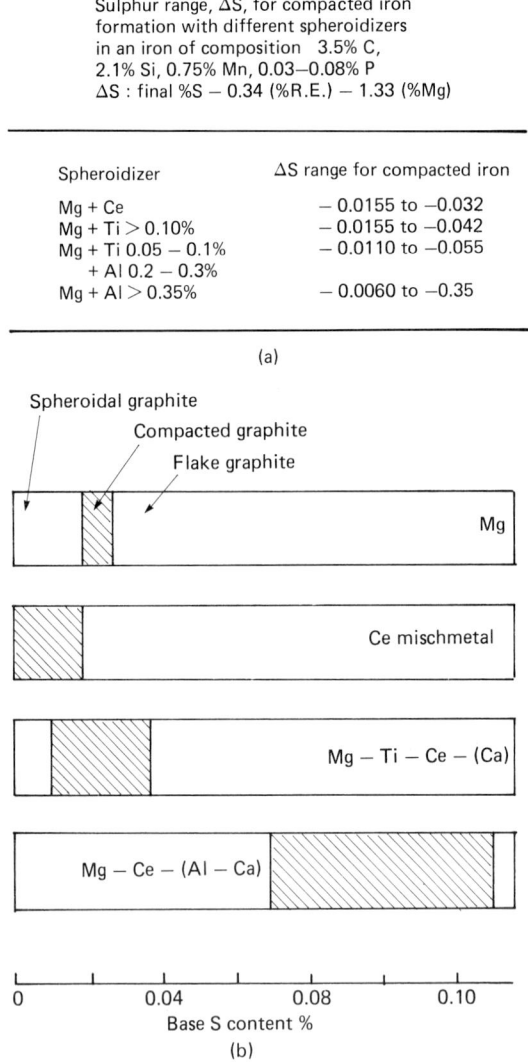

Sulphur range, ΔS, for compacted iron formation with different spheroidizers in an iron of composition 3.5% C, 2.1% Si, 0.75% Mn, 0.03–0.08% P

ΔS : final %S − 0.34 (%R.E.) − 1.33 (%Mg)

Spheroidizer	ΔS range for compacted iron
Mg + Ce	− 0.0155 to −0.032
Mg + Ti > 0.10%	− 0.0155 to −0.042
Mg + Ti 0.05 − 0.1% + Al 0.2 − 0.3%	− 0.0110 to −0.055
Mg + Al > 0.35%	− 0.0060 to −0.35

(a)

(b)

Figure 1.15 (a) Comparison of S ranges for different compacted iron production methods; (b) S ranges for the successful production of compacted irons using different compacting agents (after ref. 55)

incorporate the Ti addition. With both Mg and Ce present, Ce is the first to desulphurize, leaving the Mg free to produce compacted graphite. Ca acts in a similar manner and has been incorporated into a proprietary alloy in order to introduce a certain degree of insensitivity regarding the dependence on the Mg/S relationship. A commercially available alloy contains 4–5% Mg, 8.5–10.5% Ti, 4.0–5.5% Ca, 1.0–1.5% Al, 0.2–0.5% Ce, 48–52% Si and the balance Fe.

The iron can be treated using the techniques

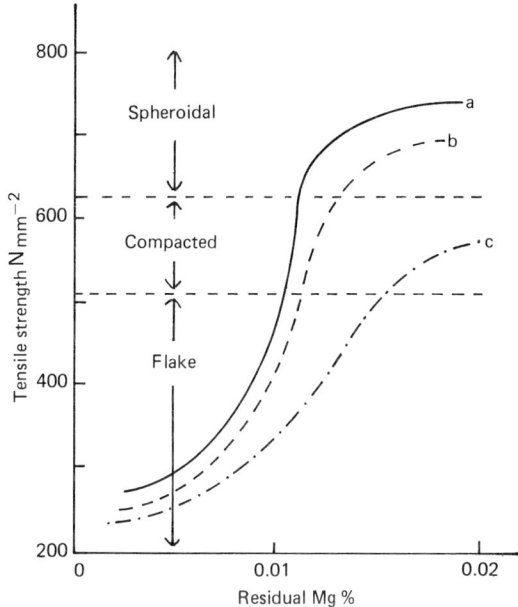

Figure 1.16 a, Changes in the structure of grey irons with Mg additions; b, Tensile strength change with Mg additions for a grey pearlitic iron; c, Change in structure with a Mg–Ti–Ce addition (after ref. 53)

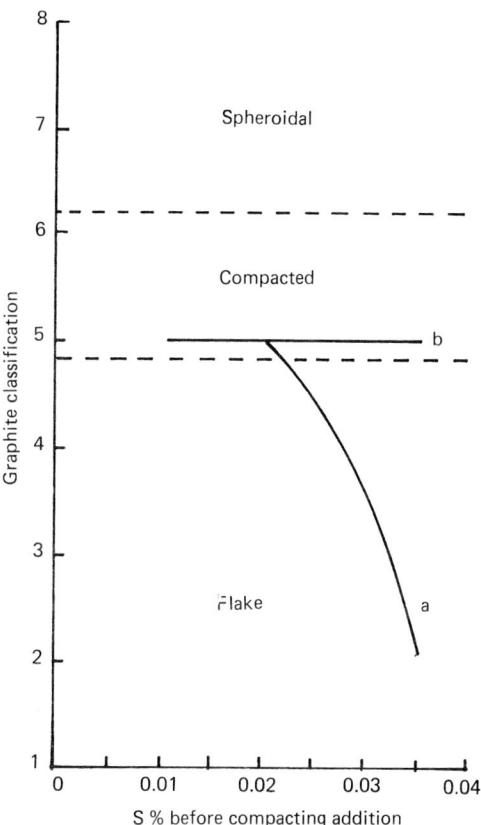

Figure 1.17 Influence of base S level on graphite structure in 10-cm diameter bars showing the improvement obtained when using a compacting agent incorporating Ca; a, early agent; b, agent containing Ca (after ref. 55)

described previously for spheroidal irons, including the in-mould technique[68,69]. A ladle inoculant is usually added and the S content should not exceed 0.035%. Between 0.6 and 1.6% of the alloy should be added depending on the addition method, S content and temperature. Although *Figure 1.17* suggests an insensitivity to S content, a S-dependent alloy addition is necessary to ensure a structure containing > 80% compacted graphite. Although this method is used extensively, particularly in the USA, it suffers from the disadvantages of section sensitivity, a tendency to form spheroids in thin sections and Ti contamination of returns.

General Motors have demonstrated that compacted irons can be produced from high S iron (0.07–0.13%) using a Mg–Ce–Ca–Al alloy[70]. Alloying additions of up to 1.8% are required depending on the S content (see *Figure 1.18*). Close control over treatment temperature (1475 to 1510 °C) is necessary. The numerous sulphide particles resulting from the high S level create graphite nuclei and inoculation is not required for unalloyed irons. However, post inoculation is recommended for low C.E.V. alloyed irons. Similar studies conducted in Austria in the 1970s did not lead to a recommendation for the use of post inoculation because of the risk of slag inclusions and graphite degeneracy in larger castings[55].

The production of compacted irons by adding only Ce mischmetal to a low S iron is used extensively in Europe and is based on original research by Morrogh

and Williams[71]. The advantages of this method are that Ce can be added easily by a single ladle addition without problems from fume and flare and without the risk of scrap contaminations. Ce in amounts between 0.02 and 0.05% promotes compacted graphite in a low S iron. Knowledge of the S level is critical because at any S level there is a narrow range of Ce that produces compacted graphite. Production is facilitated if the S content is reduced below 0.01%. Excessive Ce has a powerful chilling effect. Strong final inoculation with ferrosilicon may be used to ensure carbide-free castings and such a step is usually necessary in thinner castings. An excess of Ce in the presence of inoculants will form spheroidal graphite, which itself may be undesirable. It is difficult to obtain a fully pearlitic as-cast structure with Ce. Therefore, additions of Sn, Sb or Cu may be necessary.

The effect of the rare earth element balance in the addition has been examined in an attempt to reduce chilling tendency[72,73]. Treatment of a melt with a balanced ratio of Ce and La decreases the chilling tendency, but this benefit is restricted to a narrow

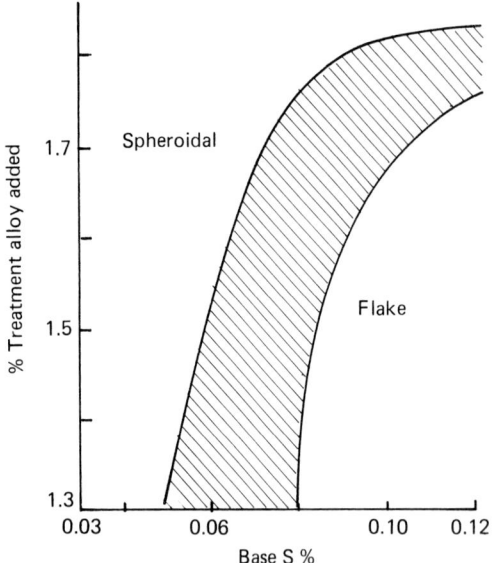

Figure 1.18 S range for high S compacted graphite iron production with a Mg–Ce–Al–Ca addition (after ref. 70)

alloy addition range. An alternative approach has been described using a Fe–Si–Ce–Ca alloy[74]. Whereas Ce is a spheroidizer and carbide-former, Ca is a spheroidizer and a graphite-promoter. This alloy addition decreases section sensitivity and the tendency to form spheroids in thin sections.

Special purpose white and alloy cast irons

These irons differ from the alloyed irons described in previous sections because they contain a greater alloy content (> 3%) and cannot be produced by making ladle additions to irons of a standard base composition. They can be divided into graphite-free and graphite-bearing alloys, as shown in *Figure 1.19* and are noted for their corrosion, elevated temperature, and wear and abrasion resistance properties.

Corrosion resistant alloys

Although unalloyed grey irons exhibit reasonable corrosion resistance, particularly in alkalies, the special duty high Ni (Ni-resist), high Ni-Si (Nicrosilal), high Si and high Cr irons offer outstanding corrosion resistance in appropriate media. The microstructures of these irons are illustrated in Chapter 6, *Figures 6.37, 6.39, 6.40* and *6.41*.

The austenitic Ni-resists are produced with flake or spheroidal graphite; the latter offer superior mechanical properties but are more expensive. The Ni (13–36%) and Cr (1.8–6%) concentrations are varied depending upon the nature of the corrosive environment. Important uses include gear pumps for the

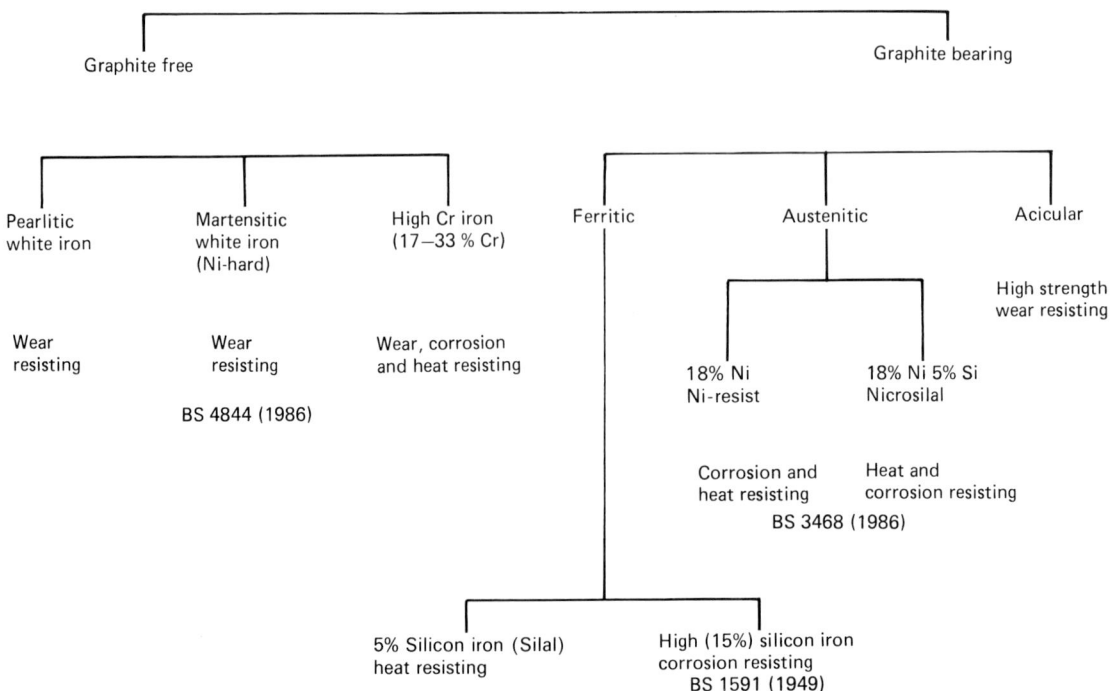

Figure 1.19 The family of special purpose white and alloy cast irons

processing and transport of sulphuric acid, pumps and valves in sea water applications, components used in steam services and in the handling of alkali, caustic and ammoniacal solutions and for the pumping and handling of sour crude oils in the petroleum industry. The spheroidal Ni-resists provide excellent resistance to cavitation erosion under severe conditions[76].

High Si irons containing between 14.20 and 14.75% Si have excellent resistance to corrosion in sulphuric and nitric acids at all temperatures and concentrations but, unlike the Ni-resists, they have the disadvantage of being brittle and can only be machined by grinding. Their corrosion resistance in hydrochloric and hydrofluoric acids is poor. Resistance to attack in hydrochloric acid can be improved by increasing the Si content to 16–18% or by adding 3–5% Cr or 3–4% Mo to the base alloy. High Cr (30%) irons are the only ones which may be considered to be stainless under atmospheric conditions. A passive film develops under oxidizing conditions and this confers resistance to nitric acid attack at all temperatures up to boiling point and all concentrations up to 70%. The alloy is machinable provided the C content is limited to 1.5%.

Heat resisting alloys

Cast irons intended for critical engineering applications at high temperatures must exhibit good high temperature mechanical properties, structural stability, oxidation resistance and resistance to growth, which is a phenomenon unique to grey cast iron that is caused by structural decomposition and internal oxidation[76-79]. *Table 1.14* shows a selection of irons used for heat resisting applications. Unalloyed grey irons give satisfactory service up to about 500 °C for applications such as fire-grates, fire-backs and grate-bars, where growth and scaling can be tolerated. Minor alloying additions such as Cr and Mo improve performance for critical engineering applications[13].

The choice of iron type and composition is usually a compromise to provide an optimum combination of properties for the particular application. Ingot moulds, cylinder heads and exhaust manifolds fall into this category[80]. Local temperatures between 800 and 900 °C are generated at the metal–mould interface during ingot casting. Temperature differences in the mould can exceed 300 °C and induce severe thermal cycling. High thermal conductivity flake grey irons (3.6–3.9% C, 1.5–2.4% Si, 0.4–0.8% Mn) minimize the thermal stresses that develop. Damage to the mould manifests itself in the form of heat checking on the inside surface followed by extensive vertical cracking. This is the result of oxidation and structural change. The ferrite-pearlite matrix is austenitized and retransformed to fine pearlite in each cycle. Oxidation occurs at the interface after stripping and is followed by decarburization and scaling.

Mould life can be extended by 20–40% by alloying with Mn. A suitable composition is 4.2–4.5% C, 0.8–1.2% Si, 1.2% Mn. The increased Mn stabilizes the matrix structure and the pearlite is spheroidized rather than ferritized. However, care is necessary in balancing the composition in order to avoid premature cracking.

An alternative approach adopted more recently is the use of compacted irons. The combination of properties indicated in *Figure 1.14* produces a stronger mould than one made of flake iron and one that is more resistant to distortion than a spheroidal iron mould. Spheroidal irons have a lower thermal conductivity, higher elastic modulus and greater tensile strength than flake irons. This means that although higher thermal gradients lead to higher induced thermal stresses, spheroidal iron moulds are more crack resistant. However, their yield and creep resistance are insufficient to resist the higher thermal stresses. Mould life is limited by bulging, particularly in larger moulds.

Similar reasoning controls the selection of irons for exhaust manifolds. Thermal stresses are induced because the manifold is bolted to the cooler engine block and is under thermal restraint as well as because of temperature differences induced by gas flow and air cooling. Structural stability and oxidation are less of a problem at temperatures below 600 °C and unalloyed or Cr-alloyed flake irons with a medium C content give satisfactory service. However, these alloys suffer from thermal cracking and oxidation with higher exhaust temperatures. This can be countered by increasing the C content to 3.5–3.6% and introducing small concentrations of Cr and Mo. Ferritic spheroidal irons have been introduced to cope with more severe conditions but distortion can lead to difficulties in remounting manifolds.

The special alloy cast irons are usually specified when service conditions are more severe. There are four main types:

1. Ni-resists which rely on Ni, in combination with Mn and Cu, to produce a stable austenitic matrix and Cr in combination with Ni to form an effective oxidation resistant scale;
2. Intermediate Si irons with a stable ferritic matrix and excellent oxidation resistance at high temperatures;
3. Al-alloyed irons which exhibit good oxidation resistance and
4. High Cr white irons in which Cr provides excellent oxidation resistance.

The Ni-resists combine corrosion resistance with heat resisting properties. American and British specifications for flake and spheroidal irons differ slightly and are given in *Tables 1.15* and *1.16*, respectively. Grade 1 Ni-resist is not available with spheroidal graphite because the high Cu content, which supple-

Table 1.14 Cast irons suitable for heat-resisting applications

Type	Typical service conditions	Total C %	Si %	Mn %	S %	P %	Ni %	Cr %	Mo %	Room T.S. $N\,mm^{-2}$	Temp. Prop. BH	Coefficient of expansion per °C × 10⁶	Stress to rupture In 10000 h 350°C	500°C	700°C
High Phosphorus grey iron	cheap: moderately scale resistant to 600°C suitable for fire bars.	3.0 3.3	2.2 3.0	0.4 0.7	<0.12	0.9 1.4				154 185	200 240	13 upto 650°C	154 232	46	2 3
Low alloy grey iron	used 400–700°C: dimensional stability and oxidation resistance better than unalloyed iron	3.0 3.4	1.6 2.8	0.4 1.0	<0.12	0.1 0.4		0.5 2.0	⩽0.5	293 386	200–260 Grey 350–450 White	11–12 air temperature 13–14 up to 650°C	232 309	46 62	2 3
Silal	Cupola melted: cheapest oxidation and growth resistant iron: up to 800°C H.F. melted: low C, high growth and oxidation resistance up to 850–900°C	up to 2.3 / 1.8 2.0	5.5 7.0 / 5.5 7.0	0.5 0.8 / 0.5 0.8	0.06 / 0.06	0.1 0.3 / 0.1 0.3				139 263 (21.875 cm diameter bar)	220 255	13 up to 500°C		31	2 3
Nicrosilal	austenitic, 650–900°C	1.5 2.0	4.5 5.0	0.6 1.0	0.1	<0.1	18 23	2.0 2.4		139 247	150 200	18	108 185	62 93	<15
Ni-resist austenitic	Very resistant 800–850°C stable 500–600°C	2.4 3.1	1.0 2.0	0.8 1.4	<0.1	0.1 1.0	12 17	2.6 3.6	Cu 5.5–8.0	124 247	120 170	18	108 185	62 93	<15
Ni-resist austenitic	can be used at temp > 800°C	1.7 2.0	4.9 5.4	0.4 0.7	<0.1	<0.1	34 36	1.8 2.0		485		stabilized at 950°C			17
Chromium iron ferritic	Scale resistant to 950–1000°C	2.0 2.8 / 1.0 1.3	1.4 1.8 / 1.0 1.3	1.0 1.6 / 0.7 1.0	0.1 / —	<0.1 / —	— / —	14.0 17.0 / 30.0 33.0		309 541 / 371 541	250 480	9–10 / 13	309 386	124 154	31 46

Table 1.15 Summary of USA specifications for flake and spheroidal high alloy cast irons

Flake Irons

Specification	Grade	Total C	Si	Mn	Ni	Cr	Cu	S	P		Uses
ASTM	1		1.0	0.5	13.5	1.5	5.5			Minimum	valve guides, insecticide
A436–78		3.0	2.0	1.5	17.5	2.5	7.5	0.12		Maximum	pumps, flood gates
	1b		1.0	0.5	13.5	2.5	5.5			Minimum	sea water valve and
		3.0	2.0	1.5	17.5	3.5	7.5	0.12		Maximum	pump bodies
	2		1.0	0.5	18.0	1.5				Minimum	pump impellers, pump
		3.0	2.0	1.5	22.0	2.5	0.5	0.12		Maximum	casings, plug valves
	2b		1.0	0.5	18.0	3.0				Minimum	caustic pump casings
		3.0	2.0	1.5	22.0	6.0	0.5	0.12		Maximum	valves
	3		1.0	0.5	28.0	2.5				Minimum	turbocharger housings
		2.0	2.0	1.5	32.0	3.5	0.5	0.12		Maximum	stove tops, caustic pumps
	4		5.0	0.5	29.0	4.5				Minimum	range tops
		2.0	6.0	1.5	32.0	5.5	0.5	0.12		Maximum	
	5		1.0	0.5	34.0					Minimum	glass rolls and moulds
		2.4	2.0	1.5	36.0	0.1	0.5	0.12		Maximum	machine tools, gauges
	6		1.5	0.5	18.0	1.0	3.5		Mo	Minimum	
		3.0	2.5	1.5	22.0	2.0	5.5	0.12	1.0 Mo	Maximum	valves
A518–80		0.7	14.2						Mo	Minimum	pumps and piping for
		1.1	14.75	1.5		0.5	0.5		0.5	Maximum	corrosive liquids

		Tensile strength N mm^{-2} minimum	B.H.	Impact J 30 mm unnotched	Expansion per °C × 10^6	Thermal conductivity W m^{-1} k^{-1}
A436–78	1	172	131–183	136	19.3	41
	1b	206	149–212	108	19.3	41
	2	172	118–174	136	18.7	41
	2b	206	171–248	81	18.7	41
	3	172	118–159	203	12.4	40.2
	4	172	149–212	108	14.6	38.8
	5	206	99–124	203	5.0	40.2
	6	138	124–174			

		Total C	Si	Mn	P	Ni	Cr		Uses
ASTM	D2		1.5	0.7		18.0	1.75	Minimum	valve components in petroleum,
A439–80		3.0	3.0	1.25	0.08	22.0	2.75	Maximum	salt water and caustic
	D2-B		1.5	0.7		18.0	2.75	Minimum	turbocharger housings
		3.0	3.0	1.25	0.08	22.0	4.0	Maximum	
	D2-C		1.0	1.0		21.0		Minimum	electrode guide rings, steam
		2.9	3.0	2.4	0.08	24.0	0.50	Maximum	turbine dubbing rings
	D3		1.0			28.0	2.5	Minimum	turbocharger nozzle and
		2.6	2.0	1.0	0.08	32.0	3.5	Maximum	housings, turbine diaphrams
	D3-A		1.0			28.0	1.0	Minimum	high temperature bearing rings
		2.6	2.8	1.0	0.08	32.0	1.5	Maximum	
	D4		5.0			28.0	4.5	Minimum	diesel engine manifolds
		2.6	6.0	1.0	0.08	32.0	5.5	Maximum	
	D5		1.0			34.0		Minimum	gas turbine shroud rings,
		2.4	2.8	1.0	0.08	36.0	0.1	Maximum	glass rolls
	D5-B		1.0			34.0	2.0	Minimum	optical system parts for
		2.4	2.8	1.0	0.08	36.0	3.0	Maximum	dimensional stability
	D5-5	1.7	4.9	0.4		34.0	1.8	Minimum	
		2.0	5.4	0.7	0.08	36.0	2.0	Maximum	
A571–71	D2-M	2.2	1.5	3.75		21.0		Minimum	pressurized components matrix
		2.7	2.5	4.5	0.08	24.0	0.2	Maximum	stable to −230 °C
US Military	MIL + 24137	2.4	1.8	0.8		18.0	1.7	Minimum	shipboard use and
		3.0	3.2	1.5	0.20	22.0	2.4	Maximum	propellers; heat, corrosion
		2.7	2.0	1.9		20.0		Minimum	shock resistance and
		3.1	3.6	2.5	0.15	23.0	0.5	Maximum	non magnetic.

Table 1.15 Continued

		Tensile strength (min) $N\,mm^{-2}$	B.H.	Elongation %	Impact J
ASTM	D2	400	139–202	8	16.3
A439–80	D2-B	400	148–211	7	13.6
	D2-C	400	121–171	20	40.0
	D3	380	139–202	6	9.5
	D3A	380	131–193	10	19.0
	D4	414	202–273	–	–
	D5	380	131–185	20	23.0
	D5-B	380	139–193	6	8.1
	D2-M	440	121–171	30	–

ments the Ni content and provides a high level of corrosion resistance interferes with spheroidization. A major use of this alloy is for piston ring inserts. Its coefficient of expansion is very similar to that of the Al alloy used for pistons. The harder, more wear resistant Ni-resist inserts withstand the constant rubbing and pounding at the elevated temperatures far better than the Al alloy. Grade 2 is the most commonly used alloy in corrosive environments and for heat and oxidation resistance up to 700 °C. Grade 3 offers good thermal shock resistance and erosion resistance in wet steam and corrosive slime. Grade 4 has higher Si and Cr contents and is superior in its resistance to erosive, corrosive and oxidizing atmospheres.

Grade 5 has minimum thermal expansion and good dimensional stability. It displays superior shock resistance up to 430 °C. The Si and Cr contents are adjusted to suit the application. For example, type D 4 contains between 5 and 6% Si and 4.5 to 5.5% Cr. It shows good oxidation resistance but relatively poor mechanical properties and, because of carbide present in the microstructure, is not structurally stable under thermal cycling conditions. On the other hand, grade D 5B does not suffer these limitations. However, it does not have adequate resistance to oxidation, growth and scaling at the highest temperatures. Consequently, a new alloy grade D 5S, based on D 5B but with an increased Si content has been introduced recently. This alloy exhibits oxidation resistance similar to that of a 25% Cr–20% Ni heat resisting cast steel. It also has good elevated temperature mechanical properties and a high resistance to thermal shock.

It is suitable for applications such as turbo-charger casings, manifolds, hot forming dies and jet engine components with operational temperatures exceeding 850 °C[16]. However, Cox has warned of a possible instability of the high Si austenite matrix at intermediate temperatures[81]. The low Mn and Cr grade 2 spheroidal irons are prone to cracking in the heat-affected zone on welding. This has been attributed to a decrease in the ductility at welding temperatures and has been overcome[82] by limiting P to 0.025% and controlling the Mg and Si contents according to

Si % + 75 Mg % $\leqslant$ 7.5.

A new iron, designated D 2W, which is less prone to cracking and which can be repair welded has been

introduced into some standards. When applications require low creep rates whilst sustaining high stress levels at higher temperatures, the standard grades of Ni-resist are alloyed with up to 1% Mo. The effect of a 1% Mo addition on stress rupture properties is shown in *Figure 1.20*. A further improvement in properties can be gained by adding 0.015% B to the Mo alloyed iron.

Flake and spheroidal ferritic irons containing 4 to 6% Si offer suitable alternatives for many high temperature applications at relatively low cost. Si increases oxidation resistance continuously up to 6%, first of all by forming a resistant oxide film and then a silicate film which provides resistance to the transport of O atoms into the metal and to the diffusion of metal atoms towards the surface. Si raises the critical temperature at which ferrite transforms to austenite ($\sim$970 °C with 6% Si) thereby expanding the useful temperature range of application with respect to resistance to growth. Si also strengthens ferrite, but in amounts exceeding 5% it reduces toughness and increases the brittle ductile transition to above room temperature. Hence, the Si content is controlled within the range 4–6% to optimize properties. As the Si content is increased from 4 to 5%, the C content is decreased from 3.8 to 2.9% in order to retain a hypoeutectic composition and avoid primary graphite formation.

The flake iron Silal was one of the earliest heat resisting alloys developed[83]. The advantages of a high critical temperature, a ferritic matrix and a fine undercooled type D graphite provide excellent growth and scaling resistance. Silal is machinable, but is brittle at room temperature. However, at temperatures exceeding 260 °C it has a toughness comparable to other cast irons.

The spheroidal grades of the 4–6% Si irons are popular because the higher strength and greater ductility qualify them for more rigorous service. They are predominantly ferritic as-cast but, on occasions, contain small quantities of eutectic carbide and pearlite. Annealing is often used to minimize any adverse effect of these constituents on mechanical properties. Some manufacturers prefer to use a sub-critical anneal at 790 °C which develops superior high temperature properties with only a small sacrifice of ductility. Si irons can be alloyed with 0.5 to 2.0% Mo to combine adequate room temperature strength and

Table 1.16 British specification, BS 3468 (1986) for flake and spheroidal high alloy cast irons

Flake Austenitic

Grade	C%	Si%	Mn%	Ni%	Cu%	Cr%	Nb%	P%	Mg%	
F1		1.5	0.5	13.5	5.5	1.0	–	–	–	minimum
	3.0	2.8	1.5	17.5	7.5	2.5	–	0.2	–	maximum
F2		1.5	0.5	18.0		1.5	–	–	–	minimum
	3.0	2.8	1.5	22.0	0.5	2.5	–	0.2	–	maximum
F3		1.5	0.5	28.0		2.5	–			minimum
	2.5	2.8	1.5	32.0	0.5	3.5	–	0.2		maximum

Grade	Tensile strength ($N\,mm^{-2}$)	B.H.	Elongation %	Thermal Conductivity $W\,m^{-1}\,K^{-1}$
F1	170–240	140–220	1–2	38–42
F2	170–240	140–220	1–3	38–42
F3	190–250	150–230	1–2	38–42

Spheroidal Austenitic

Grade	C%	Si%	Mn%	Ni%	Cu%	Cr%	Nb%	P%	Mg%	
S2		1.5	0.5	18		1.5	–			minimum
	3.0	2.8	1.5	22	0.5	2.5	–	0.08	–	maximum
S2B		1.5	0.5	18		2.5	–			minimum
	3.0	2.8	1.5	22	0.5	3.5	–	0.08	–	maximum
S2C		1.5	1.5	21			–			minimum
	3.0	2.8	2.5	24	0.5	0.5	–	0.08	–	maximum
S2M		1.5	4.0	21						minimum
	3.0	2.5	4.5	24	0.5	0.2	–	0.08	–	maximum
S2W		1.5	0.5	18		1.5	0.12			minimum
	3.0	2.2	1.5	22	0.5	2.2	0.20	0.05	0.06	maximum
S3		1.5	0.5	28		2.5	–			minimum
	2.5	2.8	1.5	32	0.5	3.5	–	0.08	–	maximum
S5S		4.8		34		1.5	–			minimum
	2.2	5.4	1.0	36	0.5	2.5	–	0.08	–	maximum
S6		1.5	6	12						minimum
	3.0	2.8	7	14	0.5	0.2		0.08	–	maximum

Grade	Tensile strength ($N\,mm^{-2}$)	B.H.	Elongation %	Impact J (10 mm specimen) minimum
S2	370–490	140–230	7–20	–
S2B	370–490	140–230	7–20	4
S2C	370 minimum	140–220	20 minimum	20
S2M	420 minimum	160–250	25 minimum	15
S2W	370–490	140–200	7–20	–
S3	370–490	140–230	7–20	–
S5S	370–470	130–180	7–20	–
S6	390 minimum	120–200	15 minimum	–

Matching specifications

BS 3468			
	F1	A436	1 and 1B
	F2		2
	F3		3
	S2	A439	D2
	S2B		D-2B
	S2C		D-2C
	S2M	A571	D-2M
	S2W	–	–
	S3	A439	D-3
	S5S		D-5S
	S6	–	–

S2W special welding grade
S2M room temp. mechanical properties retained at low temperatures

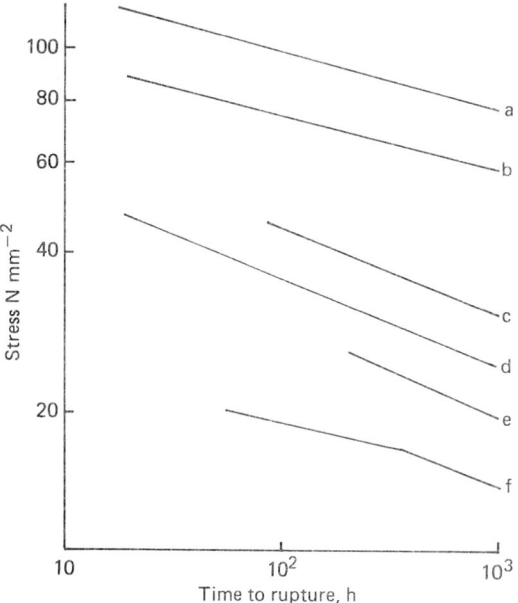

Figure 1.20 Stress rupture data at 700 °C for: a, type D-2 Ni-resist with 1% Mo; b, type D-2 Ni-resist; c, 4% Si spheroidal iron with 2.5% Mo; d, 4% Si spheroidal iron with 1.0% Mo; e, 4% Si spheroidal iron with 0.5% Mo; f, 4% Si spheroidal iron. The 4% Si irons were subcritically annealed at 850 °C (after ref. 85 and 86)

ductility with good high temperature strength and oxidation resistance. The effect of Mo on tensile properties is illustrated in *Figure 1.21*[84]. *Figure 1.20* shows that creep rupture strength increases continuously with increasing Mo addition but the greatest increase occurs with 0.5 and 1.0% additions[86,87].

Although the mechanical properties of alloyed Si

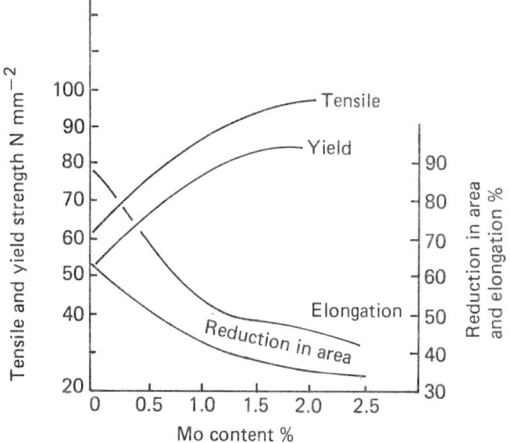

Figure 1.21 The effect of Mo on the tensile properties of 4% Si spheroidal iron at 705 °C after annealing at 790 °C (after ref. 84)

irons are inferior to those of alloyed Ni-resists, their lower cost, low creep rate in the temperature range 700–820 °C and good resistance to oxidation make them suitable for various applications below their critical temperature. The 4% Si–0.6% Mo alloy is used in turbo-charger castings and the 1% Mo alloy is used in exhaust manifolds, gas turbine components, glass moulds and industrial furnace parts.

Al alloyed cast irons of commercial interest fall into two categories, alloys containing up to 6% Al and those containing between 18 and 25% Al. Al increases the graphitization potential in both of these composition ranges such that grey iron forms on solidification. Alloys are produced with flake[87] compacted[88,89] and spheroidal[90] graphite. Al largely replaces Si in the low concentration alloys. The idea of substituting Al for Si was first conceived by DeFrancq *et al.*[91].

Advantages to be gained include higher tensile strength, better resistance to thermal shock, high graphitizing tendency and low chilling tendency, which enables thinner sections to be cast. Al stabilizes pearlite during the eutectoid transformation. The low Al irons exhibit good scaling resistance coupled with good machinability. Suggested uses[92] include exhaust manifolds, turbo-charger housings, disc brake rotors, brake drums, cylinder liners, camshafts and piston rings. Al is present in addition to the normal Si content in the higher concentration range and these alloys exhibit good mechanical properties with superior scaling resistance at high temperatures. These characteristics have been described by Petitbon and Wallace[93] and Yaker *et al.*[94]. The influence of Al on oxidation resistance is indicated in *Figure 1.22*. The influence of Al on the tensile properties of flake and spheroidal irons is given in *Table 1.17* and stress rupture curves are shown in *Figure 1.23*.

The elevated temperature properties of Al alloyed irons can be improved further by alloying with Mo. Dickenson[90] has shown the improvement obtained in 5% Al flake and spheroidal irons by adding 2% Mo (see *Table 1.18*). Sponseller *et al.*[95] have shown that a spheroidal iron alloyed with 4% Si, 0.5–1.0% Al and 2% Mo is suitable for applications in the temperature range 650–820 °C that require good oxidation resistance, good structural stability and high strength at elevated temperatures. These alloys do not suffer from some of the casting difficulties experienced with 5% Al alloys. Spheroidal irons containing 5.5% Al, 3.0% Si and 2.0% Mo have been suggested for applications at temperatures above 820 °C. A considerable reduction in oxidation resistance at 925 °C was reported for alloys containing only 2.0% Si.

Al alloyed irons have not been exploited fully because of difficulties encountered during casting. The formation of dross can be a severe problem but clean castings can be obtained[89,96] by flushing the system with inert gas prior to pouring. The careful use of slag traps and gating systems, which prevent turbulence in

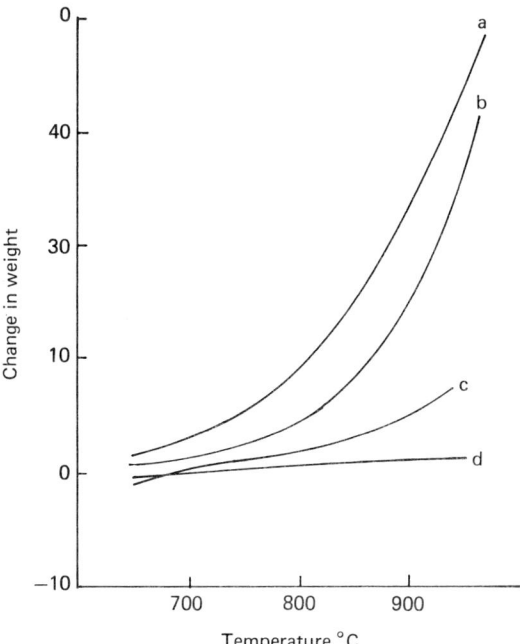

Figure 1.22 The effect of Al content on the oxidation of grey irons in air at various temperatures for 200 h; a, 2.47% Al; b, 4.28% Al; c, 5.99% Al; d, 20.79% Al (after ref. 93)

the mould and so avoid the incorporation of oxide films into the casting, also help obtain clean castings. The alloys can be susceptible to gas porosity. Dissolved Al can react with moisture or hydrocarbons in the mould to produce free hydrogen which dissolves in the molten metal and causes pinholing on solidification.

High Cr white irons offer excellent resistance to growth and oxidation at elevated temperatures and are cost effective alternatives to stainless steel in applications that are not subjected to severe impact loading. The irons fall into one of three categories;

1. martensitic irons with 12 to 28% Cr;
2. ferritic irons with 30 to 34% Cr or
3. austenitic irons with 15 to 30% Cr and 10 to 15% Ni to stabilize the austenitic matrix to low temperatures.

The classification depends on the service temperature, stressing, required life and economic factors. The beneficial effect of Cr on scaling resistance is evident from *Figure 1.24*[97]. Applications include recuperator tubes, breaker bars and trays for sinter furnaces, various furnace parts, glass bottle moulds and valve seats.

Wear resistant alloys

Cast irons are used in many wear resistant applications. Grey irons are used under lubricated and dry

Table 1.17 The influence of Al on the mechanical properties of flake and spheroidal irons (after ref. 94)

Alloy	U.T.S. $N\,mm^{-2}$	0.2% P.S. $N\,mm^{-2}$	Elongation %	Temperature °C
3.18% C, 1.93% Si, 0.95% Mn	256	235	0.5	27
flake	72	69	0.7	649
	28	27	1.6	760
2.53% C, 1.60% Si, 0.45% Mn, 5.9% Al	174	159	0.4	27
flake	54	54	2.5	649
	25	25	5.4	760
	19	18	3.5	871
	13	12	3.1	982
2.1% C, 1.06% Si, 0.40% Mn, 20.5% Al	220	220	0.1	27
flake	51	47	31.2	871
	25	23	> 14.8	982
3.5% C, 1.93% Si, 0.23% Mn	401	241	18.9	27
spheroidal	54	48	33.2	649
	25	22	43.5	760
	30	28	30.0	871
3.12% C, 2.1% Si, 0.02% Mn, 1.25% Al	383	339	2.7	27
spheroidal	77	78	12.1	649
	34	34	30.5	760
	26	25	28.5	871
3.32% C, 2.11% Si, 0.03% Mn, 3.8% Al	563	550	0.3	27
spheroidal	112	109	18.8	649
	43	39	28.8	760
	19	17	48.5	871
	22	19	20.2	982

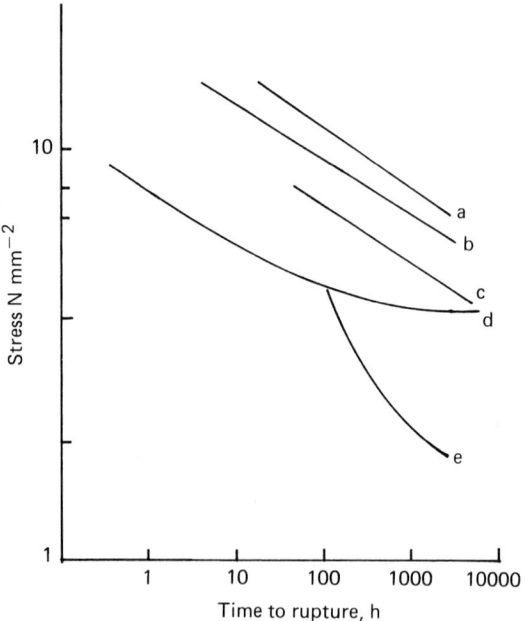

Figure 1.23 Comparison of the stress rupture curves of unalloyed and Al alloyed spheroidal iron at 650 °C; a, 3.8% Al spheroidal iron; b, 1.25% Al spheroidal iron; c, unalloyed spheroidal iron; d, 5.9% Al grey iron; e, unalloyed grey iron (after ref. 94)

sliding conditions. In the former case, movement can be continuous as in bearings or reciprocating as with pistons and cylinders. Braking systems and clutch plates operate under dry sliding conditions. During the running-in period under lubricated sliding, graphite provides a renewable solid lubricant between mating surfaces and subsequently, flake cavities provide reservoirs for oil and act as termination points for local damage by providing sinks for abrasive material that would otherwise be trapped between the sliding surfaces. The cast iron structure is of little consequence if lubrication is complete. However, with marginal or poor lubrication, wear resistance is improved if the iron is fully pearlitic with moderately coarse random flake graphite (type A) and dispersed, hard phosphide or carbide particles.

Good thermal conductivity, ability to resist galling, good frictional properties, reasonable strength, low modulus of elasticity and a facility for renewing graphitic areas make grey flake irons resistant to the development of high local temperatures during dry sliding as encountered in braking systems[98]. Such systems do not fail usually as a consequence of directly applied stress but as a result of wear or heat checking. The wear resistance of brake discs is improved appreciably by the addition of small quantities of Ti and V which produce a fine dispersion of carbides in the matrix[99]. Heat checking occurs as a result of internal oxidation and structural change as already described for ingot moulds. Frequent but not particularly fierce braking can increase the temperature of the iron sufficiently for structural deterioration to occur. This can be resolved by improving design. Additions of Cr and Mo will increase pearlite stability and the strength of the

Table 1.18 The influence of Al and Mo on the high temperature mechanical properties of flake and spheroidal irons (after ref. 90)

Alloy C%	Si%	Mn%	Mo%	Al%		U.T.S. N mm^{-2}	Temperature °C	Stress to Rupture 100 h N mm^{-2}	1000 h N mm^{-2}
3.0	1.57	0.64			flake	193	540		
						102	650	23	15
						60	760		
2.75	1.67	0.7	1.9	5.21	flake	271	540		
						155	650	37	26
						79	760		
						63	870		
3.7	2.36	0.3			spheroidal	202	540		
						105	650	35	25
						60	760		
						72	870		
3.43	2.06	0.3	2.2	6.35	spheroidal	433	540		
						228	650		
						122	760		
						77	870		
3.11	3.54	0.26	1.96	6.27	spheroidal		650	138	97

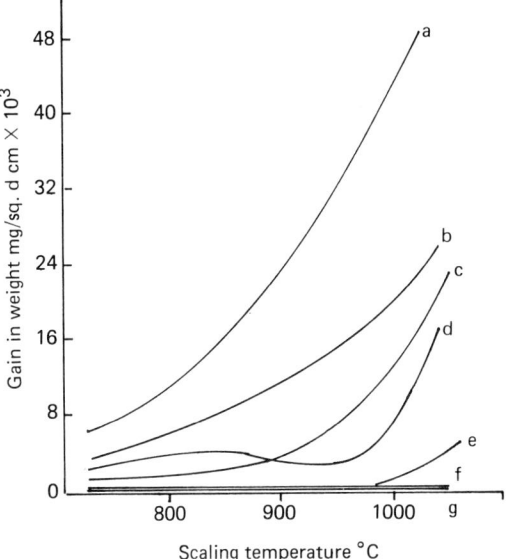

Figure 1.24 Oxidation behaviour of several irons at temperature for 200 h in air; a, 1% Cr + 1% Ni; b, Ni-resist; c, Silal; d, Nicrosilal; e, 17% Cr; f, 33% Cr; g, 30% Cr + 1.0% Ni (after ref. 97)

matrix. Heat checking may occur also as a result of the sudden application of braking loads which cause the surface temperature of the iron to increase above the critical temperature. Ferritic spheroidal irons have been used under these conditions as they are more resistant to thermal shock.

White irons are usually used to combat abrasive wear. Carbides in the microstructure, depending on their type, morphology and volume fraction, provide the hardness required for crushing materials without degradation. The supporting matrix structure can be controlled by alloy content and/or heat treatment to develop pearlitic, austenitic or martensitic structures to provide the most cost effective balance between abrasive wear resistance and toughness.

Three basic types of abrasive wear have been identified[100], namely, gouging, which is usually associated with impact, high stress or grinding abrasion and low stress or scratching abrasion. Macroscopic penetration of the working surface by coarse abrasive particles accompanies gouging, as for example, when hammer bars crush material in an impact pulverizer. Grinding abrasion occurs when abrasive particles are crushed under the grinding influence of moving metal surfaces. Scratching abrasion occurs when freely moving particles impinge on a wearing surface, as in shot blasting. Impact forces are small in scratching abrasion and cutting wear prevails for small impingement angles and deformation wear for large angles.

The nature of abrasive wear depends on many factors including microstructure of the abraded surface, type of abrasive, relative movement and type of loading, chemical action and temperature. Consequently, the relationship between microstructure and wear resistance is complex and care must be taken to ensure that any laboratory wear test reproduces the service conditions. For example, laboratory tests have underestimated the abrasion resistance of austenitic irons[101]. They rarely reproduce the continuous impact loading experienced in service and the full work hardening ability of the matrix is not developed during the laboratory test.

The complex dependence of wear on microstructure is illustrated in a recent study of the grinding abrasion resistance of Cr irons in the as-cast and heat treated conditions[102]. Martensitic matrix structures displayed higher abrasion resistance than austenitic or pearlitic structures. The 15% Cr–3% Mo martensitic iron displayed the highest resistance. The 10–15% Cr irons were more resistant than 25–30% Cr irons due to the lower C martensite produced after destabilization of the higher Cr iron. The 5% Cr irons exhibited a greater wear resistance than the 10–15% Cr irons. This was because cracking in the carbide phase did not lead to immediate removal from the structure with a 'continuous' carbide as in the 5% irons. The fractured carbide rods in the 'discontinuous' eutectic in the 15% irons became detached readily during wear.

These findings differ from observations made when impact loading accompanies wear. Under these conditions, structures with continuous carbides have insufficient toughness and suffer gross fracture. Structures with the discontinuous eutectic can withstand impact and gross failure is avoided at the expense of greater wear, particularly if the carbide is not adequately supported by a hard matrix.

Three types of cast iron are used for wear resistance. Specifications are given in *Tables 1.19* and *1.20*. The oldest class contains the unalloyed and low alloy grades which display a structure of massive 'continuous' M_3C carbides in a pearlitic matrix. Although still used, these extremely brittle alloys have been largely superseded by the tougher, alloyed white irons.

The first alloyed irons to be developed were the Ni-hards. These relatively cheap, cupola-melted martensitic white irons contained Ni to increase hardenability to ensure austenite transforms almost completely to martensite following heat treatment. They also contained Cr to increase the hardness of the eutectic carbide. A typical microstructure is shown in Chapter 6, *Figure 6.36*.

The optimum alloy composition depends on casting thickness and intended use. If abrasion resistance is the prime consideration, the C content is maintained in the range 3.2–3.6%. However, if impact loading is of concern, it is usual to restrict the C to the range 2.7–3.2%. The latest alloy in this series is Ni-hard IV which corresponds to Class 1D in *Table 1.19*

Table 1.19 Composition and mechanical properties of abrasion resistant irons according to ASTM specification A532

Class	Type	Designation	C%	Mn%	Si%	Ni%	Cr%	Mo%	P%	S%	Cu%	
I	A	Ni–Cr–HC	3.0			3.3	1.4					minimum
			3.6	1.3	0.8	5.0	4.0	1.0	0.3	0.15		maximum
I	B	Ni–Cr–LC	2.5			3.3	1.4					minimum
			3.0	1.3	0.8	5.0	4.0	1.0	0.3	0.15		maximum
I	C	Ni–Cr–GB	2.9			2.7	1.1					minimum
			3.7	1.3	0.8	4.0	1.5	1.0	0.3	0.15		maximum
I	D	Ni–HiCr	2.5		1.0	5.0	7.0					minimum
			3.6	1.3	2.2	7.0	11.0	1.0	0.10	0.15		maximum
II	A	12% Cr	2.4	0.5			11.0	0.5				minimum
			2.8	1.5	1.0	0.5	14.0	1.0	0.10	0.06	1.2	maximum
II	B	15% Cr–Mo–LC	2.4	0.5			14.0	1.0				minimum
			2.8	1.5	1.0	0.5	18.0	3.0	0.10	0.06	1.2	maximum
II	C	15% Cr–Mo–HC	2.8	0.5			14.0	2.3				minimum
			3.6	1.5	1.0	0.5	18.0	3.5	0.10	0.06	1.2	maximum
II	D	20% Cr–Mo–LC	2.0	0.5			18.0					minimum
			2.6	1.5	1.0	1.5	23.0	1.5	0.10	0.06	1.2	maximum
II	E	20% Cr–Mo–HC	2.6	0.5			18.0	1.0				minimum
			3.2	1.5	1.0	1.5	23.0	2.0	0.10	0.06	1.2	maximum
III	A	25% Cr	2.3	0.5			23.0					minimum
			3.0	1.5	1.0	1.5	28.0	1.5	0.10	0.06	1.2	maximum

Class	Type	Designation	Brinell hardness Sand cast (minimum)	Chill cast (minimum)	Hardened (minimum)	Softened (minimum)	Typical section thickness (mm)
I	A	Ni–Cr–HC	550	600			200
I	B	Ni–Cr–LC	550	600			200
I	C	Ni–Cr–GB	550	600			75 diameter ball
I	D	Ni–HiCr	550	500	600	400	300
II	A	12% Cr	550		600	400	25 diameter ball
II	B	15% Cr–Mo–LC	450		600	400	100
II	C	15% Cr–Mo–HC	550		600	400	75
II	D	20% Cr–Mo–LC	450		600	400	200
II	E	20% Cr–Mo–HC	450		600	400	300
III	A	25% Cr	450		600	400	200

and Grade 2D in *Table 1.20*. This alloy is considerably tougher than the earlier Ni-hards. Until recently, the increased toughness of Ni-hard IV and other high Cr irons was attributed to the 'discontinuous' carbide. It was generally thought that the fracture path passed through the 'continuous' carbide but with the 'discontinuous' M_7C_3 carbide this mode of fracture is less likely and results in increased toughness. However, both eutectic phases are continuous[103] and their different mechanical properties are related to the different types of anomalous eutectic structure[38].

Unalloyed irons and the low alloy Ni-hards display a quasi-regular eutectic structure with massive continuous M_3C carbides forming the major phase in the eutectic. Ni-hard IV and higher Cr irons display a eutectic structure consisting of a mixture of blades and hollow faceted rods and belong to the broken lamella class of eutectics. This structure is composite-like in nature and displays a greater toughness than the unmodified quasi-regular structure. The mechanical properties are improved by increasing the cooling rate[104,105]. Examples of this occur in hard facing operations when the proportion of rods in the structure increases or when Mo is added, as in high Cr–Mo irons, when a cellular eutectic morphology forms with rods in the centre of the cells and blades at the periphery. The faceted M_7C_3 rods are thermally stable and the morphology is not easily changed by heat treatment.

The possibility of impurity modification of the anomalous eutectic to produce a non-faceted fibrous structure as in the Al–Si eutectic or a discontinuous eutectic structure has not been explored. Ni-hards are usually stress relieved at 200–230 °C for at least 4 h prior to service in order to relieve martensitic transformation stresses and to promote the transformation

Table 1.20 British specification, BS 4844 (1986), for abrasion-resistant cast irons

Grade	C%	Mn%	Si%	Ni%	Cr%	Mo%	P%	S%	Cu%	
1A	2.4	0.2	0.5							minimum
	3.4	0.8	1.5		2.0					maximum
1B	2.4	0.2	0.5							minimum
	3.4	0.8	1.5		2.0					maximum
1C	2.4	0.2	0.5							minimum
	3.0	0.8	1.5		2.0					maximum
2A	2.7	0.2	0.3	3.0	1.5					minimum
	3.2	0.8	0.8	5.5	2.5	0.5				maximum
2B	3.2	0.2	0.3	3.0	1.5					minimum
	3.6	0.8	0.8	5.5	2.5	0.5				maximum
2C	2.4	0.2	1.5	4.0	8.0					minimum
	2.8	0.8	2.2	6.0	10.0	0.5				maximum
2D	2.8	0.2	1.5	4.0	8.0					minimum
	3.2	0.8	2.2	6.0	10.0	0.5				maximum
2E	3.2	0.2	1.5	4.0	8.0					minimum
	3.6	0.8	2.2	6.0	10.0	0.5				maximum
3A	1.8	0.5			14.0					minimum
	3.0	1.5	1.0	2.0	17.0	2.5	0.1	0.1	2.0	maximum
3B	3.0	0.5			14.0					minimum
	3.6	1.5	1.0	2.0	17.0	3.0	0.1	0.1	2.0	maximum
3C	1.8	0.5			17.0					minimum
	3.0	1.5	1.0	2.0	22.0	3.0	0.1	0.1	2.0	maximum
3D	2.0	0.5			22.0					minimum
	2.8	1.5	1.0	2.0	28.0	1.5	0.1	0.1	2.0	maximum
3E	2.8	0.5			22.0					minimum
	3.5	1.5	1.0	2.0	28.0	1.5	0.1	0.1	2.0	maximum
3F	2.0	0.5			11.0					minimum
	2.7	1.5	1.0	2.0	13.0	2.5	0.1	0.1	2.0	maximum
3G	2.7	0.5			11.0					minimum
	3.4	1.5	1.0	2.0	13.0	3.0	0.1	0.1	2.0	maximum

Grade	Hardness (minimum)	
	Thickness ≤ *50 mm*	*Thickness* > *50 mm*
1A	400	350
1B	400	350
1C	250	200
	Thickness ≤ *125 mm*	*Thickness* > *125 mm*
2A	500	450
2B	550	500
2C	500	450
2D	550	500
2E	600	530
	All thicknesses	
3A	600	
3B	650	
3C	600	
3D	600	
3E	600	
3F	600	
3G	650	

of retained austenite. This treatment improves strength and impact resistance.

The straight high Cr irons, Class III in *Table 1.19*, must have a high C content and a martensitic matrix for good wear resistance. Although these irons are cheaper than Ni-hards, limited hardenability restricts their use to moderate to thin section castings such as pump volutes and slurry pump impellers. The matrix structure is predominantly pearlitic in heavy sections and austenitic in light sections. The wear resistance of the austenitic irons is greater than that of the pearlitic irons and increases during service as a result of surface work hardening to martensite. The lower the Cr to C ratio, the greater is the tendency for pearlite formation and if pearlite is to be avoided in a 25 mm section, the C content must be restricted to 2.8%. The limitations of this class of white irons can be overcome by further alloying.

Class III of BS 4844 and Class II of ASTM A532 define the range of high Cr–Mo white irons. The recent upsurge in demand for large size equipment in the mining, coal and mineral processing industries has led to a search for alloys that combine abrasion resistance and fracture toughness[106,107]. This has led to considerable developments in the metallurgy and founding of these alloys[108-118]. The alloys are mainly hypoeutectic in composition. The predominant eutectic carbide is M_7C_3 and the matrix can be austenitic, martensitic or pearlitic according to the application. Some components are cast pearlitic to aid machining and then heat treated to produce a martensitic matrix for abrasion resistance.

Heat treated heavy castings are replacing martensitic Ni–Cr alloys, which are too brittle, and steel castings, which have good toughness but lack abrasion resistance. Chemical composition and heat treatment must be adjusted to suit section thickness and to promote the transformation of austenite to martensite and at the same time ensure sufficient hardenability to prevent pearlite formation, which reduces abrasion resistance dramatically. A martensitic matrix with secondary carbides exhibits good abrasion resistance and is normally obtained by heat treatment. The high alloy content renders the transformation behaviour more like that of a high speed tool steel rather than that of a low alloy steel or cast iron. The austenite formed on solidification is saturated with C, Cr and other alloying elements and is very stable.

Heat treatment involves a destabilizing anneal between 900 and 1000 °C prior to quenching during which secondary carbides precipitate. Quenching is done in air to avoid the risk of cracking. The composition of the iron is adjusted to permit air hardening in the section size and within the possible quenching rates for the particular casting.

The range of C and Cr compositions is defined by the need to form M_7C_3 carbide on solidification. Increasing the C content increases the volume fraction of carbide and the abrasion resistance but decreases the toughness and hardenability. High C contents require close control over cooling rate in order to avoid cracking. Alloys with the best abrasion resistance contain between 12 and 22% Cr. Lower values lead to the formation of M_3C carbide with lower abrasion resistance and toughness. When the Cr content exceeds 22%, most of the C partitions to the

carbide and the low C martensite has lower abrasion resistance.

The elements Mo, Mn, Ni and Cu promote hardenability. However, both underalloying and overalloying are undesirable. Cases of poor abrasion resistance have been traced to excessive Si contents which produce an iron of abnormally low hardenability. Hence Si levels are often restricted to 0.6%. However, if higher Mn contents are used, the Si level can be raised to 1%.

Heat treatment is necessary with heavy castings to obtain satisfactory performance under severe repeated impact conditions which occur in large ball mills, wet mills or larger roller pulverizers where uneven wear of table segments occurs. When impact conditions are less severe, as-cast irons with substantially pearlite-free microstructures are used. The advantage of as-cast irons is the elimination of costly, high temperature heat treatment. The disadvantage is the need to closely control mould cooling to avoid residual stresses or transformation induced stresses and the need for greater alloying additions to prevent pearlite formation.

The iron founding industry – past, present and future

Although many ferrous components are forged or rolled to shape following solidification, casting is the only feasible manufacturing method for components of complex shape. It is the chosen method for many components because it is the most versatile, most flexible, most economical and the shortest and fastest way to transform raw material into finished product. The many types of iron and their wide range of properties have been described in previous sections. These assured a large market demand for irons in the past. Foundry technology was dominated by the continuous need for increased productivity and profit whilst satisfying growing demands for new and improved products. Many processes, such as cupola melting, were operated inefficiently and innovations such as spheroidal irons, introduced to satisfy new markets, became established very slowly.

The 1970s was a decade of change. The market demand for irons fell against the background of a world economic recession. Social and political pressures gradually took over from market demands and enforced many changes. Our perception of the importance of energy and resource conservation changed; for example, energy costs increased ten times in the USA between 1973 and 1979. Many other factors have resulted in change, including enforced control over emissions and work place environment, the introduction of robots and automation and the rapid growth in the use of computers throughout the foundry to facilitate design and aid melting operations, as well as in production control and statistical quality control. The growth of alternative materials, particularly in the automobile industry and liability legislation changes in the USA have reduced demand for irons. All these factors have been effective to varying degrees in the different countries, as reflected in the production trends shown in *Figure 1.4*.

In the 1980s, with few exceptions, foundries cannot rely on market growth and the economies of scale it offers. These have been replaced by sounder practices and more efficient utilization of resources by the foundries that have survived. This has been achieved through a greater understanding of all aspects of the casting process. Whilst room still exists for the artisan, founding practice is no longer an art.

Casting in the 1980s

The factors to be considered in selecting an iron to satisfy component properties have been discussed in several earlier sections. However, this is only one of three important interrelated aspects that must be considered in establishing a design that will ensure that a component can be produced and function satisfactorily. The other two are its geometrical shape and the casting process. All three factors must be considered simultaneously. For example, the casting process imposes few constraints on shape but designs with complicated internal cavities may be limited by the type of iron that can be cast satisfactorily. Design has been traditionally a human responsibility but, increasingly, computers are being programmed to design castings[119,120].

Several years ago engineers at John Deere demonstrated the increased productivity resulting from computer aided design (CAD) by developing a program for designing spherical iron cotton picker cams. The design program was the result of several years cooperation between computer programmers and design engineers, but, once developed, it could design a new cam to near perfection in several hours compared to the several weeks of trial and error used previously.

Once the design and casting process have been decided a pattern must be created and the computer is also improving productivity[121]. The choice of pattern material is no longer wood or metal. However, machineable epoxy resins are capable of providing greater dimensional accuracy[122,123]. Recently a new technique for creating a complete three dimensional model/pattern has been described. A computer is programmed with x, y and z co-ordinates taken from the three views of an engineering drawing of the casting. The computer controls the movement of two laser beams scanning a container of liquid plastic. When the beams intersect they promote a chemical action which hardens the plastic. Any liquid

remaining can be removed leaving an exact copy of the image defined by the computer data.

Investment casting offers near cast-to-shape perfection and new advances have been made in replacing conventional waxes with new chemicals, especially styrenes[124-126]. The lost foam or full-mould process uses mass produced, often die-cast, polystyrene shapes that duplicate the casting shape accurately. The pattern with attached polystyrene gates and feeders is buried in dry, unbonded silica sand which is packed by vibration. The molten iron evaporates the polystyrene immediately on entry to the mould. This is a model of simplicity and economy. The need for cope and drag flasks, parting lines and cores is eliminated. Since the pattern is not removed from the mould this method offers unlimited freedom of design. Defects originating from mould binders are eliminated[127,128].

Developments have been as rapid in more conventional core and moulding materials. Prior to 1960, there was a limited choice of core making processes, oil sand, shell and CO_2-sodium silicate[129]. During the past two decades significant advances have been made in chemical sand binder technology. Furans, phenolics, aldehydes, formaldehydes and urethanes have been combined in a range of systems in which polymerization by both addition and condensation reactions can be catalysed by acids, gases, solids and/ or heat. Cold setting systems are now well established in foundries[130].

New tooling designs to limit gas emissions have resulted in a rapid increase in the use of the amine and SO_2 curing systems. *Figure 1.25*, in which casting dimensions reflect mould rigidity and casting density indicates casting soundness, shows that provided resin-bonded sands are fully cured they can produce castings as sound and as dimensionally accurate as cement sands. Curing time varies from system to system and depends on the resin and catalyst content, temperature and humidity[131]. For example, acid catalysed resin systems cure more slowly than urethane resin systems. Consequently, close process control is required during the production and storage of moulds and cores. This, coupled with the increasing costs of organic binders and increasing concern over the environment has led to a rekindling of interest in the sodium silicate binders[132]. The original CO_2 process has been improved with new chemical formulations, hot box core making and new additions to promote breakdown, latitude in gassing, improvement in bench life, strength and handling properties[133-135].

Notable advances have been made in fluid or castable sands and self-setting silicates using ester hardeners[136]. Particular advantages of sodium silicate moulds are ecological, mould rigidity, freedom from N_2, compatibility with green sand and cost. Recently, a synthetic amorphous hydrogel (SAH) binder system has been reported[137] for no-bake moulds and cores.

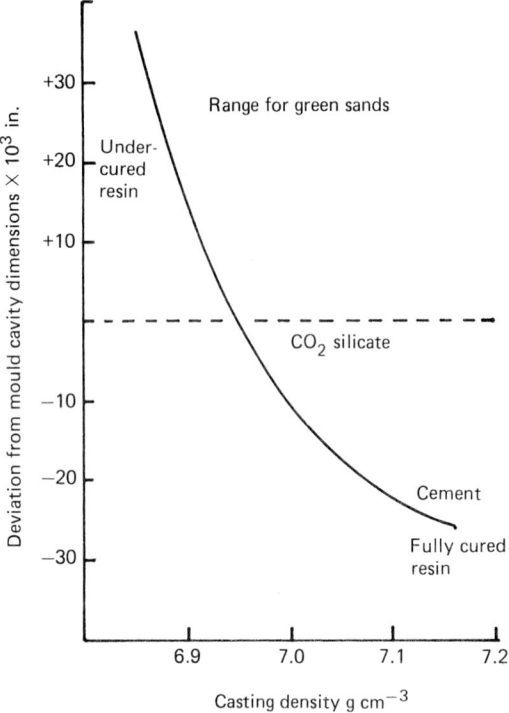

Figure 1.25 The deviation from mould cavity dimensions and the density of a spheroidal iron in spheres related to the curing of resin-bonded sands and other sands (after ref. 1)

This is an inorganic, C-free binder with set times in the range 30 s to 45 min dependent on temperature and binder ratio. This ceramic bond system is compatible with common foundry sands. It produces a rigid mould for excellent dimensional accuracy, has easy shake out, causes little environmental pollution and is easily reclaimed[138]. Economies can be made by using shell moulding to reduce the amount of sand used. This is the case with the Colshel process[139] which compacts a thin layer of sand against the pattern using a diaphragm supported by air pressure. The original process used an amine gas–low density system with a closed box gassing system and absorbing tower to prevent atmospheric pollution. However, other sand systems such as SO_2 furan or CO_2 gassed silicate can be used.

One of the oldest, yet most extensively used moulding mediums is green sand, which is silica sand containing clay and moisture. It is used in the unbaked condition. Individual moulds can be made by hand or, at the other extreme, highly mechanized and automatic machinery produces up to 300 moulds per hour, either using boxes or boxless, in foundries with a high output of long run castings. As indicated in *Figure 1.25*, green sands show varying mould rigidity. Rigidity has been improved through high

pressure moulding[140,141]. This reduces mould wall movement, particularly in heavy castings, and results in more reliable dimensions, closer tolerances, better surface finish, uniform hardness and a reduced susceptibility to shrinkage problems.

Many foundries are instituting total reclaimers rather than merely reconstituting the sand. Various systems, as well as dry attrition, and wet and thermal reclamation are being employed. Advances in understanding the factors that influence sand mouldability have emphasized the importance of removing fines during reclamation to maintain sand grain size distribution. It is also important for the reused sand to be recooled, retempered and reconstituted differently from new sand. The computer has made a major impact in this area. There is so much information to be collected, analysed and related to casting quality that without a computer giving a rapid diagnosis, control over a sand system can be lost[142,143]. Several novel techniques using vacuum have been introduced into the foundry. The Japanese 'V' vacuum moulding process holds unbonded sand in place in the mould by vacuum[144-146]. The pattern is covered by a tightly conforming thin sheet of plastic which is applied with vacuum after being heated. The flask is placed over the plastic-coated pattern and filled with sand. A second sheet of plastic is placed over the top of the sand and the flask is evacuated. The sand is hardened around the pattern which can then be withdrawn. This technique has been used in the production of grey iron piano frames. The use of vacuum has been extended to the casting process.

A process similar to the Cosworth process, which is well established for Al castings, has been patented by the Hitchiner Manufacturing Company in the USA[147]. This is an economical process for casting small thin section components. The parts are gated through the drag half of the mould directly into the cavity or a blind feeder. The cavity is filled by submerging the drag into a molten bath and attaching the cope to a vacuum chamber. Iron is drawn into the mould cavity and any decomposition gases are sucked into the vacuum system. An automatic machine is operated by a programmable controller and requires only operator surveillance. Iron components are also made by centrifugal casting, die casting, stir casting and continuous casting.

This brief consideration of the current trends in moulding demonstrates that the 1970s were not a time of stagnation. It also indicates that successful production, productivity and profitability are dependant on effective process control at each of the various stages of the casting process. In addition to pattern making and moulding these include melting, liquid metal treatment, solidification, cooling, shake out, cleaning, inspection, heat treatment, machining, repairing and final inspection. Process control is only possible with an understanding of solidification science (reactions in the liquid, nucleation and growth of the solid from the liquid, eutectic solidification, segregation during solidification, solid state transformations etc.)[148] of the chosen alloy.

Overall process control must take into consideration the interrelationship of each of the individual casting stages. For example, the volume changes that flake and spheroidal irons undergo on cooling and freezing differ from those of other casting alloys including white irons[149-150]. The important difference is that the contraction of the cooling liquid stops at or above the eutectic temperature. Subsequent cooling and freezing is accompanied by an expansion which lasts nearly to the completion of freezing. The last liquid to freeze, however, contracts, resulting in secondary shrinkage. Although conventional methods of feeding may be used, these unusual characteristics allow more economical methods to be contemplated in which the expansion is used to compensate for secondary shrinkage.

However, increasing the casting yield in this way depends on other steps in the casting process because the pattern of volume changes is not a material constant but depends on the cooling rate. This itself depends on the type of mould used, the graphite-forming tendency as influenced by composition, metallurgical quality, which depends on several process variables, and mould rigidity. Computers are being used increasingly in the calculation of feeder sizes and placement, with a significant increase in productivity[151].

The relationship between efficiency and process control and the dependence of both on solidification science is nowhere more evident than in the energy and cost intensive area of iron melting. Important changes have occurred in each of the six stages (charge selection, charging, melting, liquid metal treatment, transfer to pouring station and pouring) of this process. Advances in solidification science have exposed the considerable effect that small concentrations of impurity can have on structure and properties. They have also emphasized the need for careful control over charge constitution as high priced pig iron is replaced with cheaper steel and cast iron scrap. Advances in charge selection have been made with the introduction of linear programming to calculate optimum charge materials/material price ratios using the computer[152]. Changes in charging procedures include automatic and continuous charging[153] and charge preheating in electric melting. The old workhorse, the cupola, retains its position as the major melting unit, but only after withstanding challenges from electric melting and undergoing considerable process changes to increase its efficiency[154,155]. These changes include hot blast, divided blast, oxygen enrichment, projecting water cooled tuyeres, improved air cleaning, water cooled shell, improved refractories, increased size, improved coke quality and the use of duplexing[156].

The need for increased efficiency and the ease with which it can be achieved by applying solidification science has been demonstrated in a recent survey of 30 Canadian foundries[157]. The melting efficiency of each cupola was determined both separately and in the context of its operation within the foundry under normal working conditions. This was then related to the design and operational practice of the cupola and the constraints imposed by the foundry operation. The study shows that the commonly quoted metal/coke ratio is misleading as a measure of melting efficiency. Nominal performance metal/coke ratios varied from 15.4:1 to 3.8:1 and actual efficiencies varied from 9 to 46%. Factors influencing cupola performance and factors external to the cupola influencing melt efficiency were identified and several 'no cost' improvements were suggested for each cupola. Once a fundamentally sound, controlled cupola operation was established in this manner, increased efficiency could be realized by selectively implementing improved technology based on recent advances in solidification science.

Cupolas represent a large capital investment and often design changes to implement improved technology are difficult. Improvements made in the 1970s were achieved often as a result of long trial and error procedures. These were high risk ventures because of potential consequences in product quality and productivity. Recent advances in process modelling have eliminated these risks[158,159] by providing a means of quantitatively estimating the effects of changing process variables without risk to the actual melting unit.

In a similar vein, microcomputers are proving invaluable in promoting process control. An example is the Canmet cupola software package, a project dedicated to the assessment and improvement of cupola melting technology in Canada. The package comprises programs for melting efficiency, charge calculations, acid slag analysis, melt cost analysis, cast iron specifications and cupola trouble shooting[160,161]. Whereas net diagrams[162,163] which interrelate melt rate, C rate, blast rate and iron temperature are used to effect process control once the diagram has been established for the cupola in question, the Canmet programs can be used to reveal the relationship between a greater number of process variables for any cupola once the characteristics of the cupola have been fed into the program.

Electric melting in the form of arc melting[164], plasma arc melting, in which machine borings and turnings can form part of the charge and rod resistance furnace, and, more particularly, mains and medium frequency induction coreless melting[165,166] and channel induction melting[167,168] offers an alternative melting method that is attractive to the smaller foundry. The coreless induction furnace has less effluent and a simpler refractory system and lower power consumption than an arc furnace. The arc furnace has a major advantage in the variety of charge material that can be used. Both methods have good chemistry and temperature control, which offers rapid melting, production flexibility and consistent quality. Both can melt a variety of iron compositions. The main advantages of the channel furnace are lower power consumption and good furnace up time.

The choice of a primary iron melter represents a large capital expenditure. The changes that occurred in the 1970s have resulted in a marked increase in sensitivity to energy cost and cost structures in furnace selection. In the USA, in particular, there is a wide range of energy availabilities, energy cost rate structures, metallic charge availability and cost variation from area to area. Consequently, it is not possible to draw a generalized conclusion as to the most economical melting method. Each selection should go through a complete analysis if the final decision is to be correct for a particular foundry in a specific location[169].

Although the modern cupola is a very efficient melter, it is an expensive superheater. Duplexing, in which the iron is transferred from the primary melter into a holding furnace, which is often a channel or rod resistance furnace, is becoming more attractive, particularly with automated casting systems[170]. The holding furnace acts as a flywheel in the system and provides a ready source of metal for metering to the pouring operation. It also allows liquid metal temperature control and the superheating required for castings with thinner sections, as well as smoothing out and adjustment of iron composition. In addition it provides a buffer for variations in metal demand and availability.

Iron quality begins with the liquid metal, and the old axiom that you cannot control until you have measured has been proved with recent developments in liquid iron assessment. High speed spectrographs interfaced with computers can provide print out analysis of several elements in a matter of minutes[171]. Thermal analysis is a valuable evaluation tool for providing information concerning iron chemistry, quality and process variables. Solidification of a grey flake iron in a small Te-coated cup causes the iron to solidify white, producing sharp primary and eutectic arrests on the cooling curve which can be used to calculate C and Si contents[172-175]. This technique has been extended to spheroidal irons by using a coating containing both Te and S. The increased use of computers has made possible more detailed analysis of cooling curves without any sacrifice in speed.

Heine and coworkers[176-179] have shown that the liquidus arrest is sensitive to the melt condition. The differences, ΔCE, between the C.E.V. measured from the cooling curve and that calculated from a spectrographic analysis provides a measure of the degree of oxidation of the liquid iron. This varies because of changes in the C and Si activity in the liquid. These

changes can be related to process variables such as superheating, holding, furnace atmosphere, rusty steel in the charge and melt additions. Measurement of ΔCE has been used to monitor melt condition[180] and to indicate the susceptibility of castings to defects, for example, difficulty with feeding.

The use of the computer allows quantitative information to be derived from the cooling curve. This has been achieved in the form of a differential heat analysis as illustrated in *Figure 1.26* for a malleable iron[176]. The neutral body used in the conventional differential thermal analysis (DTA) is replaced by the portions of the cooling curve which plot as straight lines on a Newtonian cooling graph of log relative temperature vs. time, i.e. cd, de in *Figure 1.26*. The log relative temperature vs. time data is converted to heat evolved during the various stages of solidification. Heat evolved during the straight line log relative temperature curve serves as the neutral body cooling. Departures from the curve represent the thermal events of solidification. A graph of the total heat evolved relative to that of the neutral body takes the shape of the conventional DTA, but as a difference in heat evolved rather than as temperature differences between the unknown and neutral body. The final stage in the program is a mathematical normalizing procedure which permits overlapping solidification processes such as proeutectic and eutectic formation to be isolated and studied separately. The technique has been used to measure the total percentage of

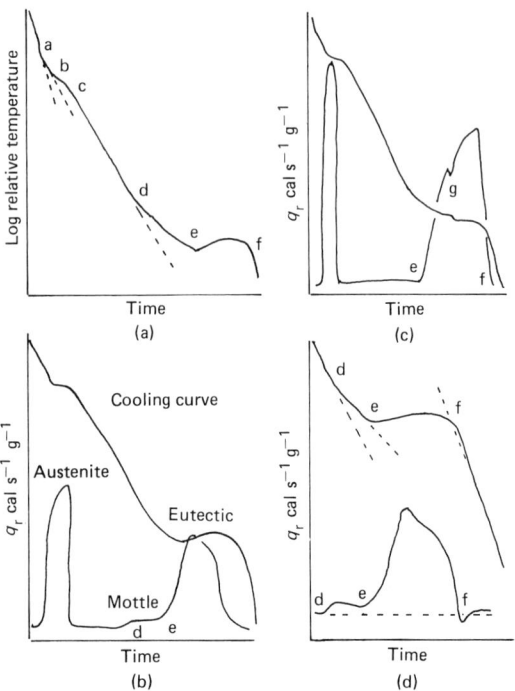

Figure 1.26 Features of the cooling curve revealed on a DHA curve (after ref. 176)

proeutectic austentite, which is the percentage austenite formed at the liquidus. This is related to the occurrence of spiking and coarse oriented dendritic grains[181]. It can also be used to measure the percentage carbide eutectic formed as related to C.E.V. and processing variables.

The occurrence of flake graphite (type D or E) mottle is registered as an inflection in the cooling curve at point d (*Figure 1.26a*) but is easily detected as a thermal event in the DTA analysis (*Figure 1.26b*). A second type of mottle, temper mottle, does not have a grey eutectic cell structure. It appears as temper carbide and shows on the cooling curve (*Figure 1.26c*) as an initially higher temperature eutectic plateau, a step and then a lower carbide eutectic temperature plateau. It is easily detected on the differential heat analysis (DHA) curve between points e and g as shown in *Figure 1.26c*. *Figure 1.26d* shows a negative peak at point f after eutectic solidification is complete and at an inflection in the cooling curve where there is a change from an accelerating rate of cooling to a decelerating rate. This peak is of considerable interest because it is related to the constituents formed late in the solidification process and provides a potential for evaluating these constituents.

The cooling curve of a grey iron shows different features (shape, arrest temperatures, degree of recalescence) depending on the graphite morphology. Several recent studies have shown how these features can be used to predict graphite morphology and the constitution of mixed structures[182-186]. During the past decade considerable advances have been made in understanding eutectic solidification and, in particular, how graphite grows in liquid iron. The influence of growth velocity, temperature gradient in the liquid, segregation effects and impurities in modifying interface kinetics and promoting constitutional undercooling is now relatively well defined[38].

Using this knowledge Minkoff has suggested[187] a computational method for the control of cast iron structure over the entire range of possible graphite morphologies. The method is based on calculation of the undercooling of the graphite-liquid interface due to solute influences in solution. The method is called 'Delta Tee'. It emphasizes that cast iron structures are a direct function of the undercooling of the interface under the conditions of graphite growth. Following analysis of the liquid iron after melting, the analytical data is fed into a computer which processes it to give corrected analysis of the liquid as required for the given structure and properties at the stated casting cross-section. The procedure corrects for solute interactions in solidification and for segregation. It provides calculation of any additions to be made, or it may reject the liquid composition as being unsuitable for the type of casting required.

This brief insight into two of the casting stages, moulding and melting, shows that there is no lack of improved processes for foundrymen to select from to improve productivity, the environment of the workplace and to reduce energy usage. Indeed, the problem is not one of finding a process to improve operations, but rather of selecting one from a wide array of process improvements which will result in better quality and reduced costs to allow them to remain competitive and to gain new customers. The aim of the following chapters is to explain various aspects of solidification science and to show how use of this knowledge leads to better cast products.

References

1. HUGHES, I.C.H., The development of iron founding. Some recent contributions and opportunities, *Brit. Foundryman*, **74**, 229 (1981)
2. MORROGH, H., Influence of some residual elements and their neutralisation in magnesium treated nodular cast iron, *A.F.S. Trans.*, **60**, 20 (1952)
3. WALLACE, J.F., The effects of elements on the structure of cast irons, *A.F.S. Trans.*, **83**, 363 (1975)
4. COLE, G.S., Solidification of ductile iron, *A.F.S.*, **80**, 172 (1972)
5. STRONG, G.R., A literature survey on nitrogen in malleable iron, *A.F.S. Trans.*, **85**, 29 (1977)
6. FARQUHAR, J.D., Nitrogen in ductile iron – a literature review, *A.F.S. Trans.*, **87**, 433 (1979)
7. ANON., Nitrogen in cast iron, *BCIRA Broadsheet*, **41**, (1971)
8. DAWSON, J.V., Influence of N and V on the soundness and strength of grey iron, *49th Int. Foundry Congress, Chicago, paper 23*, (1982)
9. HUGHES, I.C.H., The role of gases in the structure of cast iron, *BCIRA Journal*, **17**, 327 (1969)
10. GREENHILL, I.M., Some practical observations on Pb contamination of cast iron, *Brit. Foundryman*, **77**, 370 (1984)
11. JANOWAK, J.F. and GUNDLACH, R.B., Development of a ductile iron for commercial austempering, *A.F.S. Trans.*, **91**, 377 (1983)
12. DORAZIL, E., BARTA, B., MUNSTEROVA, E., STRANSKY, L. and HUVAR, A., High strength bainitic ductile cast iron, *A.F.S. International Cast Metals Journal*, **7**, 52 (1982)
13. PARKES, L.R., Austempered irons and the automotive industry, *Metals and Materials*, **1**, 53 (1985)
14. Austempered ductile irons, papers presented at BCIRA Conference, Coventry, Dec. 1984
15. COX, G.J., Developments in alloy cast irons, *Brit. Foundryman*, **76**, 129 (1983)
16. RICKARD, J., An engineering cast iron for service at temperatures exceeding 800 °C, *Brit. Foundryman*, **77**, 313 (1984)
17. CLOW, S.C., The effect and control of sulphur in iron, *A.F.S. International Cast Metals Journal*, **4**, 45 (1979)
18. ROTE, F.B. and WOOD, W.P., Segregation of molybdenum in phosphorus-bearing alloyed gray cast iron, *A.S.M. Trans.*, **35**, 402 (1945)

19. BAK, C., DEGOIS, M. and SCHISSLER, J.M., Scanning electron microscopic examination of the embrittlement of ferritic ductile cast iron, *A.F.S. Trans.*, **88**, 131 (1980)

20. CASPERS, K-H., The production of lamellar graphite cast iron engine parts, *A.F.S. International Cast Metals Journal*, **5**, 51 (1980)

21. NECHTELBERGER, E., *The properties of cast iron up to 500°C*, Technology Ltd., Stonehouse, Glos. (1980)

22. ZAKHARTCHENKO, E.V., AKIMOV, E.P. and LOPER, C.R. JR., Kish graphite in gray cast iron, *A.F.S. Trans.*, **87**, 476 (1983)

23. SUN, G.X. and LOPER, C.R. JR., The influence of hyper-eutectic graphite on the solidification of gray cast iron, *A.F.S. Trans.*, **91**, 217 (1983)

24. KHAN, M.H., The influence of titanium on the structure of gray cast iron, *A.F.S. International Cast Metals Journal.*, **3**, 35 (1978)

25. MOORE, C.T., Heat treatment of malleable cast iron, *Brit. Foundryman*, **72**, 75 (1979)

26. MILLS, K.D., Spheroidal graphite cast iron – its development and future, *Brit. Foundryman*, **65**, 34 (1972)

27. ANON., Chinese founders made spheroidal graphite castings 2000 years ago, *Foundryman Trade Journal*, March 1983

28. DIXON, R.H.T. and HINCHLEY, D., Ferrosilicon magnesium alloy development for the production of S.G. iron by various processes in 1983, *Brit. Foundryman*, **77(5)**, (1984)

29. KARSAY, S.I., *Ductile iron production*, Q.I.T. Publication, p. 93 (1976)

30. ELSE, G.E., The first two years operating experience at the Hallam plant of Stanton and Staveley, *Brit. Foundryman*, **76**, 145 (1983)

31. JACOBS, F.W., Malleable vs. nodular iron, *A.F.S. Trans.*, **83**, 263 (1975)

32. GUNDLACH, R.B., JANOWAK, J.F. and RÖHRIG, K., On the problems with carbide formation in grey cast iron, paper presented at *Third International Symposium on the Physical Metallurgy of Cast Iron*, Stockholm (1984)

33. GODSELL, B.C., Producing as-cast ductile iron in iron permanent moulds, *Modern Casting*, **72**, 35 (1982)

34. GRAHAM, P.S., Areas of consideration in the manufacture of heavy section ductile iron, *A.F.S. Trans.*, **90**, 313 (1982)

35. CAMPOMANES, E. and GOLLER, R., The effect of certain carbide promoting elements on the microstructure of ductile iron, *A.F.S. Trans.*, **87**, 619 (1979)

36. EVANS, W.J., CARTER, S.F. JR. and WALLACE, J.F., Factors influencing the occurrence of carbides in thin sections of ductile iron, *A.F.S. Trans.*, **89**, 293 (1981)

37. SUN, G.X. and LOPER, C.R. JR., Graphite flotation in cast iron, *A.F.S. Trans.*, **91**, 841 (1983)

38. ELLIOTT, R., *Eutectic solidification processing*, Butterworths, London (1983)

39. WHITE, C.V., FLINN, R.A. and TROJAN, P.K., An investigation of three different methods of nodularity evaluation in ductile cast iron, *A.F.S. Trans.*, **89**, 639 (1981)

40. TYBULCZUK, J., SAKWA, W., NAKANO, J., KAWANO, Y and MARGERIE, J.C., A study of the degenerate forms of graphite in terms of magnetic inspection of ductile iron castings, *A.F.S. International Cast Metals Journal*, **3**, 39 (1978)

41. PAPAKAKIS, E.P., BARTOSIEWICZ, L., ALSETTER, J.D. and CHAPMAN, G.B. II, Morphology severity factor for graphite shape in cast iron and its relation to ultrasonic velocity and tensile properties, *A.F.S. Trans.*, **91**, 721 (1983)

42. REMONDINO, M., PILASTRO, F., NATALE, E., COSTA, P. and PERETTI, G., Inoculation and spheroidising treatments directly inside the mould, *A.F.S. International Cast Metals Journal*, **1**, 39 (1976)

43. MINKOFF, I., Physical metallurgy of cast iron, J. Wiley, London (1983)

44. FARRELL, T.R., The influence of A.S.T.M. type V graphite form on ductile iron low cycle fatigue, *A.F.S. Trans.*, **91**, 61 (1983)

45. STRIZIK, P. and JEGLITSCH, F., Contribution to the mechanism of formation of chunky graphite, *A.F.S. International Cast Metals Journal*, **1**, 23 (1976)

46. WATMOUGH, T. *et al.*, Combined effects of selected elements on the properties of ductile irons, *A.F.S. Trans.*, **79**, 225 (1971)

47. BUHR, R.K., The effects of lead, antimony, bismuth and cerium on microstructure of heavy section nodular iron castings, *A.F.S. Trans.*, **79**, 247 (1971)

48. BOFAN, Z. and LANGER, E.W., Mechanism of interaction of Pb, Bi and Ce in ductile iron, *Scand. J. Metallurgy*, **13**, 15 (1984)

49. BOFAN, Z. and LANGER, E.W., Mechanism of interaction of Mg, Sb and Ce in ductile iron, *Scand. J. Metallurgy*, **13**, 23 (1984)

50. LIETAERT, F., HILAIRE, P. and STAROZ, C., Development of more powerful inoculants for spheroidal graphite irons, *A.F.S. International Cast Metals Journal*, **7**, 30 (1982)

51. ELLIOTT, R., Eutectic solidification, *Materials Sci. and Eng.*, **65**, 85 (1984)

52. JANOWAK, J.F. and GUNLACH, R.B., A modern approach to alloying gray irons, *A.F.S. Trans.*, **90**, 847 (1982)

53. SERGEANT, G.F. and EVANS, E.R., The production and properties of compacted graphite irons, *Brit. Foundryman*, **71**, 115 (1978)

54. ALTSTETTER, J.D. and NOWICKI, R.M., Compacted graphite iron. Its properties and automotive applications, *A.F.S. Trans.*, **90**, 959 (1982)

55. NECHTELBERGER, E. *et al.*, Cast iron with vermicular/compacted graphite – state of the art. Developments, production, properties, applications, *Proc. 49th International Foundry Congress*, Chicago, April 1982

56. POWELL, J., A review of some recent work on compacted graphite irons, *Brit. Foundryman*, **77**, 472 (1984)

57. FOSBINDER, L.L. and SAWER, S.A., Compacted graphite/vermicular cast iron – the John Deere experience, *51st International Foundry Congress*, London (1984)

nucleation is influenced by the type of electric melting. Graphite solution is easier using mains frequency, medium frequency and channel furnace in that order. The methods used to produce irons ready for pouring are described in this Chapter.

Cupola melting

The forerunner of the present day cupola was a chimney type furnace called the Kuppel. The first European cupola is accredited to an English patent granted to John Wilkinson in 1794. The patent for the first cupola in the USA was granted to Mackenzie in 1820. A successful cupola operation produces molten iron of the required composition at the correct metal or tapping temperature at a rate that slightly exceeds the demand from the moulding line. All of these requirements should be met at the lowest possible operating cost.

Although the demise of the cupola is predicted continuously, it remains the most used melting method because of its dependability and ease of operation. It can be operated in several different modes as described in the following sections. It is most economical when large, hot blast, liningless cupolas operate for long melting periods. In general, the cupola does not easily adapt to rapid, production-related fluctuations and, consequently, it is quite often used in tandem with a duplexing furnace.

General features of cupola melting

The main features of the cupola are shown in *Figure 2.1*. Cupolas have diameters ranging from 0.5 to 3.8 m and can be used to produce irons containing between 2.4 and 3.8% C at melting rates in the range 1–120 tonnes/h. A cupola is a vertical shaft furnace lined wholly or in part with refractory. It has air ports or tuyeres located around the outer periphery of the shell near the bottom and a charging door in the upper section. The well is the section below the tuyeres in which metal and slag gather. The metal is tapped from the bottom, and the slag is tapped from the top. When melting is continuous over long periods, pressurized slag separators, designed so that hot metal and slag are not retained in the well but collected in an external separator, are used[1]. Although the cupola is of simple design, attention to detail is important. The shaft must be of the correct shape to avoid bridging of the

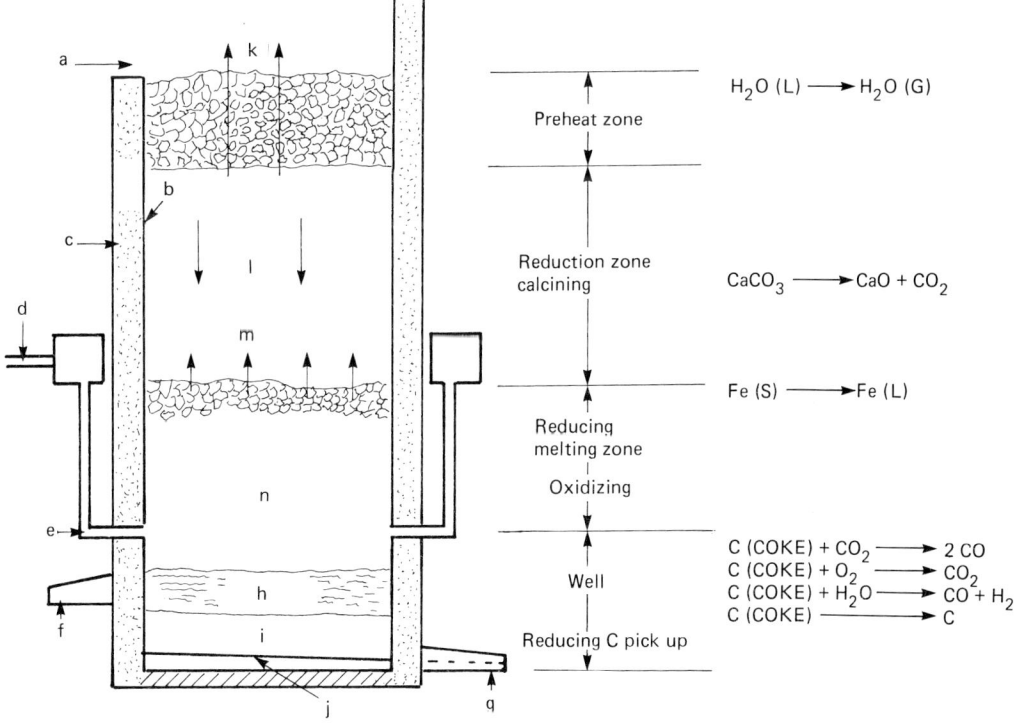

Figure 2.1 Schematic diagram of a conventional cupola and the basic reactions during melting; a, charging door, b, firebrick lining, c, steel shell, d, air blast duct to windbox, e, tuyeres, f, slag spout, g, metal spout, h, slag, i, molten metal, j, sand bottom, k, gases to exhaust, l, charge descends by gravity, materials are mixed as they descend, m, hot gases from the coke bed heat the charge above, n, incandescent coke bed; molten slag and metal are superheated as they pass through to collect in the well

Table 2.1 Factors influencing gasification (after ref. 4)

Factors controlling gasification			*Primary coke variable*	*Secondary coke variable*
< 850 °C	*850–1250 °C*	*> 1250 °C*		
X	X		Reactivity (k')	coke quality and coking conditions
	X		Effective diffusivity (D_e)	coke pore volume and pore structure
	X	X	Harmonic mean diameter (D)	coke strength and size
	X	X	Molar concentration (C_m)	coke density and fixed C
		X	Bed porosity (Σ_b)	coke size and uniformity

charge and for even distribution of blast. In hot blast cupolas, Cu water-cooled tuyeres and effective design of shell cooling equipment are important features. In all cupolas, reliable charging with even charge distribution and accurate weighing facilities are important.

The furnace operates on the principle of continuous counter current flow; solids pass down the shaft and gases rise up. The solid charge comprises metallics, coke and fluxes (usually limestone) charged as a bed of coke. On top of the solid charge are layers of iron, coke and limestone. Foundry coke is the basic fuel. Combustion in the bed provides the thermal energy required to melt and superheat the iron and for slag formation. It has a strong influence on the amount of C that dissolves in the iron[2], as well as on the yields of Si and Mn and cupola back pressure[3]. The proportion of coke in the charge can be as much as 18% of the metal addition by weight depending on the furnace operating conditions, combustion efficiency and C requirements.

Air is injected through the tuyeres. The sensible heat derived from the coke in the combustion process depends on three gas-solid reactions. The exothermic combustion reaction provides approximately 80% of the thermal energy;

$$\text{coke}\,(\sim 90\%\ \text{C}) + \text{air}\,(\sim 21\%\ O_2) \rightarrow CO_2$$

$$+\ 30\ 240\ \text{kJ/kg coke} \qquad (2.1)$$

Endothermic gasification reactions by CO_2 generated in Reaction (2.1) and H_2O derived from moisture in the blast reduce the available heat;

$$\text{coke}\,(\sim 90\%\ \text{C}) + CO_2 \rightarrow 2CO\ -$$

$$12\ 248\ \text{kJ/kg coke} \qquad (2.2)$$

$$\text{coke}\,(\sim 90\%\ \text{C}) + H_2O \rightarrow CO + H_2\ -$$

$$8900\ \text{kJ/kg coke} \qquad (2.3)$$

Reactions 2.1 and 2.3 proceed to completion under normal conditions. However, Reaction 2.2 does not, and between 15 and 75% of the CO_2 produced in combustion reacts endothermically with coke. Consequently, one way of reducing the amount of coke used in a cupola charge is to minimize the rate of CO

formation. Katz[4] has identified CO_2 gasification mechanisms at various temperatures and related these to behaviour in the cupola. Five primary variables influencing gasification and the factors to which they are related are shown in *Table 2.1*. Katz combines these factors to define two important coke quality indexes in an attempt to define the coke properties that affect cupola performance using a scientific basis. Experience has led to the suggestion that properties such as coke density, size, strength of the largest coke, reactivity, hot strength and ignition temperature are factors that influence day to day variations in cupola performance. *Table 2.2* indicates typical coke specifications.

The first index suggested by Katz is the hydraulic radius[5], which is the ratio of void volume to surface area ($R_h = \Sigma_b D/6(1 - \Sigma_b)$). This is inversely related to the rate of mass transfer of CO and CO_2 between the gas phase and the external surface of the coke which controls the intensity of gasification in the coke bed (temperature < 1250 °C). A high hydraulic radius is associated with good thermal yield. The second index suggested by Katz is related to the intermediate temperature gasification mechanism. This is considered to control the temperature at which gasification terminates. It is given by the pore diffusion limited gasification rate constant, $k_{1000} = 1/D\,(k'D_e/Cm)^{1/2}$. This index contains temperature dependent terms and is defined at 1000 °C for comparative purposes. Experimental studies[6,7] confirm that a low index is indicative of good coke quality.

Strength is a further critical parameter suggested in order to relate D and Σ_b, measured before introducing the coke into the cupola, with the actual values which exist in the coke bed[8]. The hot combustion gases rise up the stack and transfer heat to the cooler charge and the furnace walls. The CO and CO_2 content of the exit gas provide a good indicator of the combustion efficiency in the coke bed. Cupolas commonly operate with a CO level of $\sim 13\%$. Anti-pollution regulations require cleaning and CO reduction of the exit gas prior to discharge to the atmosphere.

The metallics in the charge determine the tapped iron composition and represent up to 75% of melting

Table 2.2 Specifications for foundry coke (after ref. 46)

Coke Plant	South Wales		Durham		
	Cwm	*Coedely*	*Norwood*	*Lambton*	*Derwenthaugh*
Moisture	4% maximum	5.5% maximum	3.0% maximum	3.0% maximum	3.0% maximum
Ash	9% maximum	9% maximum	9.0% maximum	9.0% maximum	9.0% maximum
Volatile matter	1% maximum	0.7% maximum	0.7% maximum	1% maximum	1% maximum
Sulphur	0.85% maximum	0.85% maximum	1% maximum	1% maximum	1% maximum
5 cm shatter Index	90 minimum	90 minimum	90 minimum	90 minimum	90 minimum
Mean size	10.7 cm minimum	10.2 cm minimum	10.7 cm minimum	10.7 cm minimum	10.2 cm minimum
Undersize	Not more than 4% less than 5.1 cm				
The above properties are requirements at the point of dispatch at the coke oven plant					

costs. Because these are variable, they offer an area in which careful selection can determine the viability of the melting operation[9]. The metallic charge contains pig iron, cast iron, steel scrap and, occasionally, iron briquettes. The proportions are determined by material availability, cost, product chemistry and cupola mode of operation. Further adjustments in iron chemistry can be made using carburizers and ferroalloys[10]. Examples of charges found technically and economically suitable for grey flake irons are given in *Table 2.3*[11].

The use of pig iron is a matter of tradition, although it is relatively expensive. It provides C, Si, Mn and P and alloying elements when required. If the liquid is not overheated, it imparts characteristics such as freedom from carbides, good response to inoculation and consistently good microstructures to the iron. These are the heredity characteristics referred to in the previous section. Blast furnace pig iron is available in various grades which differ in P content. There are different Si and Mn ranges in each grade. Refined alloy and special pig irons of various compositions are available. Silvery pig iron (10–12% Si, 12–14% Si) offers Si in more diluted form than ferrosilicon

($\sim 50\%$ Si). It can minimize Si variations in cupola melted Si irons and reduce the risk of Al contamination. Special pig irons, like Sorelmetal, which is low in Mn and residual elements, are used for spheroidal iron production.

The most acceptable source of cast iron scrap is the foundry's own scrap of known composition. Steel scrap is low in C, Si, S and P. It is used to lower C and Si contents, particularly in the production of higher strength irons. However, all bought scrap must be scrutinized carefully to avoid the introduction and build up of harmful elements. These may originate from Al components in engine scrap, Cr introduced in stainless steel mixed with mild steel scrap, Pb from leaded free-cutting steel scrap and non ferrous parts mixed with steel scrap. Their influence on structure and properties has been described in Chapter 1. Purchased iron scrap should not be used in spheroidal and malleable iron production. Only selected low P grades should be used for producing high duty engineering irons.

Alternative fuels are sought continuously due to the high price of coke[12]. One material suggested is directly reduced iron ore (sponge iron), which is lower in C

Table 2.3 Typical furnace charges and iron compositions for several grades of grey flake iron

	Typical furnace charges	
Grade 150	*Grade 300*	*Grade 400*
10% pig iron	20% refined pig iron	20% Ion C pig iron
45% foundry returns	30% Gr 300 foundry	20% Gr 400 foundry
45% bought light scrap	scrap	scrap
Ferromanganese to keep S balance	50% steel scrap	60% steel scrap
	Typical iron composition	
3.1–3.4% C	2.9–3.1% C	2.9 maximum % C
2.5–2.8% Si	1.5–1.8% Si	1.4–1.6% Si
0.5–0.7% Mn	0.5–0.7% Mn	0 6–0.75% Mn
0.15% S	0.12 maximum % S	0.12 maximum % S
0.9–1.2% P	0.15 maximum % P	0.15 maximum % P
	0.4–0.6% Cr	0.5–0.6% Mo
	0.8–1.2% Ni	1.5–2.0% Ni

and residual elements than pig iron and steel scrap. This makes it a possible charge material for ferritic spheroidal and malleable irons. However, experience shows that sponge iron must have a high degree of metallization and low gangue content if cupola performance is not to be affected adversely. The cost of sponge iron must be sufficiently attractive to offset any increased cost incurred due to extra additions of pig iron, coke or ferroalloys and increased wear on refractories.

An optimum size exists for charge materials. They should not exceed one third of the cupola diameter, otherwise scaffolding may occur leading to irregular melting performance. Light, thin and rusty scrap results in excessive Si and Mn oxidation losses and reduced C pick up. Heavy steel scrap may not have melted on reaching the coke bed. Whatever the melting rate, the charge should be fluxed with limestone. Fluxing produces a non soluble (in iron) liquid slag that absorbs coke ash, rust, dirt and sand. This aids the separation of slag and iron. Although the mechanics of cupola melting are relatively simple, control of chemistry and economical operation are complex and depend on a unique combination of physical and chemical factors as described in the following sections.

Modes of cupola operation

There are several different cupola technologies and modes of operation that influence iron chemistry and thermal efficiency.

Acid or basic operation

Cupola operation may be acidic or basic depending on the type of furnace lining and the slag composition. Acid refractories are silica-based and are reasonably inert to attack by molten iron and slag. They are less expensive and of longer life than basic refractories which are based on magnesia[13]. The slag composition exerts a strong influence on iron quality and operational conditions. Slag basicity is defined as the $(CaO\% + MgO\%)/SiO_2\%$ ratio of the liquid. If this ratio exceeds 1, the slag is basic; if it is less than 1, it is acidic.

Basic slags can be obtained by charging the cupola with sufficient fluxes (limestone and dolomite) to ensure the required Ca and Mg content. These additions result in a greater slag volume than in the acid cupola. The extent to which slag basicity can be increased in an acid-lined cupola by fluxing is limited due to attack of the lining. On the other hand, basic linings are expensive and present installation and maintenance problems. Consequently, the tendency is to water cool the melting zone, which is unlined, and to use a neutral lining in the well for slags with moderate basicity and a C lining for slags of high

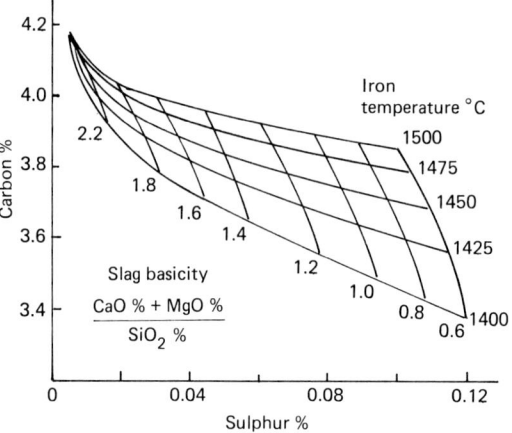

Figure 2.2 The relationship between C and S of an iron, slag basicity and iron temperature (after ref. 46)

basicity. This mode of operation is particularly cost effective for large cupolas (> 130 cm diameter). Increasing the slag basicity reduces S and O, increases C pick up and causes higher Si loss. The relationship between slag basicity and tapped C content for charges of similar composition at various tapping temperatures is shown in *Figure 2.2*[14].

Operations with an acid slag allow variation of C over a wide range by varying the operational

Table 2.4 A comparison of the features of acid slag and basic slag cupola operation

Feature	Acid slag	Basic slag
Carbon content	Easily controlled preferable	Too high for sections thicker than 20 mm
Silicon loss	About 10% preferable	Very high ~30%
Charge	Pig iron usually needed	Pig iron may be omitted, preferable
Spheroidizing	More spheroidizer may be needed	Preferable
Refractory cost	Low preferable	High price but offset in part by lower use
Coke consumption	Low preferable	High
Melt rate	High preferable	Lower
Iron cleanliness	Lower due to high S	Preferable
Tapping temperature	Higher preferable	Lower

parameters that influence tapping temperature. Low S levels may be achieved by basic slag operations but the range of C contents is limited. This characteristic, coupled with higher Si losses and the cost of fluxes and refractories, has resulted in a movement away from basic cupola melting for spheroidal irons to the use of cold or hot blast acid-lined furnaces with carburization and desulphurization performed outside the cupola. Basic and acid melting are compared in *Table 2.4.*

The cold blast cupola

This is a much used mode of operation, particularly in smaller capacity units. It offers a simple and cheap construction suitable for producing irons of limited quality at lower tapping temperatures. However, its thermal efficiency is poor at ~ 30%. The main output variables are melting rate, tapping temperature and composition. Input variables affecting them are:

1. amount of coke charge and its quality;
2. the rate at which air is blown into the cupola;
3. the composition and the nature of the metallics in the charge;
4. the quantity and composition of fluxes and
5. furnace design.

As the percentage of coke in the charge increases at constant blast rate, the melting rate is reduced and the metal temperature increased. The rate of increase of the metal temperature decreases as the coke charged increases. The coke influences composition indirectly through its effect on metal temperature. Higher temperatures promote C pick up and reduce Si and Mn losses. Smaller pieces of coke have a greater surface area:volume ratio. This makes more C available for gasification, with a consequent loss in metal temperature. Optimum cupola performance requires larger coke in larger diameter cupolas. S is transferred from coke to iron during melting. The higher the coke S content, the greater will be the iron S content.

Increasing the blast rate increases the melting rate. This increases the metal temperature to a maximum, after which it decreases with further increase in blast rate. The blast rate also influences metal composition indirectly through its influence on metal temperature. Increased blast humidity decreases metal temperature. Blast rate and blast velocity, which can be controlled by tuyere area adjustment, influence the shape and size of the combustion zone and the melting efficiency. Works[15] has described how reducing tuyere area to increase blast velocity increased melt efficiency. This consequently reduced operational costs as a result of a lower coke consumption, a less costly mix of charge metallics, a higher metal temperature and reduced lining loss due to a smaller but more centrally located melting zone.

Composition of the charged metal is the major input variable influencing the tapped metal composition. In general, an increase in the level of an element charged will result in a corresponding, but diminished, increase in its tapped value. The tapped metal composition is influenced by other elements present in the charge. For example, increases in Si, S and P and a decrease in Mn reduce the C pick up. The physical nature of the metallic charge pieces influences performance. Small light pieces increase heat transfer. This increases metal temperature and combustion efficiency, for example, when light steel scrap is substituted for heavier pig iron. However, such a change can lead to an increased S pick up.

The amount of limestone added is normally ~ 30% of the coke addition. Too little limestone reduces the slag basicity. This causes higher S pick up, reduces metal temperature and does not flux the coke ash, effectively reducing the efficiency of the coke combustion. Too much limestone increases basicity in an acid cupola without improving performance and can lead to excessive Si loss. Auxiliary fluxes[16,17] such as fluorspar, soda ash and calcium carbide have been used to supplement the fluxing action by increasing the fluidity of the slag and reducing the risk of slag build up around the tuyeres. These additions reduce the S content and either increase or decrease (soda ash) the metal C content.

An important cupola design feature is well depth. This is the vertical distance between the centre of the taphole and the centre of the tuyeres (see *Figure 2.1*). If the cupola is tapped intermittently, the well depth

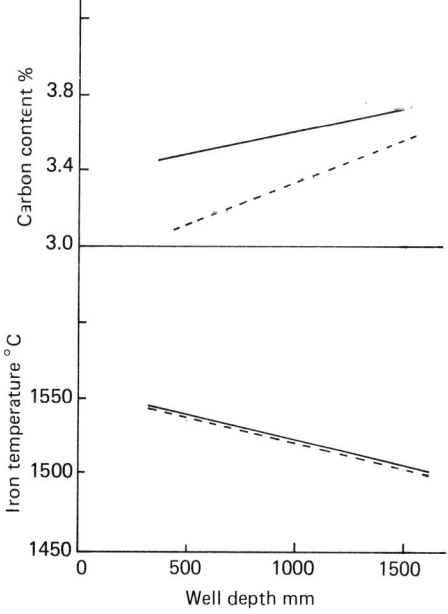

Figure 2.3 Effect of well depth of a continuously tapped cupola on the C content and iron temperature; ——, 90% pig iron, 10% steel charge; – – –, 50% pig iron, 50% steel charge (after ref. 18)

is determined by the quantity of liquid required from the furnace, each tap or that necessary to provide adequate mixing capacity plus an allowance for the slag produced between taps. If tapped continuously, less metal and slag are held in the well and the above requirement is relaxed. The well depth can then be selected either to increase thermal efficiency (shallow well) or to increase C pick up (deep well).

Experiments conducted by the British Cast Iron Research Association (BCIRA) have illustrated these effects[18]. A series of melts were performed with well depths between 46 and 152 cm. Two charges (90% pig iron: 10% steel scrap and 50% pig iron: 50% steel scrap) were melted in a 76 cm diameter cupola operated with a single blast and a coke charge of 15% by weight of metal. *Figure 2.3* shows the influence of well depth on C content and metal temperature. The C increase is equivalent to 0.13 and 0.43% for the high and low C charges, respectively, for each metre increase in well depth. Increasing the well depth did not influence the Si, Mn, S and P contents. The ability to increase C content in this way can be used for economic gain by substituting steel scrap for pig iron in the charge for a particular iron grade, although it may be necessary to increase the coke charge to maintain the desired tapping temperature.

Some European cupola manufacturers design well depth to suit the type of iron produced. Cupola tuyeres are placed just above the sand bed to restrict C pick up if a low C malleable iron is being produced. On the other hand, if a high C iron is to be produced, the tuyeres are raised to a level that will give the required degree of C pick up. The diameter or cross sectional area of a cupola influences melting performance. The smaller surface area:volume ratio of large diameter furnaces reduces the radiation losses. This results in higher metal temperatures, but the combustion ratio (CO_2%/[$CO + CO_2$]%) is lower.

The variation of metal temperature with shaft height (distance between tuyeres and charge door sill) is small, provided the shaft height exceeds that desirable for the cupola size. Undesirable fluctuations in metal temperature and C content occur if the bed height (height above the tuyeres to which the shaft is filled with coke at the start of melting) changes during melting or if the initial bed height is selected incorrectly. An optimum bed height (usually between 130 and 140 cm) must be chosen. A compromise must be made between a shallow bed, which gives metal of consistent composition but, initially, of low temperature, and a deep bed which quickly gives the correct metal temperature at the expense of variation in the tapped C content.

The above illustrates some of the many interrelationships between design features, input and output parameters that make control of the melting process in the cold blast cupola such a complicated process. Design data for cold blast cupolas that can be used to provide a specification to meet given requirements have been published by BCIRA[19]. An example of this data is given in *Table 2.5*. An example of its use to specify characteristics of an intermittently tapped cupola capable of delivering 70 tons of metal in 7 h with a coke charge of 12.5% is given below.

The well capacity requirement is 3 tons and the charge consists of 40% returns, 35% steel and 25% pig iron. The melting rate required assuming a 90% utilization, is 11.1 tons/h. Columns 1 and 2 of *Table 2.5* indicate a blowing rate of 5300 ft³ of air per minute for a metal:coke ratio of 8. This blowing rate corresponds to a melting zone area of 14.19 ft² and a zone diameter of 51 inches. Column 5 recommends a blower capacity of 6360 ft³ per minute at a discharge pressure of 52 water gauge. Column 6 recommends a well depth of 26 inches to accomodate 3 tons of liquid iron plus 6 inches for slag, making a total well depth of 32 inches. The tuyere area should be between 290 and 510 inches² distributed between 8 tuyeres. The lining thickness should be 12 inches. A charge weight of 15 cwt is recommended for the high steel scrap charge to provide a bed height of just over 5 ft. The shaft height should be 22 ft. A well capacity of 3 tons will provide a mixing facility for four charges.

Table 2.5 An example of BCIRA cold-blast cupola design data

Col. Nos.													
1	*1*	*1*	*2*	*3*	*4*	*5*	*5*	*6*	*6*	*7*	*8*	*9*	*9*
Melting rate at various metal:coke ratios ton/h			*Blowing rate Ft³/min*	*Melting zone area ft²*	*Diameter melting zone in.*	*Blower capacity*		*Well capacity*		*Total tuyere area in.²*	*Numbers of tuyeres*	*Bed weight cwt*	
						Vol. ft³/min	*Discharge pressure w.g.*	*per ft. ht.*	*per 3 in. height*			*per 5 ft. height*	*per ft. height*
10:1	*8:1*	*6:1*											
1.6	1.3	1.1	665	1.77	18	800	40	3.6	0.9	35–65	4	2.2	0.4
5.5	4.7	3.7	2230	5.94	33	2680	45	11.9	3.0	120–215	6	7.4	1.5
13.2	11.2	8.9	5300	14.19	51	6360	52	28.0	7.1	290–510	8	17.7	3.5
26.5	22.5	17.9	10650	28.27	72	12700	63	56.0	14.0	580–1020	10	35.5	7.1

The divided blast cupola

The thermal efficiency of a cold blast cupola can be improved using a divided blast[20]. Most new cold blast cupolas are built with this facility. Maximum benefits are obtained[18,21,22] when;

1. the two rows of tuyeres are 90 cm apart;
2. the blast rates and controls for each row of tuyeres are independent and
3. the blast is divided equally between the two rows.

Figure 2.4 shows that it is possible to obtain;

1. a higher tapping temperature and a higher C pick up for a given coke consumption or
2. a reduced charge coke consumption and, if required, an increased melting rate while maintaining a given metal temperature.

For example, a coke charge of 15% used in a cupola with one row of tuyeres gives a metal temperature of 1500 °C; with two rows, the temperature is 1545 °C. If the metal temperature is maintained constant (1500 °C) a coke charge of 15% is reduced to 10.8% with two rows of tuyeres. With a total blast rate of 43 Nm^3/min, the melting rate increased from 3.10 tonnes/h to 3.69 tonnes/h with the lower coke charge. Thus, conversion to divided blast allows coke consumption to be reduced by 28% and the melting rate is increased by 19%. Depending on operating conditions, coke savings were between 20 and 32% and the melting rate increase between 11 and 23%. C

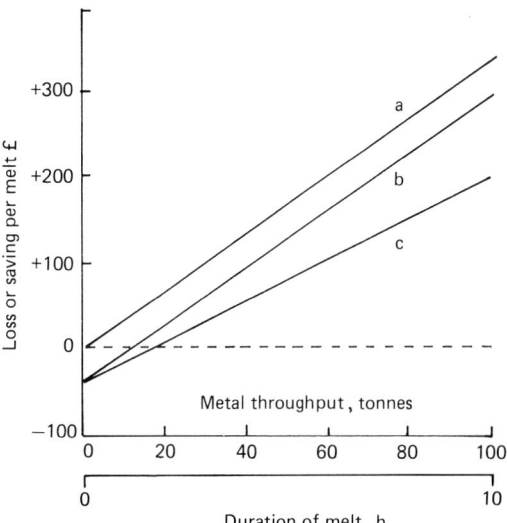

Figure 2.5 An estimate of savings as a result of changing a 10 tonnes/h cupola to divided blast operation (after ref. 18)

pick up increased by 0.2% with no other change in metal composition when operating without reducing the coke charge. When this charge was reduced to maintain a constant metal temperature, C pick up increased by 0.06% and the Si decreased by 0.18%.

Divided blast operation extends the operating height of the coke bed and increases the depth of the superheating zone. This means that the molten iron droplets pass through a greater depth of incandescent coke. Melting occurs well above the upper tuyeres in a correctly designed and operated divided blast cupola. Whether divided blast is used to increase C pick up, reduce coke consumption or for an intermediate combination of these benefits is determined by the balance between the savings that can be made in coke and charge material costs.

For example, *Figure 2.5* shows estimated operational savings as a result of converting a 137 cm diameter cupola producing 10 tonnes/h to divided blast operation. Line a shows savings in coke consumption based on *Figure 2.4* assuming that the coke charge was 12.5%. Actual savings are less. With divided blast, the initial coke bed height should extend 90 cm above that required for normal operation. Extra coke is required for this and the reduction in coke saving is indicated by line b. For short melts, the saving in charge coke does not compensate for the additional requirement for bed coke. The charged Si level must be increased to compensate for Si loss. Line c describes the overall saving when this is taken into account and represents minimum savings.

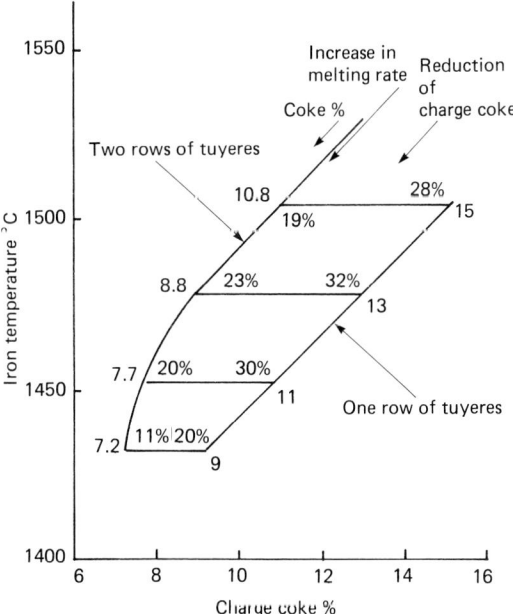

Figure 2.4 A comparison of conventional and divided blast operation indicating the benefits of increased tapping temperature or reduced coke consumption and increased melting rate gained from divided blast operation

Oxygen enrichment

Oxygen enrichment of the blast is an alternative

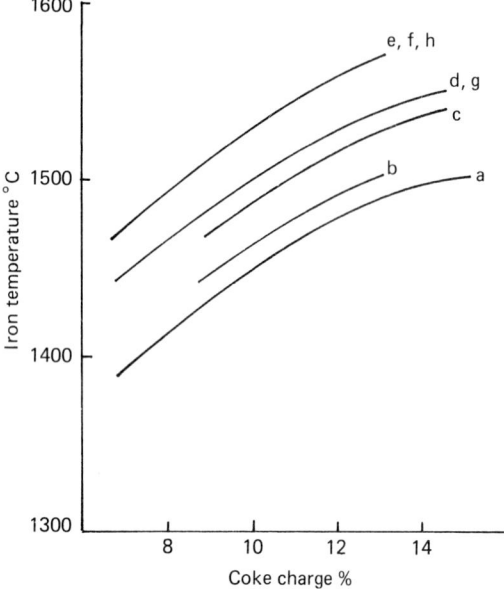

Figure 2.6 The relationship between iron temperature and coke charge for conventional and divided blast cupola operation and the effect of various methods of using O_2 (4% enrichment level). Conventional operation; a, without O; b, blast enrichment ($+15\,°C$); c, tuyere injection ($+40\,°C$); d, injection 23 cm below tuyeres ($+50\,°C$); e, injection 61 cm below tuyeres ($+85\,°C$); f, injection 91 cm below tuyeres ($+85\,°C$). Divided blast operation; g, without O ($+50\,°C$); h, blast enrichment ($+85\,°C$)(after ref. 27)

method of increasing thermal efficiency[23-26]. Compared with normal operation, the use of oxygen:

1. results in a higher metal temperature, a higher C pick up and a lower Si melting loss for the same coke consumption. The higher C pick up allows metallic charge costs to be reduced by partial replacement of pig iron with cast iron or steel scrap and the smaller Si loss reduces charge costs;
2. allows coke consumption and cost to be reduced for a constant metal temperature with no increase of C pick up and no reduction in Si melting loss and
3. permits more rapid attainment of metal temperature at the start of melting or following a blast cut off period and so minimizes cost penalties or casting defects due to the use of cold metal.

The effectiveness of oxygen depends on the method of injection either:

1. blast enrichment achieved by feeding oxygen into the main blast where it is mixed with air before entering through the tuyeres;
2. injection at the tuyeres; entry is through stainless steel injectors inserted in the tuyeres or
3. injection into the well; oxygen is injected into the coke bed beneath the tuyeres using water cooled Cu injectors.

Measurements of the influence of oxygen enrichment in a 76 cm diameter acid cold blast and divided blast cupola are shown in *Figure 2.6*[27]. Maximum increase in metal temperature occurred with well injection for single blast operation and for injection at the lower tuyeres for divided blast operation. Coke savings for constant metal temperature can be deduced from *Figure 2.6*. These results show that divided blast operation without oxygen provided a higher metal temperature than oxygen enrichment, except for injection into the well which poses a number of problems that have curtailed its use. Indeed, it is generally accepted that the first consideration in reducing melting costs in cold blast operation is the introduction of divided blast.

Further benefit may be derived from using oxygen with divided blast. This combination provides an attractive alternative to hot blast operation for cupolas producing up to 10 tonnes/h. An important use of oxygen enrichment has been to increase the melting rate of an established cupola to well beyond its normal rate[18,27]. *Figure 2.7* shows the improvements possible in the cupola referred to in *Figure 2.6*. If the metal temperature is not increased, the coke

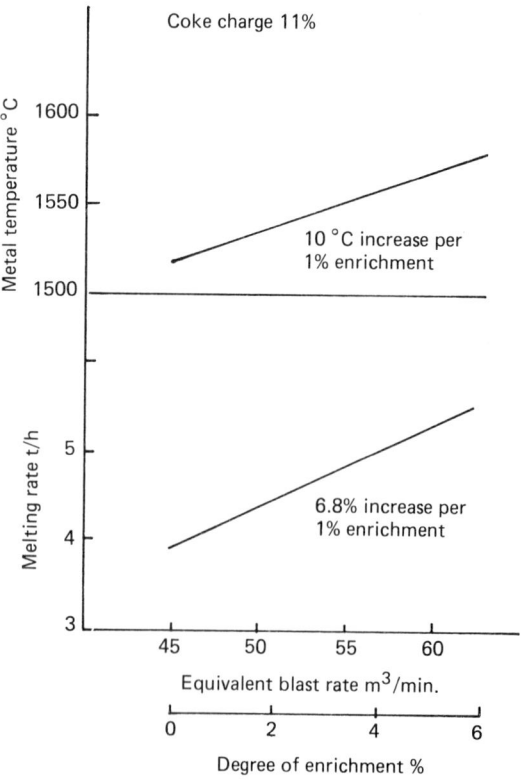

Figure 2.7 The effect of increasing the degree of O enrichment by supplementing the blast of a divided blast cupola. O is used to enrich the blast supply to the lower row of tuyeres (after ref. 18)

cooling water and by radiation to the lining. Equation (2.4) is the basis of furnace classification:

mains, normal or line frequency   50 Hz,
triple frequency   150 Hz and
medium frequency   > 200 Hz.

It also determines the minimum size of the individual components of the furnace charge. Medium frequency furnaces can melt fine charges but melting from cold in a mains frequency furnace requires a solid 'plug' of metal which is almost the dimensions of the internal diameter and occupies a third of the crucible volume.

The turbulence generated in the melt is porportional to the furnace power and inversely proportional to the square root of the frequency. This imposes an upper limit to the amount of power that can be applied to a given size of furnace at a particular frequency. For example, it is unusual to apply more than 200–250 kW per tonne of furnace capacity at mains frequency. This means that the effective maximum hourly melting rate is about a third of the total capacity. Turbulence is beneficial for the rapid assimulation of additions but can be a disadvantage when the charge is prone to oxidation or gas pick up. The crucible can be preformed in smaller furnaces but is usually formed by ramming a refractory lining between a steel former and the furnace coil. The former remains in position during the slow heating of the first charge and supports the lining until sintering occurs.

Mains frequency induction furnaces

This is the most popular type of electric furnace with capacities up to 40 tonnes. The need for a plug when melting from cold and the slow melting rate until the charge is 70% liquid means that this furnace is most economically operated when a liquid heel of about 70% of the furnace capacity is maintained at all times. Mains frequency furnaces are suited to continuous production runs on similar grades of iron.

The charge must be dry. This can be achieved by preheating, which can also reduce overall energy costs and increase the melting rate[66]. One of the advantages of electric melting is flexibility of charge selection and sizes. However, extremely dirty and rusty scrap generates excessive slag which increases deslagging time and lining thickness which results in an inability to draw full power. Induction furnaces are usually lined with SiO_2 with ~ 0.8% boric oxide[67]. Both the design and the thickness of the lining influence the performance. The thinner the lining, the better the coupling, which means that more power can be obtained with a constant voltage. However, the thinner the lining, the shorter is its useful life. The average lining thickness can be related to electrical impedance;

impedance $\propto$ average lining thickness $\propto V^2/P$

where V is voltage and P is the power. Measurements of V^2/P combined with true average lining thicknesses can be used to construct a graph relating V^2/P and lining thickness[66]. In this way V^2/P measurements can be used to determine relining time and the ideal average lining thickness to maximize power input. Lining design can be used to maximize refractory performance by increasing the thickness in areas of maximum wear but still retaining the average thickness over the coil length.

Medium frequency induction furnaces

The medium frequency furnace has become increasingly popular during the past decade, particularly since solid state frequency converters replaced rotary generators[68]. Melt capacity varies from 2 to 20 tonnes. Power is up to 750 kW per tonne capacity using frequencies in the range 200 to 1000 Hz. The charge to tap time is about 1–2 h and the melt rate 2–10 tonnes/h.

Like triple frequency, medium frequency furnaces offer the small to medium foundry a very flexible melting facility. It can be started with a cold charge. This gives an efficiency comparable to that of melting with a liquid heel and allows the composition to be changed from charge to charge. However, when melting from cold there is a greater risk of charge bridging resulting in a molten pool forming in the bottom of the crucible and overheating. To avoid this, the maximum size of any piece of the charge should not exceed one third of the bath diameter. Bridging may occur also when using borings, so their proportion should be limited to 25%.

Wilford[69] has made a detailed comparison of mains frequency and high power density medium frequency furnaces. The total cost of a medium frequency installation can be less than a mains frequency alternative because of lower installation costs. *Table 2.7* compares the characteristics, operating cycles and energy requirements of mains and medium frequency furnaces each with a metal production capability of 4.7 tonnes/h.

The lower energy consumption of the medium frequency furnace stems from the lower heat losses from a smaller capacity furnace. Although it is emptied after each melt and then charged during a significant part of the power on time, the heat losses are less than during the shorter charging time for the mains frequency furnace because the cold layer of charge on top of the melt is preheated and carries a large proportion of escaping heat energy back into the melt.

The analysis in *Table 2.7* applies to furnaces which

Table 2.7 Comparison of mains and medium frequency furnaces each with a metal production capability of 4.7 tonnes/h (after ref. 69)

Characteristic	Medium frequency	Mains frequency
Furnace capacity	4 tonnes	12 tonnes
Installed power	3000 kW	3000 kW
Frequency	500 Hz	50 Hz
Bath diameter	93 cm	114 cm
Power supply efficiency to coil	96%	97%
Coil efficiency	80%	80%
Melting power on	40 min	39 min
Time lost for charging	4 min	4 min
Deslagging, temperature and analysis time	5 min	5.5 min
Tapping time	2 min	2.5 min
Cycle time	51 min	51 min
Energy to melt 4 tonnes to 1500 °C	1480 kWh	1480 kWh
Energy consumed for this	1927 kWh	1907 kWh
Furnace loss per cycle with lid on	40 kWh	136 kWh
Energy loss during charging with lid off	10 kWh	19 kWh
Energy loss during deslagging	16 kWh	26 kWh
Extra energy used to replace energy losses with lid off	34 kWh	58 kWh
Total energy consumed	2001 kWh	2101 kWh

are producing metal continuously at maximum output. The medium frequency furnace is particularly attractive when output falls below those levels. This can occur because of:

1. operating restriction to one or two shifts per 24 h.
2. reduced melting rate because of reduced metal demand and
3. reduced melting rate because the furnace is being used as a metal dispenser as well as a melter.

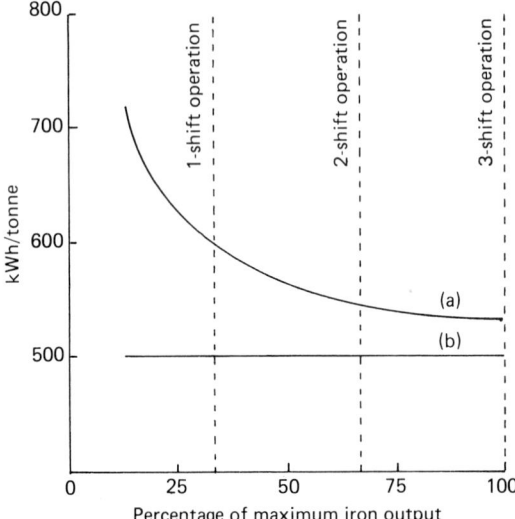

Figure 2.14 Calculated energy required to produce various daily throughputs of molten iron from a mains frequency and a medium frequency furnace; (a) mains frequency furnace, (b) medium frequency furnace (after ref. 69)

The reduced efficiency of the mains frequency furnace as productivity falls is evident from *Figure 2.14*. An important charge constituent in electric melting is alloy and C additions. Wilford[69] reports trials which showed that the recovery of C ($\sim 84\%$) was less than that obtained ($\sim 95\%$) with additions to a mains frequency heel. Si recovery was less than that of C which is contrary to general experience. However, the recovery increased to 98% when ferrosilicon was added to a deslagged bath at 1500 °C. Final compositional adjustments depend for their success on having sufficient metal movement at the melt interface. Wilford defines a characteristic metal surface velocity which governs the rate of dissolution. For the melting furnaces discussed, the metal surface velocity is considered to be proportional to the height of the surface meniscus given by:

$$M = \frac{P_L \times k}{DL^2\sqrt{f}} \qquad (2.5)$$

where M is the meniscus height as % of L,
P_L is the induced power,
k is a constant,
D is the bath diameter,
L is the coil length and
f is the frequency.

Thus, in addition to the maximum power:weight ratio discussed earlier there exists a minimum ratio necessary for adequate stirring to dissolve C additions efficiently. The ratio increases as the frequency increases, as shown in *Table 2.8*. Melting rate is not the only factor to be considered in power rating selection. If high C.E.V. irons are being produced at

Table 2.8 Calculated power levels at 200 and 600 Hz to provide similar metal flow velocities as obtained with mains frequency furnaces of three capacities (after ref. 69)

Furnace capacity tonnes	*Frequency*		
	50 Hz	*200 Hz*	*600 Hz*
2	400 kW	600 kW	1000 kW
4	800 kW	1200 kW	2000 kW
6	1200 kW	1800 kW	3000 kW

low tapping temperatures, power ratings at least equal to those in *Table 2.8* will be required. However, lower ratings which will prolong refractory life can be used if low C.E.V. irons are being produced at high tapping temperatures.

A careful approach to final tapping temperature is necessary if a high power density is used. A medium frequency furnace powered at 875 kW per tonne at 600 Hz requires only 2 min to superheat a melt from 1400 to 1500 °C. This time may be insufficient for solution of final C additions. The problem may be overcome by reducing the power input to lengthen the superheating time or making additions of a solid charge during the superheating period.

Channel furnaces

The main features of a channel furnace are shown in *Figure 2.15*. The metal ring or channel occupies about 7% of the total melt volume and is a single coil of low impedance acting as the secondary of the inductor

transformer. A high current is induced in the channel, resulting in convection currents which push hot metal through the throat into the bath and draw cold metal down into the channel. This stirring action allows cold charge additions to be taken into the melt quickly. This furnace operates at mains frequency and is primed with molten metal. A liquid heel must be maintained at all times and power drawn throughout a melting or holding campaign.

Developments such as multi-inductors, the quick change inductor, selected use of refractories, the permanent roof and improved throat and inductor design have made channel furnaces very popular. Large capacity vertical and horizontal drum furnaces with inductor ratings up to 3000 kW are used[70]. Increased power ratings have made the vertical channel furnace a useful melting unit. Its attractions are an ability to melt during off peak tariff periods and to pour a range of casting sizes up to the useful capacity (total − liquid heel) of the furnace.

An energy balance for a 40-tonne capacity furnace is shown in *Figure 2.16*[71,72]. The holding power requirement to combat energy losses from a full furnace at 1450 °C is 265 kW from *Figuree 2.16*. During melting, 19 tonnes are charged during one night resulting in an energy loss of 36 kW/tonne charged during lid off time. In addition, there is a deslagging loss of 5 kWh/tonne. A power input of 1060 kW was required to melt 1.75 tonnes/h to 1500 °C.

This corresponds to a melting efficiency of ∼ 65%, which is less than usual for a mains frequency coreless furnace (70–75%). However, the channel furnace

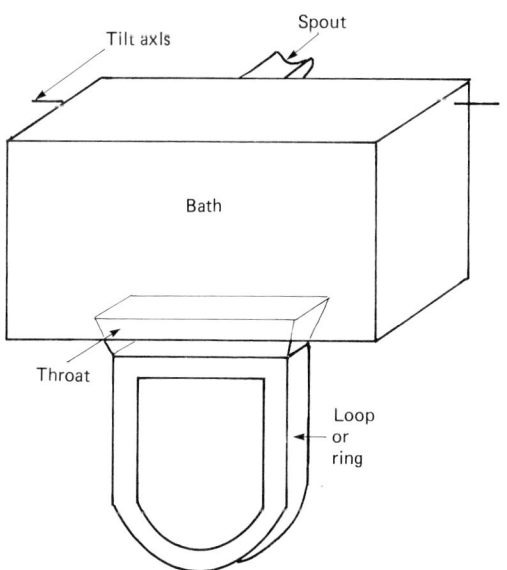

Figure 2.15 Schematic representation of a channel furnace

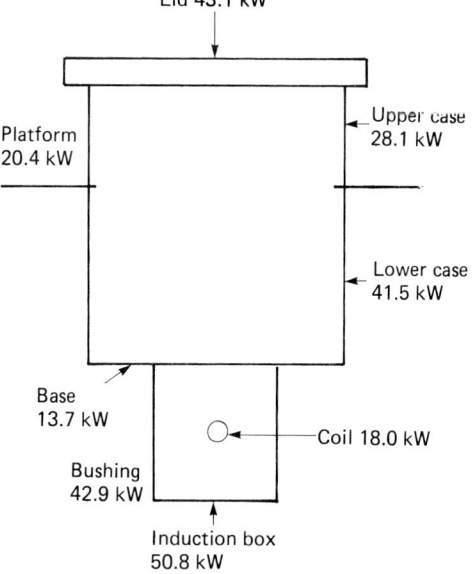

Figure 2.16 Energy balance for a 40 tonne channel furnace (after ref. 72)

provides a holding capacity. Other advantages are a consistent and repeatable chemistry and accurately controlled temperature. However, lack of surface stirring hinders rapid alloy assimilation. Also, the necessity of maintaining a liquid heel reduces flexibility in changing iron grades. As with any furnace maintaining a liquid heel, reduced daily demands cause rapid increases in the energy required per tonne of metal cast.

Horizontal channel furnaces are used extensively as holding furnaces in duplexing actions. Wilford and Langman[73] have discussed the selection of the optimum channel furnace storage capacity from a knowledge of the rate of supply of metal from the melting furnace and the demand from the moulding line. Operating experience shows that the power required per tonne to hold liquid iron is independent of furnace capacity above 25 tonnes. Hence, the power required for larger capacity holding furnaces can be evaluated from the energy balance in terms of that required to overcome circuit inefficiencies and standing heat losses, that is, 265 kW for the furnace considered in *Figure 2.16*. Any power input above that required to balance the heat losses superheats the iron with an efficiency of ∼ 95%. This energy transfer is almost independent of holding temperature. It explains why the induction furnace is more economical than the cupola for superheating.

During production, the holding temperature will fluctuate depending on the temperature of the incoming iron, the bath temperature and the relative volumes of metal introduced and that in the bulk. The fluctuations will be minimal if the volume charged is small and the holding furnace is full. The temperature of the incoming metal should be restricted to ∼ 30 °C above the bath temperature if excessive superheating is to be avoided. This is because the rate of loss of heat through the furnace cannot compensate for the high temperature input, especially if significant furnace power has to be maintained to keep the induction loop open. If the metal charged is at a lower temperature than the bath, superheating can occur if there is a delay in the supply of liquid iron or if a higher tapping rate is required, as shown in *Figure. 2.17*. Many of these fluctuations can be avoided if the furnace power is regulated.

The temperature and time of holding, type and size of furnace and the manner in which it is operated influence the iron composition and the graphitization potential of the liquid. The C and Si concentrations are controlled by the equation;

$$SiO_2 + 2C \rightleftarrows Si + 2CO \qquad (2.6)$$

If the furnace temperature is below the equilibrium temperature of this reaction the C and Si losses will be small unless the atmosphere is oxidizing. In an oxidizing atmosphere loss will occur until the surface is covered with slag. However, above the equilibrium temperature, C is lost from the iron by reaction with oxygen in the slag to form CO. If the atmosphere is oxidizing Si is subjected to simultaneous oxidation and reduction, whilst C is continuously oxidized by SiO_2 and the furnace atmosphere. A shielding atmosphere limits the oxidation reactions.

Losses are minimal in well sealed furnaces. For example, a C loss of only 0.03% was reported[73] during holding for 14 days in an alumina lined 90-tonne furnace with a teapot inlet and outlet spout. Losses are related to the exposed surface/volume ratio, the degree of stirring and the degree of turbulence caused by tapping in unsealed furnaces. Although compositional changes can be small in large furnaces, it can be difficult to prevent losses in small furnaces.

An attraction to spheroidal iron producers is the storage of several Mg-treated batches in a single furnace and the tapping of iron with a controlled Mg analysis at a controlled temperature. Different approaches have been described. One uses a pressure tight channel furnace with argon gas to force metal up a teapot outlet spout for tapping[74] and another uses a bath-tub type which is open to the atmosphere, with a wet rammed high alumina refractory in the inductor and furnace body[75]. Mg loss occurs in a controlled manner in both instances[76,77] but the formation of Mg compounds causes the channel's electrical resistance to increase[77].

In general, the nucleation level is lower in electric melted irons because the longer residual furnace time allows for graphite dissolution and non metallics, which may increase the nucleation, to be taken up by

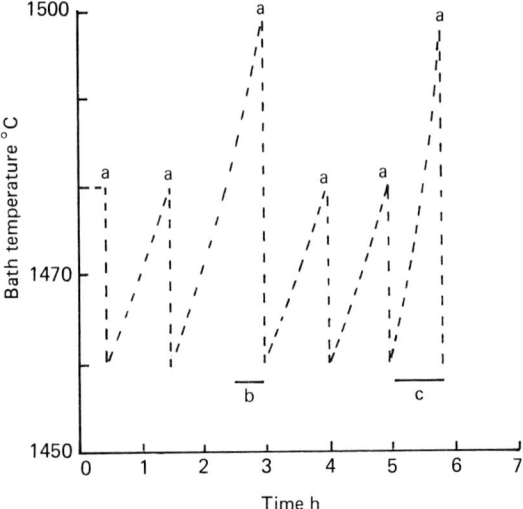

Figure 2.17 Influence of a delay in metal supply and a doubled rate of metal take off on the temperature of a 20-tonne capacity channel furnace a, 10 tonnes of metal added at 1430 °C; b, delay in metal supply; c, doubled rate of metal take off. The power input is sufficient to return bath to 1480 °C one hour after charging. The metal is charged hourly and 1 tonne is tapped every 6 minutes

the slag. Iron in a holding furnace prior to foundry start up is likely to have achieved the base level of nucleation, which depends primarily on the C.E.V. and alloy content of the iron[78]. During the working shift, movement in the liquid graphitization potential will reflect that of the added metal, diluted by the stored volume and any change due to C and Si loss during holding. The potential will return to the base level on holding to the next casting operation. Loss of graphitization potential is not a problem during holding, provided chemistry is maintained. Conditions for this have been described above.

Arc furnaces

Arc furnaces are used to melt about 5% of irons produced and although used occasionally as a holding furnace, they are not as suitable for this purpose as induction furnaces. Energy transfer in the arc furnace occurs mainly by radiation from the arc produced by current transfer from graphite electrodes to the charge material and/or melting bath. Radiation conditions are particularly favourable because of the high arc temperature. Only about 5% of energy transfer occurs by direct current flow. The furnace is a flat hearth type with a large area:volume ratio. This is partly to satisfy design requirements and to compensate for poor stirring in the liquid.

The energy supplied per unit time is the main factor influencing melt rate. It is also influenced by the proportion of borings and turnings charged, the amount of oil and water present on the charge material, charge density and power input after meltdown. Full power can be applied only during meltdown because application after melting can lead to refractory melting and local overheating of the iron. The energy consumption for melting falls in the range 520–580 kWh/tonne. Melt rates are up to 25 tonnes/h and tapping times are from 1–2 h. Arc furnaces are very reliable and robust and versatile with respect to the type and condition of the charge. A typical charge consists of 40% returns, 40% steel scrap and 20% borings. Major problems include frequent refractory repair, high noise and effluent levels, and difficulty in producing a consistent analysis for irons with > 3.2% C from low-C raw materials in a satisfactory tap-to-tap time. This is a consequence of the low level of agitation in the liquid that restricts efficient and reliable C pick up.

Refractory materials are selected to be compatible with the mode of operation. Roofs are constructed independently of the furnace from 60–95% Al_2O_3 brick. Sidewalls have to withstand the arc temperature, slag and iron splash and slag and metal at the slag-metal interface. Various grades of MgO brick are used by spheroidal iron producers to withstand basic slag operation. Grey and malleable iron producers use Al_2O_3 or SiO_2 for acid slags for both sidewalls and furnace bottoms. The standard repair practice is gunning. Areas requiring the most attention are sidewalls and upper portions of furnace bottoms. The introduction of water-cooled furnace panels has reduced the extent of damage to the sidewalls, permitting the use of full power after meltdown, and thus increasing productivity.

Resistance furnaces

This type of furnace is used in automatic pouring systems. It provides up to 30 tonnes of iron in the line to cushion demand. Energy transfer is by radiation from graphite elements situated above the bath. Energy must pass through the slag to reach the metal. This becomes the limiting factor for heat transfer because most conductive slags are good insulators and maintain large temperature gradients. The heating source is exposed to the environment, which increases heat radiation losses. Furnaces are shallow, particularly in areas distant from the heat source, in order to increase heat transfer efficiency and uniformity. Advantages of the resistance furnace include durability and reliability, superior refractory life and the ability to switch furnace power off without damage to the heating element.

Compositional control during liquid metal preparation

Composition not only defines the grade of iron but the structure, properties and soundness of castings within any grade. Compositional control is achieved during primary melting and in separate treatment processes prior to casting.

Gases in cast irons

The influence of gases on the structure and properties of cast iron has been described by Hughes[79]. Significant gases are oxygen, nitrogen and hydrogen. Typical concentration ranges are: oxygen, 0.005–0.01%, nitrogen, 0.0015–0.015% and hydrogen 0.00005–0.000025%. Actual values depend on the type of iron and method of melting as illustrated in *Table 2.9* for nitrogen. Oxygen and nitrogen are picked up mainly during melting, although nitrogen may be gained from binders used in core and moulding sands. Abnormally high hydrogen contents are not associated with melting, with the possible exception of the first metal produced from the cupola. Instead they are associated with the reaction of liquid iron with moisture in the mould promoted by reactive minor elements Mg and Al. Oxygen influences structure through both nucleation and growth. Hydrogen is surface active with graphite. It is additive in its action with elements like S, such that the combined effect is to reduce S adsorption on graphite and promote white iron solidification.

Table 2.9 Factors influencing the N content of cast irons

Typical nitrogen contents of cast irons

Malleable iron	0.005–0.014% N
Grey iron	0.004–0.007% N
Untreated spheroidal iron	0.004–0.012% N
Treated spheroidal iron	0.003–0.008% N

Nitrogen contents of malleable iron produced by different primary melting methods

Arc melting	130–190 p.p.m.
Cupola	110–145 p.p.m.
Coreless induction	80–130 p.p.m.
Channel induction	75–110 p.p.m.

Influence of the quantity of steel scrap on the nitrogen content of cupola melted iron

25% steel	0.011% N
80% steel	0.015% N
100% steel	0.017% N

Oxygen absorption in cast irons

The absorption of oxygen during the melting of different types of iron under various conditions has been measured and related to the equilibrium oxygen concentrations predicted from C, Si and Al deoxidation reactions[80–82] and to graphite nucleation[83]. Neumann and Dötsch suggest that oxygen solubility is controlled by the Si–O equilibrium in the temperature range 1400–1450 °C, but that the C–O equilibrium is followed at higher temperatures. However, oxygen probe measurements on grey, base-malleable and base- and treated-spheroidal irons suggest that Si controls the oxygen level in all melts at temperatures up to 1600 °C. It appears that C content in the range 2.5–4.25% exerts little influence. This is illustrated in *Figure 2.18*, which shows the calculated equilibrium oxygen content (calculations are detailed in ref. 81) in a Fe-4.28% C alloy due to Si and C deoxidation. *Figure 2.18* also illustrates a selection of experimental measurements and the influence of Si content on the equilibrium oxygen solubility.

Increasing Si content decreases oxygen solubility, but changes due to temperature are greater. Both C and Al are stronger deoxidizers than Si at high temperatures but neither exerted a controlling influence over oxygen solubility in the alloys studied. This is attributed to slow kinetics which prevented attainment of equilibrium. Difficulties in the nucleation and growth of CO bubbles are responsible in the C deoxidation reaction. The main source of Al is in ferrosilicon additions. *Figure 2.18b* shows the powerful deoxidation effect of Al if equilibrium is attained. Both the grey and malleable iron melts contained about 0.01% Al which showed little influence on the oxygen solubility. The spheroidal iron contained 0.02% and although the measured

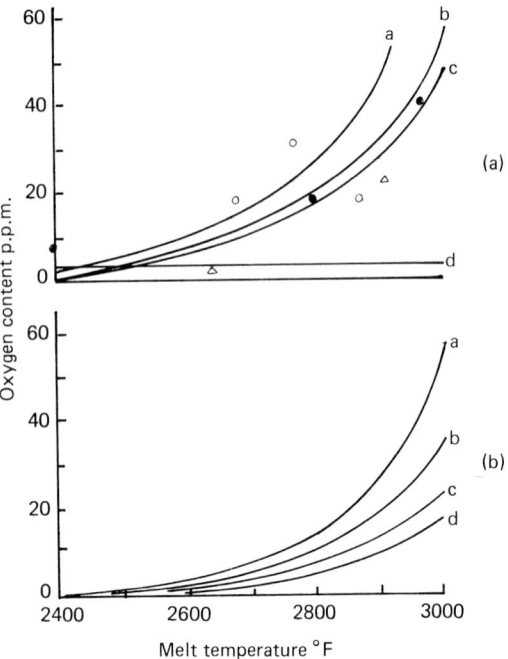

Figure 2.18 (a) the effect of Si content on the equilibrium oxygen–temperature relationship in a Fe – 4.28% C iron; a, 1.0% Si; b, 2.0%; c, 3.0% Si; d, 4.28% Si. Selected experimental measurements, △ spheroidal; ○ malleable; ● grey iron. (b) the effect of Al content on the equilibrium oxygen–temperature relationship in a 4.28% C alloy; a, 0.005% Al; b, 0.01% Al; c, 0.02% Al; d, 0.03% Al (after ref. 81)

oxygen levels fell slightly lower than the Si–O equilibrium values, they were far above the Al–O values. The later measurements by Katz *et al.*[82] confirmed that a critical level of Al must the exceeded before the oxygen level is controlled by Al deoxidation. This value increases with increasing Si content from ~0.015% for malleable irons (~1.4% Si) to ~0.025% for spheroidal irons (~2.5% Si).

Nitrogen absorption in cast iron

Control over nitrogen content is important to take advantage of its beneficial effects and to avoid its undesirable characteristics. Nitrogen in solution can be determined by the Kjeldall method and insoluble nitrogen by the Beeghly method[84]. Nitrogen solubility in Fe can be calculated using Sievert's Law. Nitrogen gas dissolves in Fe by the reaction:

$$N_2 \rightleftarrows 2N$$

where N_2 is in g.

The equilibrium constant for this reaction is:

$$K = \frac{a_N}{(P_{N_2})^{1/2}}$$

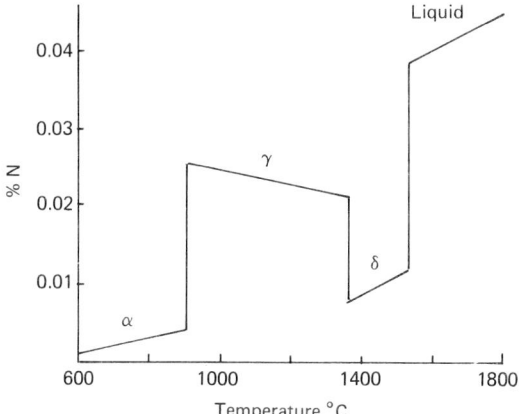

Figure 2.19 Effect of temperature on the solubility of nitrogen at one atmosphere pressure in Fe

where a_N is the activity of nitrogen and P_{N_2} the partial pressure of nitrogen gas in equilibrium with the Fe phase. The standard free energy of solution of nitrogen in iron, $\Delta G°$ is given by

$$\Delta G° = RT \ln K = \Delta H° - T\Delta S°$$

where $\Delta H°$ is the standard enthalpy of solution and $\Delta S°$ is the standard entropy of solution.

For dilute solutions a_N equals the percentage nitrogen (%N) dissolved in Fe. Consequently, assuming $\Delta H°$ and $\Delta S°$ to be temperature independent:

$$\log (\text{wt\% N})_{1 \text{ atms. } N_2} = -\frac{A}{T} - B \qquad (2.7)$$

where A and B are constants and T is the temperature in K.

The solubility of nitrogen in pure Fe is shown in *Figure 2.19*. Equilibrium solubility in the liquid phase has been described by the equation[85]:

$$\log (\text{wt\% N})_{Fe} = -\frac{247}{T} - 1.22 \qquad (2.8)$$

This solubility is much reduced in commercial irons[86-89]. The influence of various solute elements is shown in *Figure 2.20*.

The equilibrium solubility in a liquid iron of known composition can be estimated using activity interaction parameters and solute concentrations in the relationship:

$$\log (\text{wt\% N}) = \log (\text{wt\% N})_{Fe} - \sum_i e_N^i (\text{wt\% i})$$

$$- \frac{1}{2} \sum_i e_N^{ii} (\text{wt\% i})^2$$

$$- \sum_{ij}^{i \neq j} e_N^{ij} (\text{wt\% i})(\text{wt\% j}) \qquad (2.9)$$

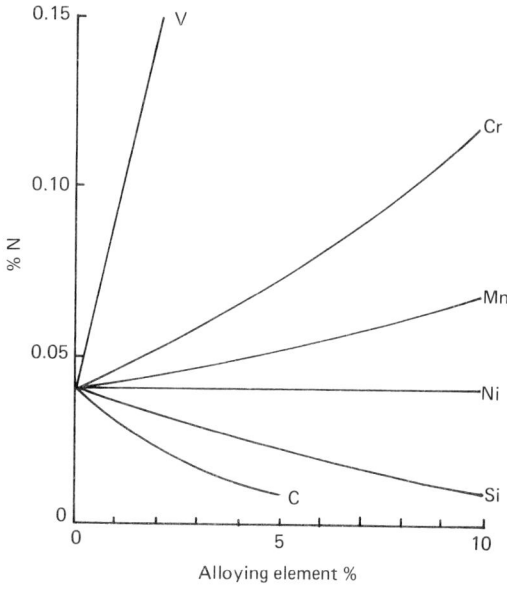

Figure 2.20 Effect of alloying elements on the solubility of nitrogen at one atmosphere pressure in Fe at 1600 °C

where e_N^i, e_N^{ii} and e_N^{ij} are the first order, second order and cross interaction parameters between nitrogen and alloying elements i and j, respectively and where $\log (\text{wt\% N})$ and $\log (\text{wt\% N})_{Fe}$ are the equilibrium nitrogen solubility in the cast iron and in liquid pure Fe, respectively. The equilibrium solubility in an iron under 1 atmosphere N_2 and containing 3.8% C and 2.5% Si (typical of a spheroidal iron) at 1600 °C is reduced from ∼0.04% to below 0.01%. Malleable irons have a lower C content and a higher solubility as shown in *Table 2.9*.

The solution of nitrogen in pure Fe–C–Si alloys is diffusion-controlled, but in the presence of surface active elements such as O and S, absorption is controlled by a surface reaction and the rate of solution is reduced although the solubility limit remains the same[90]. Sources of nitrogen include the melting atmosphere, charge, C additives and core and sand binders.

Table 2.9 also shows that the melting process influences the nitrogen level. The highest levels are found with arc furnaces. This is associated with the high percentage of steel scrap and the high ionizing potential of the arc which exposes the iron to a high nitrogen partial pressure. The partial pressure is high in cupola melting, but is reduced by oxygen enrichment and supplementary natural gas injection. The nitrogen content increases as the percentage of steel scrap increases. The nitrogen level in coreless induction furnaces is usually low. This is attributed to the vigorous stirring action in these furnaces which flushes nitrogen from the melt. The nitrogen content

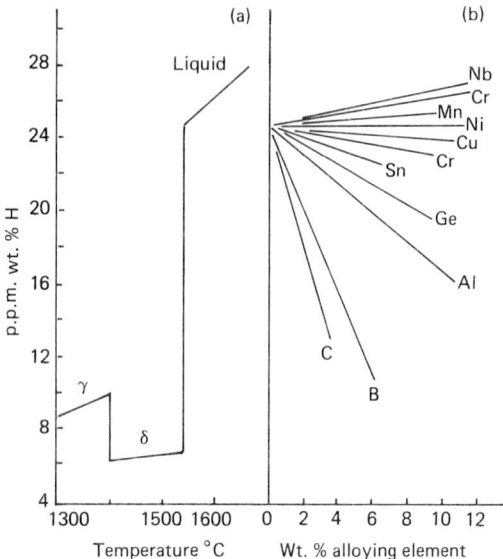

Figure 2.21 (a) The effect of temperature on the solubility of hydrogen at one atmosphere pressure in Fe. (b) The effect of alloying elements on the solubility of hydrogen at one atmosphere pressure in Fe at 1865 K

of base irons is reduced in duplexing practices involving cupolas or arc furnaces and channel furnaces[91]. The vigorous stirring action during Mg treatment reduces the nitrogen content of spheroidal irons.

Hydrogen absorption in cast irons

Figure 2.21 shows the solubility of hydrogen in pure Fe and the influence of solutes on the solubility[92,93]. Hydrogen solubility at 1865 K is decreased by Al, B, C, Co, Cu, Ge, P, S and Sn and increased by Nb, Cr, Mn and Ni.

Carbon control in cast irons

Control of C pick-up is a major factor influencing both the quality and cost of cupola melted iron. The conditions in the combustion zone are favourable for C pick-up because as the charge descends to the well, metal droplets are in intimate contact with incandescent coke. Carburization of the solid material by hot gases is not expected because of the short time that any unit volume of gas spends in the zone and the relatively low temperature of the metal. The solution of C from coke by liquid iron depends on the composition and temperature of the iron, coke-iron interface conditions and the coke quality. Although all the factors controlling the rate of solution are not fully understood, it is well established that the maximum solubility obtained under equilibrium conditions is not achieved in practice. Control over C

pick-up is important because it is necessary to avoid excessive pick-up in malleable iron melting, whereas maximum pick-up is desired in grey flake and spheroidal iron melting in order to reduce melting costs.

Equilibrium solubility data in pure Fe and in multi-component systems has been reviewd by Neumann[94]. The solubility in pure liquid Fe at different temperatures is given by:

$$\log N_{c_{max}} = -\frac{12.7276}{T} + 0.7266 \log T - 3.0486$$

$$(2.10)$$

where N_c is the mole fraction of C and T is the absolute temperature. Alternatively,

$$\%C_{max} = 1.3 + 2.57 \times 10^{-3} T(^\circ C) \qquad (2.11)$$

The influence of small concentrations of solute on this solubility has been expressed in terms of a solubility factor, m (the slope of the relationship in *Figure 2.22*) such that:

$$\Delta N_c^x = m N_x$$

where x is the added solute. Using this approach, the C solubility in a multicomponent system was expressed as:

$$\%C_{max} = 1.3 + 2.57 \times 10^{-3} T + 0.027\% \text{ Mn}$$
$$- 0.31\% \text{ Si} - 0.33\% \text{ P} - 0.4\% \text{ S} \qquad (2.12)$$

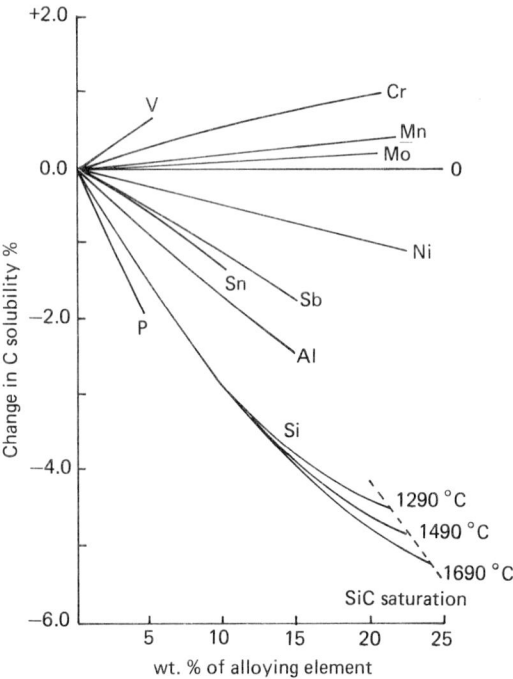

Figure 2.22 The influence of elements on the solubility of C in liquid Fe (after ref. 94)

ogeneous, were examined over the normal cupola operating range of 1450 to 1550 °C. In fact, the slags were heterogeneous, reducing the effective basicities to 1.3–1.6 with a_{CaO} values between 0.03 and 0.06. The equilibrium constant K_{23} less than doubles over the temperature range 1450 to 1550 °C. The studies suggest that desulphurization in the cupola is limited by the absence of C–O equilibrium and heterogeneous slags.

Figure 2.27 shows the desulphurization possible for iron compositions and temperatures typical of basic practice if C–O equilibrium is achieved and if Si–O equilibrium occurs. The calculations assume a heterogeneous 5% Al_2O_3 slag, a 0.75 wt% S level in the coke and weight ratios of iron, coke and slag of 10:1.8:1 respectively. Effective desulphurization is achieved only by melting high S irons at high temperatures or, alternatively, by using more powerful deoxidizers or desulphurizers or agents that suppress the formation of heterogeneous slags.

Al is a powerful deoxidizer and its role in cupola desulphurization has been described by Katz and Spironella[105]. Results are presented which show that S levels as low as those achieved in basic practice can be realized with acid practice using charges containing Al. The analysis suggests that Equation (2.26) controls the rate of desulphurization in the presence of Al. After reducing the FeO in the slag and affecting Si recovery, any remaining Al reduces the MnO level

in the slag resulting in improved desulphurization by the reaction:

$$CaO + S + Mn \rightarrow CaS + MnO.$$

As the Al charge content increases, the MnO level decreases more rapidly in acid slags. This affects a better desulphurization. Consequently, higher Al levels are required to achieve the same improvement in desulphurization in basic practice. It is suggested that Al additions might make it possible to produce irons using < 0.01% S without incurring Si loss.

The attractiveness of the cupola as a melting unit is tempered by the complexity of its operation. This has led to the development of processes in which cupola melting is performed under oxidizing conditions for cost effectiveness. Compositional control is then achieved in an external reaction vessel[109]. The mode of operation can be cold blast, hot blast or oxygen-enriched. The charge consists of steel and cast iron returns only. Melting under oxidizing conditions saves coke, increases combustion efficiency by lowering the $CO:CO_2$ ratio in the stack effluent, increases the melt rate and increases the metal loss by oxidation. This can be of benefit to ferritic spheroidal iron producers with elements Al, Ti, Cr etc.

Recently the concept of a basicity-dependent zero loss level for Si and Mn under oxidizing conditions has been suggested[110]. The Si content of cupola iron tends to an equilibrium value determined mainly by slag composition a characteristic oxygen activity and attempts to add Si to the charge in excess of the zero loss level result in most of the addition being oxidized. Consequently, little or no additions of ferrosilicon are made to the charge. A melt with a low Si level is very C hungry and 3.0% C is readily achieved during melting. Basic composition is achieved with controlled additions of SiC, ferroalloys or C additions in the external reaction vessel. In the studies described[109] this was a multi-porous plug type. Desulphurization falls into this category. S can be controlled at or below 0.015% using a basic slag as described above.

However, this mode of cupola operation presents operational difficulties and, as *Figure 2.2* shows, produces irons of high C content. Spheroidal iron pipe manufacturers find such irons difficult to centrifugally cast. This has led to a movement away from basic cupola melting to the more controllable acid melting with external desulphurization.

The standard free energy of formation of metal sulphides at 1500 °C, the sulphide melting temperature and specific gravity are shown in *Table 2.11*. Clow[111] has used this data to discuss possible desulphurizing methods. The S content of flake irons is usually limited to 0.15%. Mn added as ferromanganese, silicomanganese, spiegeleison or silvery iron can be used to control the effect of S with the aid of late inoculation in low S irons. Without Mn, S forms FeS late in the solidification sequence and this can

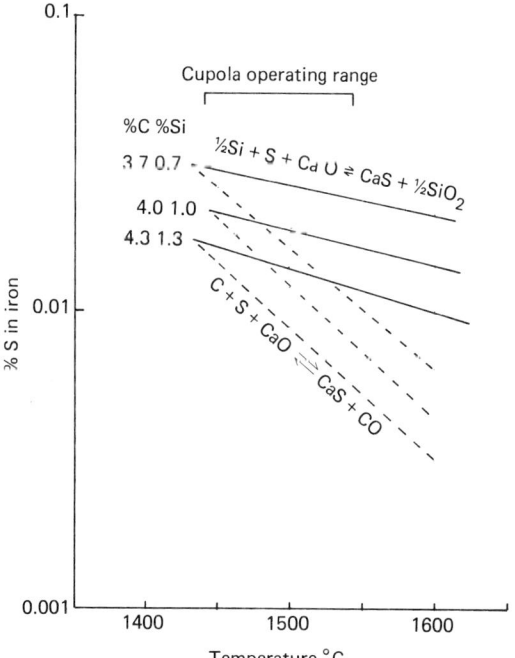

Figure 2.27 Comparison of optimum cupola desulphurization performance based on C–O equilibrium and Si–O equilibrium (after ref. 108)

Table 2.11 Free energy of formation of metal sulphides at 1482 °C with melting point and specific gravity of the sulphides (after ref. 111)

Sulphide	Free energy kcal/gm mole	Melting point °C	Specific gravity
FeS	−25	1190	4.74
MnS	−71	1620	3.99
Na$_2$S	−80 (estimated)	1175	1.85
MgS	−95	2000	2.84
CaS	−172	2450	2.50
CeS	−195	2100	5.00

restrict eutectic graphite cell growth which leads to increased undercooling and even white iron formation.

Deliberate use is made of this effect in heavy section malleable iron castings to encourage white iron formation. Mn, added to balance S, according to the relationship:

$$Mn\% = S\% \times 1.7 + 0.3\%$$

causes the S to partition between Fe and Mn. This forms inclusions predominantly of MnS early in the solidification sequence. This means they are distributed uniformly through the structure and can act as nuclei for eutectic graphite. If the S < 0.03%, although balanced with Mn, the number of MnS inclusions is too small to produce effective nucleation. Unless late inoculation is used, type D graphite may form. There is an upper limit to which Mn may be used to control S because at high S levels it is not combined completely with Mn[112] which leads to Type D graphite formation and, with low pouring temperatures, sub surface blowhole defects associated with manganese iron silicate slags may occur[113].

The active agent in Na desulphurization, Na$_2$O, is provided by soda ash or caustic soda. A high degree of desulphurization is achieved even at low temperatures. This was the standard method for treating grey flake irons for many years. However, the need to attain lower S levels for spheroidal and compacted irons, environmental problems associated with the production and use of Na compounds and difficulties with a corrosive, fluid slag product have led to its almost total abandonment.

The Ca compounds used for desulphurization are CaC$_2$ and CaO. CaC$_2$ is the more common agent. The desulphurization reaction occurs between solid CaC$_2$ and liquid Fe at the surface of the CaC$_2$ releasing C and producing solid CaS. The slag is granular, easily removed from the metal surface and does not attack ladle refractories extensively. The reaction is exothermic and results in smaller temperature losses than with other desulphurizers. Only small changes in C, Si and Mn accompany desulphurization. The

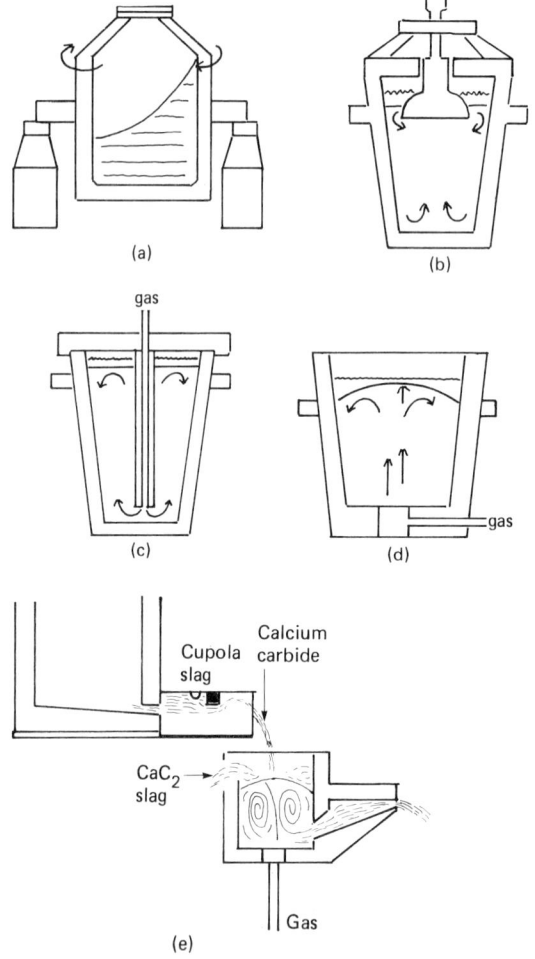

Figure 2.28 Desulphurization methods and typical operation (a) shaking ladle; 0.5% CaC$_2$; 10 minutes; S level 0.075 to 0.017 (b) Rheinstahl stirrer; 0.65% CaC$_2$; 3 minutes; S level 0.08 to 0.009 (c) Volianik process; 1% CaC$_2$; 3–4 minutes; S level 0.1 to 0.01 (d) Porous plug or Gazal process; 0.75% CaC$_2$; 3–4 minutes; S level 0.10 to 0.01 (e) arrangement of treatment vessel for continuous desulphurization of cupola iron

efficiency of S removal depends on the total CaC$_2$ surface area presented to the liquid Fe and the rate of S transport to the surface. The latter depends on the S content of the iron. Practical parameters influencing efficiency are degree of stirring, dwell time, injection rate (when applicable) and CaC$_2$ particle size.

Several methods are available for stirring. The shaking ladle provides stirring by vessel movement (see *Figure 2.28*). The eccentric motion to the vertical axis produces a wave on the bath surface, the crest of which breaks continuously. This folds CaC$_2$ under the metal surface and brings fresh metal to the surface which folds over and covers the carbide[114]. Experience with the shaking ladle[115] has shown it to be a relatively slow process which incurs temperature loss. This makes it unsuitable for the treatment of small quantities of iron that have to be poured from a high temperature. More efficient stirring is achieved in the duortical converter[116], which uses a similar action but reverses its direction every 15 seconds.

Other methods achieve metal slag contact without ladle movement. In the Rheinstahl stirrer a refractory quirl or arm rotates at about 100 r.p.m. at the iron–carbide interface producing a predominantly circular motion with a radial component[117]. The radially flowing stream of Fe carries CaC_2 particles on its surface. These move more slowly than the iron and do not circulate back to the ladle bottom as does the desulphurized iron. Temperature losses are small and little metal splashing occurs. This method is popular in the USA.

Other methods use gas (compressed air, N_2 or natural gas) to stir the liquid iron. Bubbles are forced to rise from the bottom of the ladle. This creates a metal circulation in the ladle that carries unreacted metal upwards with sufficient force to cause it to splash over the top floating CaC_2. Injection is through a graphite lance or tube in the Volianik process[118].

The porous plug process[119,120] uses a porous corundum refractory plug in the bottom of the ladle. The texture of the refractory is such that it produces very fine bubbles which result in a pronounced meniscus at the top of the ladle without serious splashing. The folding and washing action of the rising iron column is very effective and produces good desulphurization. The porous plug was developed for batch processing but has been used extensively for continuous desulphurizing[121-123] as illustrated in *Figure 2.28*. The teapot ladle configuration retains unspent slag and the wide and carefully positioned slag notch allows the spent slag to be pushed out by the action of the gas.

Continuous desulphurization is now used extensively with large tonnage continuously tapped cupolas. The major disadvantage of using CaC_2 is slag disposal. It has an obnoxious smell and can contain up to 3% unreacted carbide. If this becomes damp during storage or makes contact with water after disposal, impure acetylene, which is inflammable and explosive in certain mixtures with air, is liberated. Burnt lime (CaO) is readily available, cheap and has a theoretical capacity for desulphurization. The slag is granular and easily removed from the metal and easily disposed. However, *Figure 2.29* shows that desulphurization occurs at a much slower rate. Coon[124] has shown that additions of fluorspar of up to 10% greatly increase the reaction rate. Although twice as much CaO is required and C and Si losses are greater,

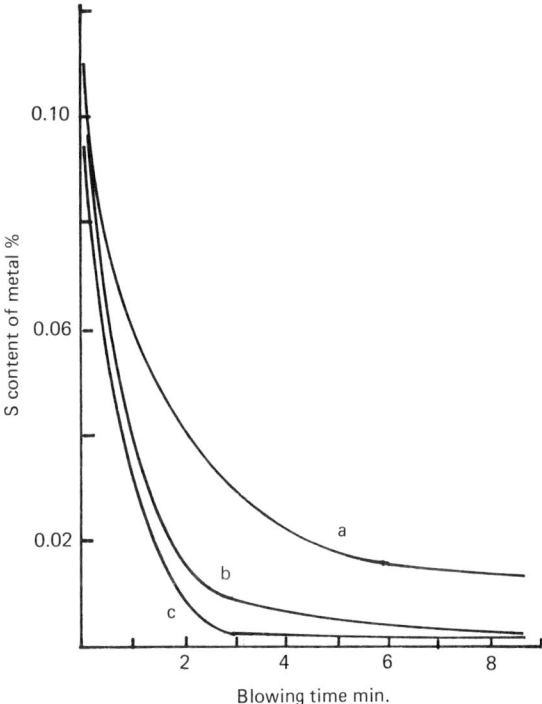

Figure 2.29 Effect of type of desulphurizing agent on the rate of desulphurization in a porous plug. Temperature at start of treatment 1520 °C; air supply 0.23 m³ min⁻¹; 2% addition of desulphurizer a, lime b, lime + 10% fluorspar c, calcium carbide (after ref. 124)

CaO is much cheaper. When used with fluorspar it is a viable alternative to CaC_2.

Mg and Ce are good desulphurizers but are rarely used for this purpose. When used as spheroidizers they will reduce the S content of the iron. For reasons of economy and to avoid incidence of dross defects the S content should be at a low level before spheroidization. However, it has been claimed[125,126] that with processes such as the magnesium converter, the Mg recovery is so high that it is economical to use high S irons so that the Mg first desulphurizes and then spheroidizes. However, there will always be a possibility of dross defects[127].

Recarburization practice

Recarburizers are used primarily for economic reasons to allow various grades of iron to be produced from lower priced or more readily available raw materials, often to the exclusion of pig iron. The main areas of use include:

1. C raising in electric furnaces;
2. external carburization of cupola melted iron;

Table 2.12 Typical analyses of recarburizing materials (after ref. 128)

Type of recarburizer	Fixed carbon	Ash %	Moisture %	Volatile matter %	Sulphur %	Nitrogen %	Hydrogen %	Approximate density kg/m³
Synthetic graphite	99.30	0.40	0.20	0.10	0.05	0.005	–	840
Crystalline natural graphite	86.30	13.20	0.06	0.44	0.35	0.06	–	–
Medium S calcined petroleum coke	98.90	0.40	0.40	0.30	1.50	0.60	0.15	770
Regular Al grade low S calcined petroleum coke	99.30	0.40	0.10	0.20	0.30	0.08	0.04	800
Needle grade dried metallurgical coke	89.70	9.00	0.30	1.00	1.00	1.00	–	640
Brown coal char ex Australia	92.00	2.50	2.00	3.50	0.25	0.60	1.10	640
Pitch coke	98.00	0.50	0.50	0.50	0.40	0.70	0.20	550

3. injection into cokeless and conventional cupolas and
4. injection into arc furnaces.

Properties of carburizing materials are given in Table *2.12*. Synthetic graphite is the most versatile because of high purity and a rapid solution rate. Its main source is from machinings and scrap from primary graphite products and from desulphurizing petroleum coke. Natural graphite is used extensively in the USA because it is plentiful, but disadvantages include a high ash content and low C content.

A popular carburizer is calcined petroleum coke, manufactured by calcining green petroleum coke. Metallurgical coke and brown coal char are derived from coal. Foundry or blast furnace coke is produced by carbonizing coking coals with blended additions of green petroleum coke or breeze at temperatures in the range 1000–1300 °C for up to 36 h followed by quenching into water, drying and crushing. Brown coal char is produced by carbonizing brown coal at about 800 °C. It has the advantage of being a primary product, which is relatively cheap and has a low S content.

SiC is an effective carbonizer although used primarily as an alternative to ferrosilicon for introducing Si. A typical composition for metallurgical grade SiC is: Si – 64%; C – 30%; Al – 0.35%; S – 0.02–0.04% and N – 0.025–0.04%. The addition of SiC provides Si low in Al and C low in S and N. A 1% SiC addition to an induction furnace produces a 0.56% Si and a 0.27% C increment. In a cupola, C is raised by 0.1–0.15% with a 1% addition.

Several factors influence the selection of a recarburizer. The maximum amount of C that can be introduced into liquid iron has been defined in Equations (2.10)–(2.19) and *Figure 2.23*. The rate of solution in practices external to the cupola increases:

1. as the difference between the saturation level and the C content of the iron being treated increases.

Thus, a malleable iron is easier to carburize than a spheroidal iron;
2. as the degree of turbulence in the liquid increases;
3. as the particle size decreases subject to limitations of the particles not becoming airborne. This would produce a deterioration in the environment, a reduced C recovery and electrical shorting. Fine materials (0–3 mm) are used:
 a. with irons of C.E.V. > 4.3%;
 b. for recarburization of cupola melted iron in a porous plug;
 c. for injection into convential and cokeless cupolas and arc furnaces;
 d. for carburization in medium frequency, channel and low-power-to-weight ratio mains frequency furnaces;
 e. as C corrective additions immediately prior to tapping.
4. if a synthetic graphite carburizer is used, this can be of significance in making final C adjustments in high C.E.V. iron at relatively low temperatures.

Economies play an important role in recarburizer selection. It is important to consider the effective cost and not only the cost per tonne of carburizer. This includes additional cost of any alloy required to neutralize S that may be introduced, as well as problems associated with higher ash contents which cause increased slag and dross formation. This increases the tap to tap time and increases refractory wear. Power consumption is a consideration. Complete solution of graphite (90% recovery) can be achieved in electric melting below 1420 °C in a 3.8% C iron. With petroleum coke, an 85% recovery requires a temperature of 1475–1500 °C and a 75% recovery with metallurgical coke requires a temperature in excess of 1500 °C.

Metallurgical considerations include effects due to moisture and H pick up, S and N pick up (see *Figure 2.30*), volatile impurities and the inoculating effect of

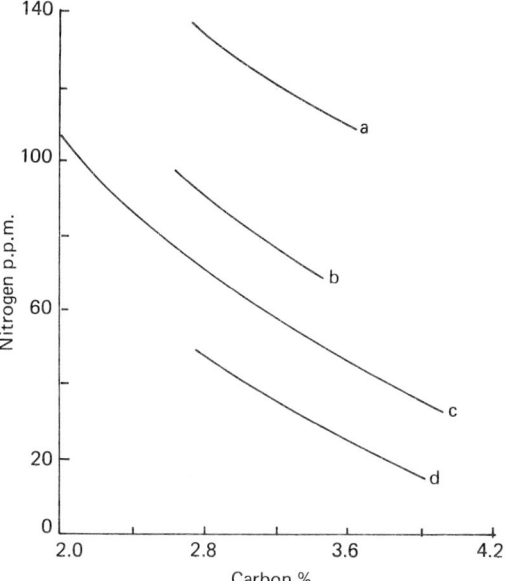

Figure 2.30 Effect of type of carburizer on the nitrogen content of iron at 1500 °C at different levels of carburization a, petroleum coke (N recovery 30–50%) b, metallurgical coke (N recovery 15–25%) c, equilibrium curve d, high purity graphite (after ref. 128)

frequency melting using petroleum coke as a recarburizer to produce two grades of iron suitable for automobile components (3.55–3.60% C, 2.55–2.65% Si and 3.45–3.50% C, 2.4–2.5% Si, 0.20–0.25% Cr). The furnace is of 31 tonnes capacity and is operated on a 4 tonne transfer tapping and charge cycle in conjunction with a 20 tonne capacity channel holding furnace. Once a liquid heel has been established, 4 tonnes of molten iron at 1480–1500 °C are tapped after deslagging and transferred to the holding furnace at 1380 to 1415 °C. The petroleum coke and SiC additions are made to the remaining heel and incorporated by the stirring action. The charge make-up is given in *Table 2.13*. The preheated (300 °C) 4 tonne charge is then added. Ladles holding 150 kg are filled from the holding furnace and inoculated with 0.1–0.3% ferrosilicon prior to casting. Constant analysis monitors the C and Si contents and corrective additions of SiC and coke are made to subsequent charges. Ferrochrome and ferromanganese are used to achieve the Cr and 0.6% Mn level.

This practice has replaced a cupola operation. The S and N levels have been reduced but compared to cupola melted iron when 40% pig iron was used, it is necessary to produce an iron with a 0.2% higher C.E.V. to give the same machinability. This is due to the increased tramp element introduced by using more steel scrap in the charge.

Spheroidization

The practical execution of spheroidization is considered in this section. The ladle transfer method is simple and popular. Mg master alloy is placed in a pocket in the bottom of the ladle (*Figure 2.31*) and liquid iron is poured quickly in a direction away from the alloy in order to reduce the tendency for flotation and burning. Several variations of this method are practised. In the sandwich technique, the spheroidizer is covered with small pieces of steel to delay the onset of the reaction and to lower the temperature of the

graphite recarburizers. The latter effect is one of the reasons why many grey and malleable iron producers recarburize with C recarburizers. In addition to being cheaper, an iron of low but consistent nucleation is produced.

A detailed account of recarburizing practice covering method of addition, melting procedure and recoveries achieved for direct addition or injection to cupola, induction and arc furnaces and for external treatments including ladle additions, porous plug and shaking ladle treatments has been presented by Coates[128]. One of the practices described is mains

Table 2.13 Typical furnace charge and C, Si, N and S balance for 3.55–3.60% C and 2.55–2.65% Si iron. Calculations based on 85% C recovery from petroleum coke, 26% C recovery from SiC and 56% Si recovery from SiC (after ref. 128)

Item	Furnace charge		Carbon		Silicon		Sulphur		Nitrogen	
	Weight (kg)	%	% in material	Contribution charge %	% in material	Contribution charge %	% in material	Contribution charge %	% in material	Contribution charge %
Foundry returns	1150	27.77	3.58	0.9942	2.6	0.722	0.055	0.015	0.0055	0.0015
Pig iron	600	14.49	3.80	0.5506	2.4	0.347	0.050	0.007	0.0050	0.0008
Steel scrap	2250	54.33	0.20	0.1087	0.1	0.054	0.030	0.016	0.0040	0.0020
Silicon carbide sized 0–9 mm	91	2.20	30.00	0.5720	63.00	1.232	0.025	0.001	0.0300	0.0006
Petroleum coke sized 3–9 mm	50	1.21	98.00	1.0285	–	–	1.500	0.005	0.6000	0.0036
Total	4141	100.00		3.2540		2.355		0.054		0.0085

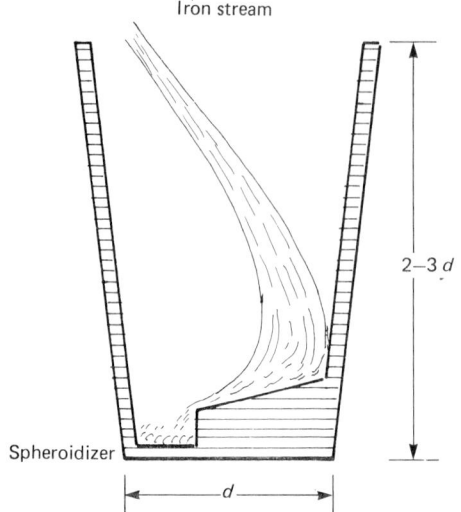

Figure 2.31 Schematic diagram of the open ladle transfer method of treating spheroidal iron

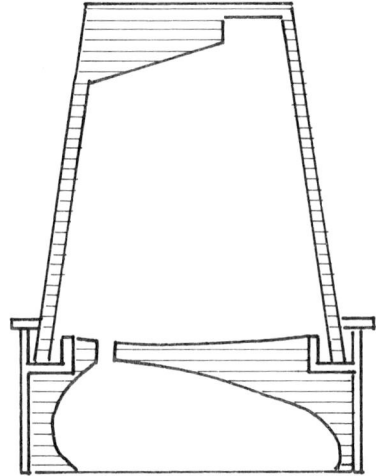

Figure 2.32 Schematic diagram of the covered ladle method for the treatment of spheroidal iron. P is present but not specified (after ref. 129)

liquid surrounding the master alloy. An alternative procedure uses shell sand to protect the master alloy from reaction until it is 'triggered' by breaking the shell with a plunger. Open ladle treatment is a violent process accompanied by considerable smoke fume emission. Although a large height/diameter ladle ratio and use of only two thirds of the holding capacity improves the Mg recovery and reduces spillage,

greater improvements have been achieved using tundish covers[129,130] (*Figure 2.32*).

The use of tundish covers offers the flexibility of a ladle treatment coupled with increased Mg recovery and freedom from smoke emission. A fixed or moveable top cover provides a low pressurized system which prevents entry of outside air without developing substantial in-ladle pressure. Alternatively, the ladle can be placed in a pressurized chamber to prevent Mg from boiling.

Table 2.14 Chemical composition of commonly used Mg-ferrosilicon alloys (after ref. 96, p. 106)

Alloy	Mg	Ce	R.E.	Ba	Ca	Li	B	N	Si	Cu	Al	Fe	La	Comment
1	2.8		3.0	3.1		3.1			48		1.0	Bal		
2	3.0				1.0				46		1.0	Bal		
3	3.0	0.4	1.0		1.0				46		1.0	Bal	0.27	
4	3.0	1.75	2.2		0.4				46		1.0	Bal		Ladle treatment
5	4.5		4.5						42		1.0	Bal		
6	5.5				1.0				46		1.0	Bal		In-mould
7	5.5	0.3	0.75		1.0				46		1.0	Bal	0.20	
8	5.5	0.65	1.25		1.0				46		1.0	Bal	0.45	
9	5.5	1.10	2.0		1.0				46		1.0	Bal	0.75	
10	5.5	0.9	1.8		1.0				46		1.0	Bal		
11	5.5	0.4	0.75		1.0				46		1.0	Bal		
12	5.5	0.5	1.0		1.0				46		1.0	Bal		
13	6.0		11.0		22.0				45		1.0	Bal		High Ca
14	8.0	2.0			10.0				50		1.0	Bal		High Ca
15	9.0				1.25				46		1.0	Bal		
16	9.0	0.35	1.0		1.25				46		1.0	Bal		
17	9.0	0.60	1.0		1.25				46		1.0	Bal		
18	9.25	0.4	0.7		1.0				46		1.0	Bal	0.26	
19	8.0		1.25		8.0				38		1.0	Bal		High Ca
20	9.0			5.0		0.5	0.4		48		1.0	Bal		Self inoculating
21	12.0								40	18		Bal		pearlitic grades
22	30.0		2.0		4.5				50		1.0	Bal		plunging
23	5–10		2.0	1.3	1.0				48		1.0	Bal		

Table 2.15 Chemical composition of some rare earth silicides and other Mg-free treatment additives (after ref. 96, p. 107). P is present but the concentration is not specified

Type	Ce	R.E.	Y	Ca	Al	Si	Fe
S1	P	30.0	P	P	P	P	P
S2				6.15	10.3	75	Bal
S3	16	33		1	1	33	Bal
S4	10	13		1	1	38	Bal
S5	16	23		1	1	30	Bal
S6	23	50		5	1	38	Bal

The stirring methods used for desulphurization can be used for spheroidization. The porous plug is the most popular method. Plunging using a refractory bell to lower the spheroidizer, which is contained in a can, to the bottom of the ladle quickly is a well established technique. Mg recoveries are good but temperature losses are greater than with ladle treatments. In-mould treatments are performed in a specially designed chamber which forms part of the gating system. This method offers high Mg recovery, minimal fading of Mg and no inoculation requirement other than from the ingredients of the ferroalloy spheroidizer.

Mg is the most popular spheroidizing agent. The metal has been used[125,126] but is usually added in alloy form. The alloys are multicomponent. They contain additions to reduce reaction violence, control spheroidization for compacted iron formation, neutralize the effect of impurities on graphite morphology and to control matrix structure. Spheroidizing alloys have been divided into three groups[131]:

1. Ni Mg, Ni-Si-Mg, Ni-Cu-Si-Mg, Cu-Mg etc.,
2. ferrosilicon containing between 3 and 30% Mg with other minor additions (FSM alloys) as illustrated in *Table 2.14* and
3. other alloys – high Ce, CaSi, Ca etc. as indicated in *Table 2.15*.

Alloys in the first group were amongst the first to be used. They have relative high densities resulting in efficient and quiet treatment. They are expensive and present day usage is restricted to irons requiring Ni and/or Cu to achieve matrix structures.

The second group, FSM alloys, are the most used. Factors influencing Mg recovery are the treatment process, temperature, base iron composition and the spheroidizing alloy composition. Higher recoveries occur with FSM alloys low in Mg and high in Si. The reactivity of these alloys depends on the Mg/Ca ratio for a constant Si content, spheroidizing alloy size and temperature. Higher Mg FSM alloys are favoured for plunging. Alloys containing 5–12% Mg are used in tundish ladle treatments and 3–7% Mg alloys are used in in-mould treatment. The spheroidizing effect fades with time. The effect is increased

1. in alloys with higher Mg contents;
2. at higher temperatures;
3. with increased deslagging time and
4. with silica furnace lining.

The graphitization potential of liquid iron

The importance of a reproducible level of base nucleation or graphitization potential of liquid iron prior to its inoculation and casting has been emphasized in several previous sections. We have seen how the primary melting method can influence the potential. Another important factor is composition. Si, P, Ni, Cu and Al increase the potential and Mn, Mo, Cr and V decrease it. These influences have been discussed in terms of electronic, thermodynamic and kinetic factors by Zhukov and co-workers[132-135].

Fe, as a transitional metal, has an electron configuration of incomplete d-orbitals with uncompensated spins. This leads to an exchange equilibrium between s- and d-valency electrons. The C atoms in graphite are bonded by covalent bonds formed by an sp^2 hybrid. However, in the fourth electron, the π-electron, the bond is weaker and joins the c-plane of the hexagonal structure. These electrons are labile and can pass to the s-orbitals of Fe. This moves the $s \rightleftharpoons d$ equilibrium to the right with the result that Fe–C bonding is reinforced and the reaction of $mC + Fe = C_mFe$ tends to the right. Zhukov has suggested that the stability of C–Fe complexes is conditioned by the grouping of six valency electrons of Fe with twelve π-electrons of two benzene rings, forming an eighteen-electron configuration of an arenic complex.

The graphitizing ability of an element in liquid iron is related to the formation of π-electrons and is measured in thermodynamic terms by the C potential π_C ($\pi_C = RT \ln a_C$ where a_C is the C activity) in the corresponding ungraphitized alloy at the eutectic temperature. When $\pi_C > 0$, the alloy has the potential to graphitize, although it may solidify white at small potentials for kinetic reasons i.e. at fast cooling rates. When $\pi_C < 0$, the alloy is stable white and will not solidify grey. Si and Al are competitors to C as valency electron donors. They inject valency electrons into the s- and d-bands of Fe and increase the concentration of π-electrons in the liquid, as indicated by the broken lines in *Figure 2.33*.

The π-electrons are responsible for the bonding in the graphitoid arenic clusters and, consequently, their number and size increase, enhancing the submicroheterogeneity of the liquid. Si does not behave in this manner at all concentrations. As its concentra-

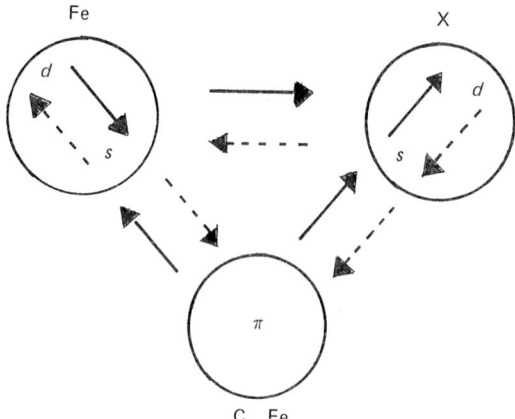

Figure 2.33 The electron interchange between Fe, alloying element X and graphitoid C_mFe clusters in liquid cast iron (after ref. 135)

tion increases Si transforms from a donor to actively interacting with C to become a carbide stabilizer, forming a silicocarbide and then, SiC. This behaviour is reflected in the variation of π_C with composition shown in *Figure 2.34*.

Al enhances graphitization, inhibits it, promotes it and finally suppresses it as its concentration increases. Cr and Mn are electron acceptors, strong carbide stabilizers and dissolve in cementite substitutionally. Electron transitions during alloying with these elements are indicated by the full lines in *Figure 2.33*. They remove electrons from the s- and d-bands of Fe, increasing the density of d-electrons with unpaired spins and also lower the number of π-electrons in the system. Hence, the number and size of the graphitoid arenic clusters in the liquid is reduced. Thus π_C is reduced until it becomes negative, as shown in *Figure 2.34*.

Zhukov has extended the analysis to the calculation of the temperature of the eutectic gutters in the equilibrium and metastable systems of several ternary alloys as shown in *Figure 2.34*. During solidification Cr segregates in the first areas of cementite to form, whereas Mn segregates progressively. Hence, Cr increases and Mn decreases the metastable eutectic temperature. This explains why a greater addition of Mn is necessary to induce chilling.

Electron transitions are not the only consideration, otherwise the chilling effect would increase progressively along the 3-d transition row – Mn, Cr, V, T, Sc – whereas it decreases after Cr. This is due, in part, to the decreasing solubility of the elements in cementite, which prevents them from exercising their acceptor properties. Sc is a very strong desulphurizer and deoxidizer and increases the graphitizing potential by removing O and S from solution. Ti removes N from solution promoting a similar effect. V- and Ti-containing irons precipitate their own carbides,

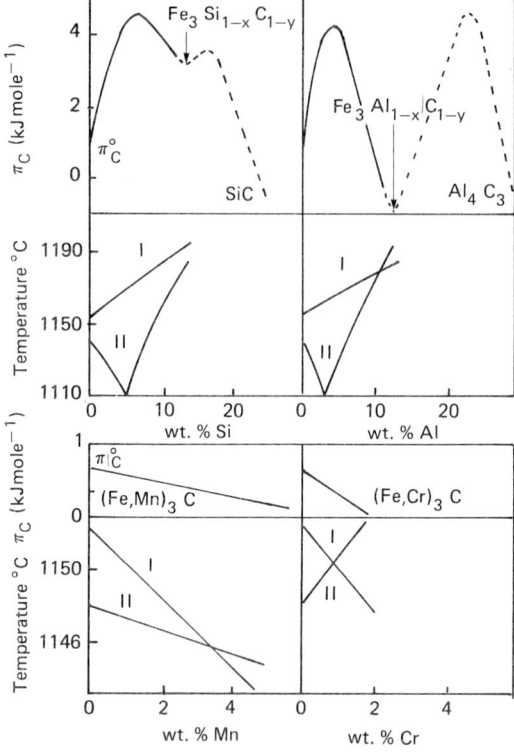

Figure 2.34 The concentration dependence of carbon potential π_c and the profile of eutectic gutters in stable (I) and metastable (II) eutectic systems Fe-C-Si, Fe-C-Al, Fe-C-Mn and Fe-C-Cr. π_C^0 is the carbon potential in unalloyed Fe-Fe$_3$C (after ref. 135)

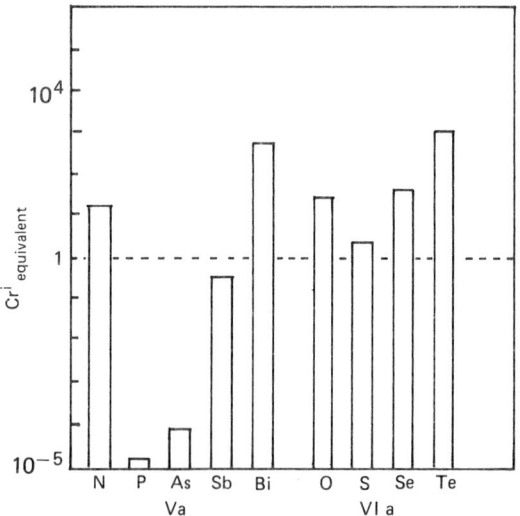

Figure 2.35 Relative chilling action of elements of Va and VIa subgroups in the periodic table. Cr_{equiv}^i is the ratio of weight % Cr/weight % i where weight % i is the weight of element i required to produce the same chilling effect as the Cr addition (after ref. 135)

Table 2.16 The graphitizing and carbide stabilizing effect of elements relative to Si

Graphitizers	Carbide stabiliziers
C + 3.0	Mn − 0.25
Ni + 0.3	Mo − 0.35
P + 1.0	Cr − 1.20
Cu + 0.3	V − 1.0 to 3.0
Al + 0.5	

initially decreasing the C activity in the liquid. However, subsequent precipitation of excess austenite can result in a net increase in the C activity in the remaining liquid to the extent of promoting austenite-graphite eutectic formation.

The interplay of these various factors has been used to account for the chilling effect of Va and VIa subgroup elements which is shown relative to that of Cr in *Figure 2.35*. Based on this reasoning, Zhukov has developed[136] a high S iron (3.2–3.4% C; 1.9% Si; up to 0.5% Mn; 0.3–0.5% S) which resembles malleable iron, but differs from it in that solid state graphitization results in a large number of compact graphite clusters which confer high anti-friction and good machinability.

It is common practice in iron production to adjust the graphitization potential by controlling the Si content. However, the effect of other elements must by considered in alloy irons. The effect of various common alloying elements relative to Si for concentrations normally found in cast irons is given in *Table 2.16*. The effect of 1% Al is approximately equivalent to the graphitizing power of 0.5% Si. One per cent Cr will neutralize the effect of 1.2% Si.

Inoculation of liquid iron

Inoculation of liquid iron is the practice of adding small quantities ($\sim 0.5\%$ for flake irons and $\sim 1.0\%$ for spheroidal irons) of alloys which induce eutectic graphite nucleation and thereby realize the graphitization potential of the liquid. It is a routine step in the production of all grey irons and is often combined with a spheroidization treatment.

There are two main methods of inoculation, ladle and late inoculation. The former describes methods in which the inoculant is added either as the liquid iron stream enters the ladle or just afterwards. Late inoculation refers to any method of treatment after the metal has left the ladle, for example, as it enters the mould (stream inoculation) or in the mould (in-mould inoculation).

The effectiveness of an inoculant is measured by its initial nucleation potency and its ability to maintain its effectiveness during the time interval between treatment and the completion of eutectic solidifica-

tion. Inoculants are rarely effective for more than twenty minutes and it is well established that inoculant composition and melting conditions (i.e. temperature) exert a considerable influence on initial potency and fading behaviour. Inoculant performance can be assessed by observing the amount of chill on the fracture surface of a small wedge test casting, by making eutectic cell counts on a standard test casting microsection or by measuring the degree of undercooling at various cooling rates.

It is important to remember that although a high eutectic cell count usually indicates a low chilling tendency, there is no standard relationship between cell number and chilling tendency. For example, calcium silicide produces a high cell count but does not eliminate chill and Sr-ferrosilicon produces a low chilling tendency without a substantial increase in the cell count. These characteristics are particularly useful in castings prone to porosity[137]. The small increase in cell count restricts the increase in expansive forces exerted on the mould during solidification thus minimizing the risk of porosity.

The benefits of inoculation include a reduced tendency for chill formation in thin sections and greater uniformity of structure, avoidance of very fine undercooled graphite with associated ferrite in thin sections and coarse flakes in thick sections. High strength, low C.E.V. flake irons can be cast without risk of cementite formation. Spheroidal irons are less prone to intercellular segregation and carbide formation. This promotes ductility and reduces the danger of cracking during knockout and fettling and the need for heat treatment to remove carbides.

Traditionally inoculants have been based on graphite, ferrosilicon or calcium silicide. Their development has been traced by Patterson and Lalich[138] and present usage described by Hughes[140]. Some typical inoculating materials are listed in *Table 2.17*. C is only effective in the graphitic form[141]. It is a very effective ladle inoculant for normal flake irons. It is so potent that it is usually used mixed with crushed ferrosilicon-based inoculants to produce a range of proprietary inoculants. It will only inoculate low S flake irons if added late, for example, by stream inoculation. Graphite is not effective as a ladle inoculant for spheroidal irons and its performance as a late inoculant is erratic.

Table 2.18 illustrates the effect of Mn:S ratio on the cell count and the chilling tendency in uninoculated iron[142]. These results show that the iron structure changes progressively from high chilling propensity, low cell count and Type D graphite at low S levels to low chilling propensity, medium cell count and Type A graphite at moderate S levels and high chilling tendency, high cell count and Type D graphite at very high S levels.

These structural features arise as a result of a dual action of S. It provides nucleation sites and, when present in excess, it restricts eutectic cell growth.

Table 2.17 Typical composition of common inoculants

Inoculant	Si%	Al%	Ca%	Ba%	Sr%	Zr%	Mn%	Mg%	Ti%	R.E.%	C%
Normal FeSi	75–80	1.2–2	0.3–1.2								
FeSi–Mn–Zr (SMZ)	60–65	1.2	1–3			5.6	5.6				
FeSi–Ba	60–65	1.0	0.8	0.8		6	6				
FeSi–Ba	60–65	0.5–1.7	1.0	9–11							
FeSi–Ba	60–65	1.5	2.0	5–6			9–10				
FeSi–Zr	80	1.5–2.5	2.5			1.5					
FeSi–Sr	75	< 0.5	< 0.1		0.8						
FeSi–Sr	45–50	< 0.5	< 0.1		0.8						
FeSi–Ti	45–50	1.5	6						10		
FeSi–Ce	45	0.5	0.5							13	
Ca–Si	60	1–2	30								
Low cost 45% FeSi	45–50	0.8	0.8								
45% FeSi–Mg	45–50	0.8	0.8					1.25			
FeSi–La	75	1.5								2 La	
Graphite											99
FeSi + graphite	40–50	1.0	1.5								45

Many foundries, melting electrically, find it necessary to add FeS in order to increase the S level to 0.05% to achieve effective inoculation. This influence of S persists to a lesser extent in inoculated irons.

The most popular inoculant is ferrosilicon. Pure Si and pure Fe-Si alloy are not effective. Effective inoculation depends on the presence of minor elements. Specially formulated inoculants containing combinations of minor elements are more effective than normal ferrosilicon[143-148]. The behaviour of inoculants listed in *Table 2.17* is illustrated in *Figure 2.36*[149].

The inoculation effect is at a maximum immediately after addition and then fades in flake and spheroidal irons[143,150,151]. A general trend is for inoculants producing a high initial cell count to fade rapidly in the first few minutes and those giving a lower initial count to fade less quickly. The Ba-containing inoculants are used with flake and spheroidal irons. Mn lowers the melting point, thus increasing their solution rate. Ba and Ca reduce chill and Ba increases the resistance to fade. *Figure 2.36* shows how Sr ferrosilicon reduces chill with a smaller increase in cell count than normal ferrosilicon. The Ti-bearing inoculant is a very effective chill reducer in thin

sections and helps to prevent N pin-holing. Ce-bearing ferrosilicon is a popular inoculant, particularly for electrically melted flake irons[152]. It is effective in combating the carbide forming tendencies of residual elements such as Cr and its low Al content helps to prevent pin-hole porosity in thin section, high C.E.V. irons. The presence of up to 0.01% Ce in most inoculants, except Sr ferrosilicon, increases spheroid count, decreases chilling tendency and reduces fading rate in spheroidal irons. However, excess additions (> 0.01%) promote carbide formation. The presence of 0.02% Bi in Ce-containing irons increases spheroid count as shown in *Figure 2.36*.

The use of late inoculation techniques leads to the virtual elimination of fading. This permits a reduction in the amount of inoculant used with cost savings. It also means that the inoculant produces a smaller change in iron composition. This can lead to further savings, for example, by enabling foundries producing spheroidal irons by Mg ferrosilicon treatment to use a greater proportion of returns in the charge. Late inoculation is more effective than ladle inoculation in preventing carbide formation in thin sections and there is considerable reduction in smoke and glare.

Mould inoculation is practised in several forms. Powdered inoculant is placed in the pouring bush or in the top of the sprue. Crushed and graded (30–300 mesh) inoculant can be placed in the bottom of the sprue. Alternatively, precast slugs can be used. The in-mould process can be used for inoculating flake irons, the simultaneous inoculation and modification of spheroidal irons and inoculation and controlled modification to produce compacted irons.

The main feature of the process is the provision of a reaction chamber in the flask as shown in *Figure 2.37*. The chamber holds the granulated inoculant and allows a regulated flow of liquid iron over it to

Table 2.18 Effect of Mn:S ratio on cell count and chilling tendency in an iron of composition 3.5% C, 1.9% Si, 0.07% P (after ref. 142)

Mn%	S%	Mn:S	Cell count cells/inch	Clear chill ($\frac{1}{32}$ inch)
0.8	0.002	66.6	362	16
1.0	0.022	45.5	362	13
0.8	0.065	12.3	517	9
0.28	0.20	1.4	1723	29

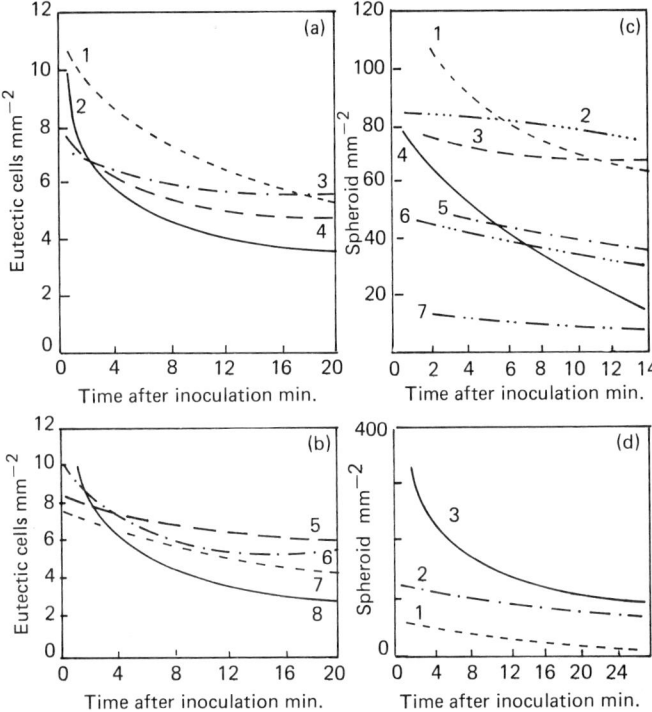

Figure 2.36 (a) and (b) Fading curves for various inoculants added to flake iron (1) Fe-Si-Ba (2) normal Fe-Si (3) Fe-Si-Ce (4) Fe-Si-Sr (5) graphite (6) Fe-Si-graphite (7) Fe-Si-Mn-Zr (8) Ca-Si (c) Influence of various inoculants on spheroid count in spheroidal irons (1) Fe-Si-Sr (2) Fe-Si-Ce (3) Fe-Si-Ba (4) normal Fe-Si (5) Fe-Si-Mn-Zr (6) Fe-Si-graph-ite (7) CaSi (d) Influence of Bi on inoculation fade (1) Fe-Si (2) Fe-Si-mischmetal (3) Fe-Si-mischmetal + Bi

promote immediate and uniform dissolution of the inoculant from the beginning to end of pouring. A parameter referred to as the alloy solution factor is used to characterize the process requirements[153-156]. This factor is defined as the ratio of the pouring rate

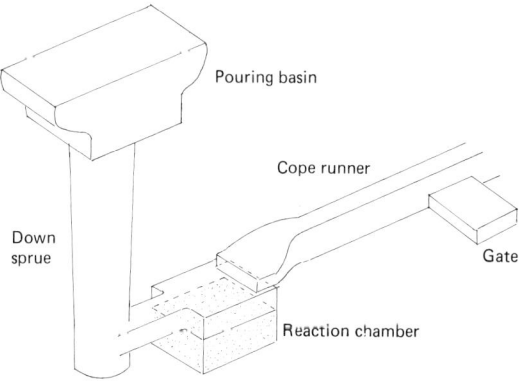

Figure 2.37 An example of an in-mould spheroidizing, gating arrangement for spheroidal irons

to the horizontal cross-sectional area of the reaction chamber. It determines the velocity of the iron as it passes over the inoculant in the reaction chamber. The amount of inoculant alloy that enters into solution in the iron depends on this velocity.

Numerous investigators have shown that there is a linear relationship between alloy solution factor and the residual Mg content in the iron as shown in *Figure 2.38*. With reference to this figure, if the casting weighs 113 kg, is poured in 50 s and a 0.06% residual Mg content is required, 1.5% or 1700 g of 5% Mg ferrosilicon inoculant must be used in a chamber with a cross-sectional area of 60.5 cm^2. This linear relation-ship allows compacted irons to be produced in a controlled manner[157,158].

The use of this technique is more difficult with vertically parted moulds than with horizontally parted moulds. Vertically parted moulds require more ingeneous designs because the alloy chamber is on the parting line. The inoculant may be added after the mould is closed[159,160] or a solid insert may be used[161]. The chamber must have a larger surface area because vertically parted moulds are usually poured faster.

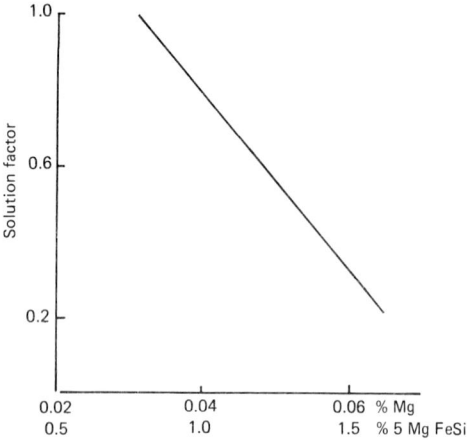

Figure 2.38 The relationship between solution factor, residual Mg content and amount of 5 Mg-Fe-Si inoculant

Cluster castings, in which identical castings are solidified sequentially from one melt, allows an examination of the benefits of the technique and also detection of any variation in properties over the treatment period. Results from this type of examination for a flake and a spheroidal iron using casting

procedures close to standard production methods for cylinder blocks and crankshafts are shown in *Figure 2.39* and *2.40*[162].

Figure 2.39 shows how the chilling tendency varies with casting sequence 1 to 5 for several inoculants in flake iron. The chill depth without inoculation exceeds 10 mm and 3.7 mm is acceptable for the casting. Differences in inoculation behaviour are evident and good ladle inoculants are not necessarily the best mould inoculants. Inoculant 5 displays a strong, well maintained action throughout the treatment period, but normal 75% ferrosilicon was only effective in the first half of the period.

Although inoculant composition influences performances[163], it must be remembered that these results relate to the casting conditions used and do not establish unique relationships. For example, ferrosilicon would be satisfactory if the casting weight was reduced by 50%, or if, for the present casting, the inoculant dissolution rate is reduced by decreasing the cross-sectional area of the reaction chamber.

Figure 2.39 shows that pouring temperature influences inoculant performance and suggests that an optimum inoculant size (1–4 mm) exists. If the grain size is too fine, the initial dissolution is too rapid. If it is too coarse, an initial delay occurs. A potential

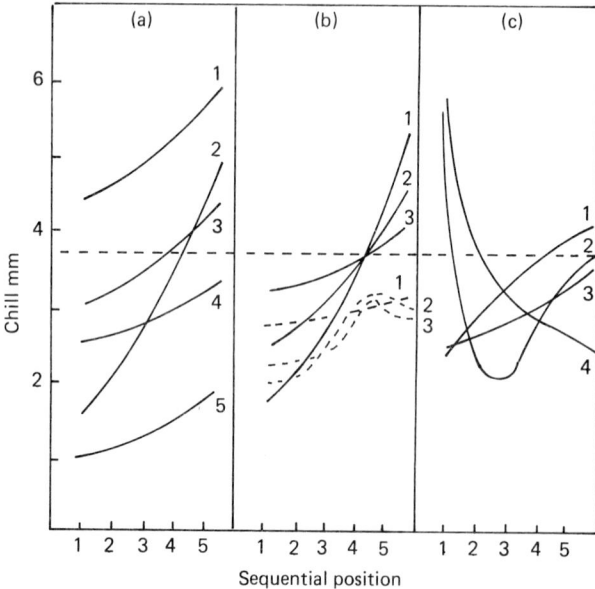

Figure 2.39 Variation of the amount of chill in a mould-inoculated iron with sequential position in a cluster casting (a) for different inoculants (1) 70% Si, 2.2% Ca, 1.2% Al, 1.2% Mn, 1.3% Zr, 1% Mg (2) 75% Si, 0.8% Ca, 1.1% Al (3) 55% Si, 20% Ca, 7% Al, 1% Mn, 4% C, 0.5% Ba (4) 50% Si, 2.5% Ca, 6% Ti, 4% Ce (5) 38% Si, 0.5% Ca, 0.5% Al, 15% Ce (b) for different pouring temperatures (1) 1440 °C (2) 1410 °C (3) 1330 °C; —— with inoculant; ––– without inoculant (c) for different inoculant sizes (1) < 0.6 mm (2) > 2 mm (3) 2–4 mm (4) a single lump (after ref. 162)

45. CARLSON, R., Renaissance of the cupola? *Foundry Facts*, No. 39, p. 11, Amer. Coke and Coal Institute, Washington DC, (April 1976)

46. LEYSHON, H.J., Developments in cupola meling, in *Proc. A.F.S.–C.M.I. Conf. on Cupola Operation, Rosemant*, p. 237 (1980)

47. EPPICH, P.E. and CORBETT, J.L., Operating experience with a 90 inch unlined divided hot blast cupola, *A.F.S. Trans.*, **84**, 427 (1976)

48. HACHTMANN, R.L. Jr. and DRAPER, A.B., Combustion control – a key to better cupola efficiency, *A.F.S. Trans.*, **88**, 489 (1980)

49. TAFT, R.T., The first twelve months operation of a totally gas fired cupola, *Brit. Foundryman*, **65**, 321 (1972)

50. TAFT, R.T., The economics of cokeless cupola melting, *Brit. Foundryman*, **78**, 484 (1986)

51. TAFT, R.T., Ten years of cokeless melting, *Brit. Foundryman*, **72**, 105 (1979)

52. JUNGBLUTH, H. and KORSCHAN, H., Melting in the cupola, *Technische Mittellurigen Knipp Forschungsbereichte*, **5**, 79 (1938)

53. MASARI, S.C. and LINDSAY, R.W., Melting in the cupola, *A.F.S. Trans.*, **49**, 94 (1941)

54. *Cupola Handbook*, 4th Ed., p. 301, A.F.S., Des Plaines, Illinois (1975)

55. WRIGHT, C.C., Anthracite as cupola fuel, *A.F.S. Trans.*, **85**, 40 (1977)

56. PATTERSON, W., SIEPMANN, H. and PACYNA, H., Materials balance and thermal equilibrium of a cold blast cupola. *Giesserei*, **13**, 1 (1961) and **14**, 1 (1962)

57. ANON. Obtaining adequate iron tapping temperature, *Foundry Management and Technology*, **105**, 105 (1977)

58. CREESE, R.C. and BURHANUDDIN, S., Net diagrams – a review and new view, *A.F.S. Trans.*, **90**, 665 (1982)

59. BRIGGS, J., Experiences in the development of a mathematical model of cupola operation, *Brit. Foundryman*, **65**, 340 (1972)

60. MEYSSON, N., Mathematical model simulating cupola operation, *Fonderie*, **294**, 165 (1975)

61. EVANS, W.J., HURLEY, R.G. and CREESE, R.C., A process model of cupola melting, *A.F.S. Trans.*, **88**, 411 (1980)

62. WARDA, R.D. and WHITING, L.V., Microcomputers in the melt shop – four programs to aid cupola operation, *A.F.S. Trans.*, **90**, 883 (1982)

63. WARDA, R.D., DARKE, E.F. and GUINDON, H.P., The true thermal efficiences of thirty cupolas, *A.F.S. Trans.*, **89**, 719 (1981)

64. WARDA, R.D. and WHITING, L.V., A study of cupola melting at 22 Canadian foundries, *A.F.S. Trans.*, **91**, 729 (1983)

65. WOLVERSON, T., The adventure of going electric, *Brit. Foundryman*, **79**, 24 (1986)

66. MIKKOLA, P.H. and WIGHT, T.R., Effectiveness of coreless induction melting, *A.F.S. Trans.*, **83**, 493 (1975)

67. STEFFORA, T.J., Induction melting for ductile iron production, *Modern Castings*, **58**, 70 (1968)

68. BATTEY, J., Induction melting and holding, *Brit. Foundryman*, **71**, p. vii (1978)

69. WILFORD, C.F., The use of large high power density medium frequency coreless furnaces for melting iron, *Brit. Foundryman*, **74**, 153 (1981)

70. SHULHOF, W.P. and COLTHURST, H.L., An American view on electric melting and holding, *A.F.S. International Cast Metals Journal*, **5**, 8 (1980)

71. WALKER, F.W., Some experiences in the use of vertical channel induction furnaces for the melting of grey cast iron, *Brit. Foundryman*, **68**, 209 (1975)

72. EDGERLEY, C.J., The performance of a 40 tonne vertical channel furnace melting grey iron, European Community Research Council Report N 1036 (March 1977)

73. WILFORD, C.F. and LANGMAN, R.D., The channel furnace and its application in the iron foundry, *Brit. Foundryman*, **71**, 196 (1978)

74. BYLAND, G. *et al.*, Holding nodular iron in a channel induction furnace, *A.F.S. Trans.*, **83**, 385 (1975)

75. PACHKIS, V. and PERSSON, J., Industrial Electric Furnaces and Appliances, Industrial Science Publications Inc., New York, (1960)

76. EVANS, W.J. and HETKE, A., Metallurgical aspects of the automatic pouring of gray and ductile irons, *A.F.S. Trans.*, **89**, 277 (1981)

77. WILFORD, C.F., An investigation into the use of electric furnaces to hold magnesium treated iron for the production of S G iron castings, European Community Research Council Report R 963 (July 1976)

78. SWINDEN, D.J., and WILFORD, C.F., The nucleation of graphite from liquid iron: a phenomenological approach, *Brit. Foundryman*, **69**, 118 (1976)

79. HUGHES, I.C.H., Role of gases in cast iron, *A.F.S. Trans.*, **77**, 121 (1969)

80. OSTBERG, G., On the occurrence of oxygen in cast iron, *Trans. A.I.M.E.*, **212**, 678 (1958)

81. GHORPADE, S.C., HEINE, R.W. and LOPER, C.R. Jr., Oxygen probe measurements in cast irons, *A.F.S. Trans.*, **83**, 193 (1975)

82. KATZ, S., McINNES, D.E., BRINK, D.L. and WILKENSON, G.A., Determination of aluminium in malleable iron from measured oxygen, *A.F.S. Trans.*, **88**, 835 (1980)

83. NEUMANN, F. and DÖTSCH, E., Thermodynamics of Fe-C-Si melts with particular emphasis on the behaviour of carbon and silicon, in *Metallurgy of cast irons*, Georgi Publ. Co., St Saphorin, Switzerland, p. 31 (1975)

84. WARDA, H. and PEHLKE, R.D., Nitride formation in solid cast iron, *A.F.S. Trans.*, **81**, 482 (1973)

85. WARDA, H. and PEHLKE, R.C., Nitrogen solution and titanium nitride precipitation in liquid Fe-Cr-Ni alloys, *Met. Trans.*, **8B**, 441 (1977)

86. PEHLKE, R.D., OPRAVIL, O. and BURGESS, P., Effect of sulphur on nitrogen content of liquid cast iron, *A.F.S. Trans.*, **84**, 452 (1976)

87. MORITA, S. and INOVAMA, N., Behaviour of nitrogen in cast iron, *A.F.S. Cast Metals Journal*, **3**, 109 (1969)

88. RAO, Y.K. and LEE, H.G., Rate of nitrogen absorption in molten cast iron, *Ironmaking and Steelmaking*, **12**, 209 and 221 (1985)

89. FARAQUHAR, J.D., Nitrogen in ductile iron – a literature review, *A.F.S. Trans.*, **87**, 433 (1979)

90. PEHLKE, R.D. and ELLIOTT, J.F., Solubility of nitrogen in liquid iron alloys: I. Thermodynamics, *Trans. Met. Soc. A.I.M.E.*, **218**, 1088 (1960)

91. STRONG, G.R., A literature survey on nitrogen in malleable iron, *A.F.S. Trans.*, **77**, 29 (1977)

92. FLINN, R.A., *Fundamentals of metal casting*, London, p. 209 (1962)

93. MINKOFF, I., *The physical metallurgy of cast iron*, J. Wiley and Sons, London (1983)

94. NEUMANN, F., The influence of additional elements on the physico-chemical behaviour of carbon saturated molten iron, in *Recent Research in Cast Iron*, H.D. Merchant, Ed., Gordon and Breach, New York, p. 659 (1968)

95. HEINE, R.W., The Fe-C-Si solidification diagram for cast irons, *A.F.S. Trans.*, **94**, 391 (1986)

96. KARSAY, S.L., *Ductile iron state of the art*, Q.I.T. Fer. Titane Inc., Montreal, Canada, p. 47 (1980)

97. BOOTH, M., Thermal analysis for composition determination of grey cast iron, *Brit. Foundryman*, **76**, 35 (1983)

98. CREESE, R.C. and HEALY, G.W., Metallurgical thermodynamics and the carbon equivalent equation, *Met. Trans. B*, **16B**, 169 (1985)

99. COON, P.M., Control of carbon pick up in the cupola, *Brit. Foundryman*, **76**, 216 (1983)

100. LEYSHON, H.J. and THIBAULT, M.R., Recent work on improved cupola performance, *A.F.S. International Cast Metal Journal*, **3**, 25 (1978)

101. LEVI, W.W., Controlling carbon in the cupola, *American Foundryman*, **12**, 28 (1947)

102. LEYSHON, H.J. and SELBY, M.J., Acid-lined cupola operation with cold blast, cold blast with oxygen and hot blast, *BCIRA Journal*, **20**, 21 (1972)

103. KATZ, S. and LANDEFELD, C.F., A kinetic model for carbon pick up in the cupola, a step beyond the Levi equation, *A.F.S. Trans.*, **93**, 209 (1985)

104. JUNGBLUTH, G.A.H. and STOCKKAMP, K., Chemical reactions in the cupola, *Foundry Trade Journal*, **99**, 377 and 405 (1955)

105. KATZ, S. and SPIRONELLO, V.R., Effect of charged aluminium on iron temperature, silicon recovery and desulphurisation in an iron producing cupola, *A.F.S. Trans.*, **92**, 161 (1984)

106. NEUMANN, F., Comparison of the metallurgical possibilities of hot and cold blast operations in the cupola: Part I. Iron temperature and melting rate, *Giesserei*, **51**, 538 (1964)

107. NEUMANN, F., Comparison of the metallurgical possibilities of hot and cold blast operations in the cupola. Part II. Relationships in regard to oxidation loss and pick up by reduction of iron and its companion elements, *Giesserei*, **51**, 697 (1964)

108. KATZ, S. and REZEAU, H.C., The cupola desulphurisation process, *A.F.S. Trans.*, **87**, 367 (1979)

109. DOELMAN, R.L., BEYERSTEDT, R.J. and SPAULDING, A.E., The reaction vessel process for alloying and inoculating cupola melted cast iron, *A.F.S. Trans.*, **88**, 123 (1980)

110. LANDEFELD, C.F. and PECK, W.J., The relation between silicon loss and metallic silicon in the cupola charge, *A.F.S. Trans.*, **91**, 1 (1983)

111. CLOW, S.C., The effect and control of sulphur in iron, *A.F.S. International Cast Metals Journal*, **4**, 45 (1979)

112. WALLACE, J.F., Effect of minor elements on the structure of cast irons, *A.F.S. Trans.*, **83**, 363 (1975)

113. TONKS, W.G., Subsurface blowholes in grey iron and their association with manganese sulphide segregation, *A.F.S. Trans.*, **64**, 557 (1956)

114. HÖHLE, L., The hot blast cupola with shaking ladle: a melting process for iron foundries, *Brit. Foundryman*, **58**, 335 (1965)

115. CURRY, T., The shaking ladle: a process to desulphurise irons, *The Iron Worker* **31**, 21 (1965)

116. OKURO, T., D.M. Converter, *Modern Casting* **52**, 73 (1967)

117. SCHULZ, H. and GILMORE, E., Reinstahl Quirl, paper presented at BCIRA Conference, Keele, 1972

118. VOLIANIK, M.N., Desulphurisation by the CTIF Volianik Process, *Modern Casting*, **52**, 502 (1967)

119. DUNCAN, J.A., SKIFFINGTON, G. and CARDEN, R.L., BCIRA Morganite porous plug process, *Foundry Trade Journal*, **135**, 225 (1973)

120. GALEY, J., FOULARD, J., LUTGEN, N. and TOMAN, W., Treatment of molten metal by agitation with neutral purge gases (Gazal Process), *A.F.S. International Cast Metals Journal*, **5**, 16 (1980)

121. SHOLL, S.A., Method for continuous desulphurising and recarburising by use of the Gazal porous plug method, *A.F.S. Trans.*, **76**, 123 (1968)

122. SHOLL, S.A., Method for continuous desulphurising and recarburising by use of the Gazal porous plug method, *Modern Casting*, **53**, 73 (1968)

123. McGLOTHLIN, G., Continuous desulphurisation of iron in high production foundries by the porous plug process, *A.F.S. Trans.*, **85**, 5 (1977)

124. COON, P.M., The development and industrial application of lime-fluorspar mixtures for the desulphurisation of cast iron, *A.F.S. Trans.*, **88**, 471 (1980)

125. ALT, A., GUT, K., LUSTENBERGER, H. and TRAPP, H.G., New method of treatment with pure magnesium to produce nodular iron, *A.F.S. Trans.*, **80**, 167 (1972)

126. LUSTENBERGER, H.A. and McCAIN, D.O., Findings from treatment of high sulphur ductile irons, *A.F.S. Trans.*, **85**, 333 (1977)

127. BARTON, R. and FULLER, A.G., Importance of process control producing dross free nodular iron castings, *BCIRA Journal*, **9**, 406 (1961)

128. COATES, R.B., Types, selection and applications of recarburisers in the UK foundry industry, *Brit. Foundryman*, **72**, 178 (1979)

129. FORREST, R.D. and WOLFENSBERGER, H., Improved ladle treatment of ductile iron by means of the tundish cover, *A.F.S. Trans.*, **88**, 421 (1980)

130. FORSHEY, T.L., ISENBERG, G.E., KELLER, R.D. Jr. and LOPER, C.R. Jr., Modification of and production experience with, the tundish cover for ductile iron treatment, *A.F.S. Trans.*, **91**, 53 (1983)

131. DIXON, R.H.T. and HINCHLEY, D., Ferrosilicon magnesium

alloy development for the production of S.G. iron by various processes in 1983, *Brit. Foundryman*, **76**, (5) xvi (1984)

132. ZHUKOV, A.A., LEVI, L.I., KLOITSKIN, Ya. G. and YAREMENKO, G.P., in *Metallurgy of Cast Iron*, Georgi Publ. Co., St Saphorin, Switzerland, p. 97 (1975)

133. ZHUKOV, A.A., Thermodynamics of structure formation in cast iron alloyed with graphitising elements, *Metal Forum*, **2**, 127 (1979)

134. ZHUKOV, A.A., Thermodynamics of microsegregation and the influence of elements on structure of unalloyed and alloyed cast iron, *Metal Sci. Journal*, **12**, 521 (1978)

135. ZHUKOV, A.A. and VASHUKOV, I.A., Influence of elements with different electron configurations on graphitisation during the solidification of cast iron, in *Proc. Solidification Technology in the Foundry and Casthouse*, Inst. Of Metals, Warwick (1980)

136. ZHUKOV, A.A. and VASHUKOV, I.A., Influence of elements with different electron configurations on graphitisation during the solidification of cast iron, *Indian Foundry Journal*, **27**, 1 (1981)

137. CLARK, R.A. and McCLUHAN, T.K., American experience with a new strontium inoculant in grey iron, *Modern Casting* **50**, 88 (1966)

138. PATTERSON, V.H. and LALICH, M.J., Fifty years of progress in cast iron inoculation, *A.F.S. Trans.*, **86**, 33 (1978)

139. MORTON, D.O. and BRYANT, M.D.B., 75 years of inoculation techniques, *Brit. Foundryman*, **72**, 183 (1979)

140. HUGHES, I.C.H., The importance and practice of inoculation in iron castings production, in *Proc. Solidification Technology in the Foundry and Casthouse*, Inst. of Metals, Warwick (1980)

141. MOORE, A., Some factors infuencing inoculation and inoculant fade in flake and nodular graphite irons, *A.F.S. Trans.*, **81**, 268 (1973)

142. WALLACE, J.F., Influence of minor elements incuding sulphur on the morphology of cast irons, in *Metallurgy of Cast Iron*, Georgi Publ. Co., St Saphorin, Switzerland, p. 583 (1975)

143. DAWSON, J.V., Factors influencing the inoculation of cast irons, *BCIRA Journal*, **9**, 199 (1961)

144. LOWNIE, H.W., Barium inoculants resist fading, *Foundry*, **91**, 66 (1963)

145. DAWSON, J.V., The stimulating effect of strontium on ferrosilicon and other silicon containing inoculants, *Modern Castings*, **49**, 171 (1966)

146. McCLURE, N.C., KHAN, A.V., McCRADY, D. and WOMOCHEL, H.L., Inoculation of grey cast iron: relative effectiveness of some silicon alloys and active metals, *A.F.S. Trans.*, **65**, 340 (1957)

147. MICKELSON, R.L., Cerium – silicon alloy reduces chill in grey iron, *Foundry*, **95**, 145 (1967)

148. KANETKAR, C.S., CARNELL, H.H. and STEFANESCU, D.M., The influence of some rare earths (Ce-Ca-Pr-Nd) and Yttrium in the magnesium ferrosilicon alloy on the structure of spheroidal graphite cast iron, *A.F.S. Trans.*, **92**, 417 (1984)

149. MORGAN, H., Inoculation of cast iron, paper presented at 33rd Annual Convention of Inst. Indian Foundrymen, New Delhi, March 1984

150. DAWSON, J.V. and MAITRA, S., Recent research on the inoculation of cast iron, *Brit. Foundryman*, **60**, 117 (1967)

151. FULLER, A.G., Fading of inoculants, paper presented at A.F.S. Conference on modern inoculating practices for grey and ductile irons. Illinois, 1979

152. GUDGING, X., ZONGSEN, Y. and MOBLEY, C.E., Solidification and structures in rare earth inoculated gray iron, *A.F.S. Trans.*, **90**, 943 (1982)

153. DUNKS, C.M., HOBMAN, G. and MANNION, G., Mould nodularising and continuous stream treatment techniques operated in Europe, *A.F.S. Trans.*, **82**, 125 (1974)

154. DUNKS, C.M. and RACE, B., Stream treatment for processing nodular iron, *Brit. Foundryman*, **74**, p. xv (1981)

155. PRUCHA, T.E., Gating for the in-mould process, *Modern Casting*, **72**, 38 (1982)

156. SHEA, M.M. and HOLTAN, S.T., In the mould treatment using elemental magnesium to produce ductile iron, *A.F.S. Trans.*, **86**, 13 (1978)

157. MICHIE, D., The production of compacted graphite iron using the in-mould treatment process, *A.F.S. Trans.*, **91**, 855 (1983)

158. FOWLER, J., STEFANESCU, D.M. and PRUCHA, T., Production of ferrite and pearlite grades of compacted graphite cast iron by the in-mould process, *A.F.S. Trans.*, **92**, 361 (1984)

159. SILLEN, R., In-mould nodulization with delayed pouring in vertically parted moulds, *A.F.S. Trans.*, **87**, 191 (1979)

160. MATHER, D.S., Production experience with automated mold nodulization, *A.F.S. Trans.*, **87**, 513 (1979)

161. MEDANA, R., NATALE, E., PRATO, A. and REMONDINO, M., Use of solid inserts of spheroidising alloys for production of in-mould ductile iron castings, *A.F.S. Trans.*, **87**, 349 (1979)

162. REMONDINO, M., PILASTRO, F., NATALE, E., COSTA, P. and PERETI, G., Inoculation and spheroidising treatment directly inside the mould, *A.F.S. International Cast Metals Journal*, **1**, 39 (1976)

163. DREMANN, C.E., New alloys for making ductile iron in the mould, *A.F.S. Trans.*, **91**, 263 (1983)

164. NIEMAN, J.R., Development of precision inoculation to control the microstructure of cast irons, *A.F.S. Trans.*, **84**, 175 (1976)

165. NEUMANN, F., Improvement in metallurgical results by introducing new inoculation methods combined with pouring and melting equipment, *Electrowarm International*, **36**, 200 (1978)

166. SANDERS, S.D., WEISS, B.R. and NIEMAN, J.R., Wire inoculation – a state of art report, paper presented at Conference on modern inoculating practices for grey and ductile iron, A.F.S./Cast Metals Institute, Illinois, Feb. 1979

167. HUGHES, I.C.H., Importance and practice of inoculation in iron castings production, in *Proceedings of Solidif-*

ication Technology in the Foundry and Casthouse, Metals Society, London (1980)

168. SERGEANT, G.F., Late metal stream inoculation: BCIRA developments, in *Proceedings A.F.S.C.M.I. Conference on Modern Inoculating Practices for Grey and Ductile Irons,* Des Plaines, p. 237 (1979)

169. EASH, J.T., Effect of ladle inoculation on the solidification of grey cast iron, *A.F.S. Trans.,* **49**, 887 (1941)

170. WANG, C.H. and FREDRIKSSON, H., On the mechanism of inoculation of cast iron melts, in *Proc. 48th Int. Foundry Congress,* Varna (1981)

171. FREDRIKSSON, H., Inoculation of iron base alloys, *Mat. Sci. and Eng.,* **65**, 137 (1984)

172. DHINDAW, B. and VERHOEVEN, J.C., Nodular graphite formation in vacuum melted high purity Fe-C-Si alloys, *Met. Trans., A.I.M.E.,* **11A**, 1049 (1980)

173. FEEST, G.A., McHUGH, G., MORTON, D.O., WELCH, L.S. and COOK, I.A., The inoculation of grey cast iron in *Proc. solidifiction technology in the foundry and casthouse,* Metals Society, London (1980)

174. KAYAMA, N. and SUZUKI, K., Influence of S on the dissolving behaviour of graphite in molten cast iron, *Report Castings Research Lab.,* Waseda University, **30**, 61 (1979)

175. LUX, B., Nucleation of eutectic graphite in inoculated gray iron by salt like carbides, *Modern Castings* **45**, 222 (1964)

176. JACOBS, M.H., LAW, T.J., MELFORD, D.A., and STOWELL, M.J. Basic process controlling the nucleation of graphite nodules in chill cast iron, *Metals Technology,* **1**, 490 (1974)

177. GADD, M.A. and BENNETT, G.H.J., Physical chemistry of inoculation in cast iron in *Third International Symposium on the Physical Metallurgy of Cast Iron,* Stockholm (1984)

178. SUN, G.X. and LOPER, C.R. Jr., Titanium carbonitrides in cast iron, *A.F.S. Trans.,* **91**, 639 (1983)

179. MUZUMDAR, K.M. and WALLACE, J.F., Inoculation – sulphur relationship of cast iron, *A.F.S. Trans.,* **80**, 317 (1972)

180. MUZUMDAR, K. and WALLACE, J.F., Effects of sulphur in cast iron, *A.F.S. Trans.,* **81**, 412 (1973)

181. NARO, R. and WALLACE, J.F., Minor elements in gray iron, *A.F.S. Trans.,* **78**, 229 (1970)

182. FRANCIS, B., Heterogeneous nuclei and graphite chemistry in flake and nodular cast iron. *Met. Trans.,* **10A**, 21 (1979)

183. YAMAMOTO, S., CHANG, B., KAWANO, Y., OZAKI, R. and MURAKAMI, Y., Mechanism of nodularisation of graphite in cast irons treated with magnesium, *Metal Sci. Journal,* **12**, 239 (1978)

184. ITOFUJI, H. *et al.*, Comparison of substructure of compacted/vermicular graphite with other types of graphite, *A.F.S. Trans.,* **91**, 313 (1983)

185. LIETAERT, F., HILAIRE, P. and STAROZ, C., Development of more powerful inoculants for spheroidal graphite irons, *A.F.S. International Cast Metals Journal,* **7**, 30 (1982)

Chapter 3

Solidification of cast irons

Introduction

Cast irons solidify by two consecutive processes, primary and eutectic phase solidification. Each of these processes occurs by a nucleation and growth process which can be monitored on a cooling curve. A typical cooling curve for a flake iron of hypoeutectic composition is shown in *Figure 3.1*. The cooling curve recorded during solidification is a direct response to changes in temperature of the thermocouple bead as affected by the latent heat evolved during solidification and the heat dissipated by the mould. Changes in the mode of solidification are reflected in changes in the cooling curve.

The liquid cools at a rate determined primarily by the component design and casting method selected. Some control is possible by selection of pouring temperature, mould materials and the use of chills. However, there are other, and often overriding, considerations in the selection of these variables. Once the liquid iron has been poured into the mould it cools continuously until the temperature TAL is reached when austenite nucleates and grows into a dendritic array during further cooling. The latent heat released slows the rate of cooling as shown in *Figure 3.1*. The liquid continues cooling until temperature TES is reached when graphite nucleation commences the formation of the eutectic cells in the C-enriched liquid between dendrite arms. Nucleation continues, accompanied by increasing latent heat liberation, until the cooling is arrested at temperature TEU. Thereafter, eutectic cell growth becomes established and occurs, initially, during recalescence. It is accompanied by a reducing driving force (undercooling) until a steady state growth temperature TER is established as a result of a balance between heat evolved and heat removed by cooling. As eutectic solidification nears completion the latent heat liberated reduces

gradually and the temperature falls. Eutectic solidification is complete at temperature TEE.

The location of these various temperatures and, hence, the structural features of the iron are determined by various particles present in the liquid which act as nucleants for primary and eutectic solidification. The structural features are also determined by the influence of primary phase formation on eutectic formation and the influence of various solute elements on the nucleation as well as growth process together with their interactions with cooling rate and their joint effect on nucleation and growth.

The foundrymen have at their disposal variations in:

1. the major element content, i.e. C, Si, Mn, S and P;
2. the minor element content, i.e. Al, Cu, Zr, Mg, Ce, Sr, Ti, Bi etc. and
3. several processing variables, including melting conditions, pouring temperature, inoculation, graphitization potential, cooling rate etc.

These can be used to exercise control over solidification events in order to obtain the iron structure necessary to satisfy the specified component properties.

The important structural features that result from these solidification events are the primary phase network which forms the matrix, the eutectic cell and the eutectic morphology. The iron properties are determined by the interaction of these structural units and any structure modifications resulting from solid state transformations during cooling or heat treatment. For example, hypoeutectic flake irons can be likened to a composite. Structural reinforcement is provided by the primary phase dendrites depending on their volume fraction (C.E.V.), structure, fineness

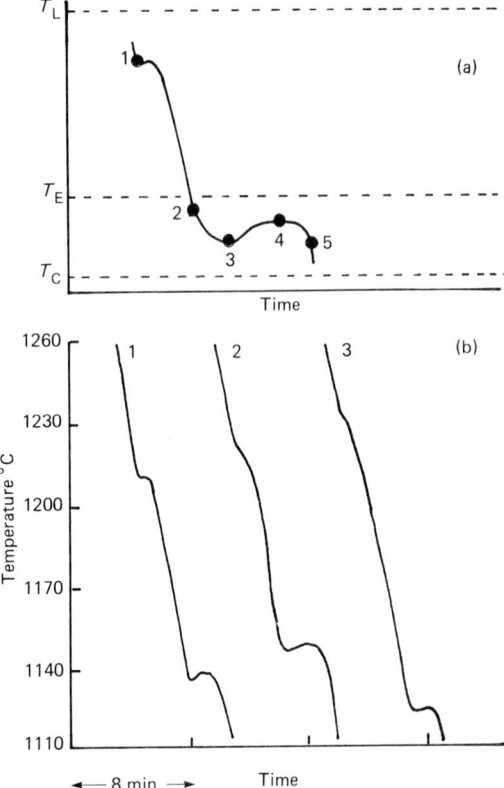

Figure 3.1 (a) Typical cooling curve of a hypoeutectic grey cast iron; T_L is the austenite liquidus temperature; T_E is the graphite eutectic equilibrium temperature; T_C is the carbide eutectic equilibrium temperature; (1) TAL is the temperature of the liquidus arrest; (2) TES is the temperature of eutectic nucleation; (3) TEU is the temperature of eutectic undercooling; (4) TER is the temperature of eutectic recalescence; (5) TEE is the end of eutectic solidification.
(b) Examples of cooling curves (1) standard cooling curve; (2) curve for a Ca–Si inoculated iron; (3) curve for an oxidized iron (after ref. 7)

and continuity. The eutectic fills the spaces between the dendrite arms. The graphite skeleton within each cell imparts a structural weakening depending on the cell size, solidification mode and graphite morphology. The eutectic austenite fills the spaces between the graphite flakes in the cell. It is continuous with the primary austenite and provides a strengthening effect within the cell.

Consequently, an understanding of the solidification process is imperative for the control of iron properties. Methods of monitoring the solidification process are an essential feature of cast iron technology. These aspects are considered in this chapter.

Solidification science

Solidification science has been described elsewhere in the present series. The purpose of this section is to briefly describe our understanding of those events that occur during the solidification of cast irons.

The nucleation process

Primary phase and eutectic cell nucleation are both heterogeneous nucleation events. Our understanding of this process is limited. Available models are extensions of simple homogeneous nucleation theory. Homogeneous nucleation theory considers a given volume of liquid of free energy G_1 at a temperature T below the solid–liquid equilibrium temperature, T_E, within which a stable cluster of atoms forms a solid nucleus. The free energy of this new configuration is given by:

$$G_2 = V_S G_V^S + V_L G_V^L + A_{SL}\gamma_{SL}$$

where V_S and V_L are the volume of solid and liquid, respectively; A_{SL} is the area of the liquid–solid interface; γ_{SL} is the liquid–solid interfacial free energy per unit area and G_V^L and G_V^S are the free energies per unit volume of liquid and solid, respectively.

As $G_1 = (V_S + V_L)G_V^L$, the change in free energy accompanying the formation of the cluster in the liquid is:

$$\Delta G = G_2 - G_1$$
$$= V_s\Delta G_V + A_{SL}\gamma_{SL} \quad (3.1)$$

where

$$\Delta G_V = G_V^L - G_V^S = \Delta H(T_E - T)/T_E$$
$$= \Delta H \times \Delta T/T_E$$

ΔH is the latent heat of fusion. The first term in Equation (3.1) represents a driving force for nucleation (undercooling ΔT). The second term represents an energy barrier. If the interfacial free energy is isotropic, minimization of the surface energy will produce a spherical cluster for which:

$$\Delta G = -\tfrac{4}{3}\pi r^3 \Delta G_V + 4\pi r^2 \gamma_{SL} \quad (3.2)$$

which is represented in *Figure 3.2*. This figure shows that there is a critical free energy barrier, ΔG^*, and a critical radius, r^*, to be exceeded before the cluster of atoms is stable and can be considered to be a nucleus.

Differentiating Equation (3.2) with respect to r gives

$$\Delta G^* = \frac{16\pi\gamma_{SL}^3}{3(\Delta G_V)^2} = \frac{16\pi\gamma_{SL}^3 T_E^2}{3\Delta H^2} \times \frac{1}{\Delta T^2}$$

and

$$r^* = \frac{2\gamma_{SL}}{\Delta G_V} = \frac{2\gamma_{SL} T_E}{\Delta H} \times \frac{1}{\Delta T}$$

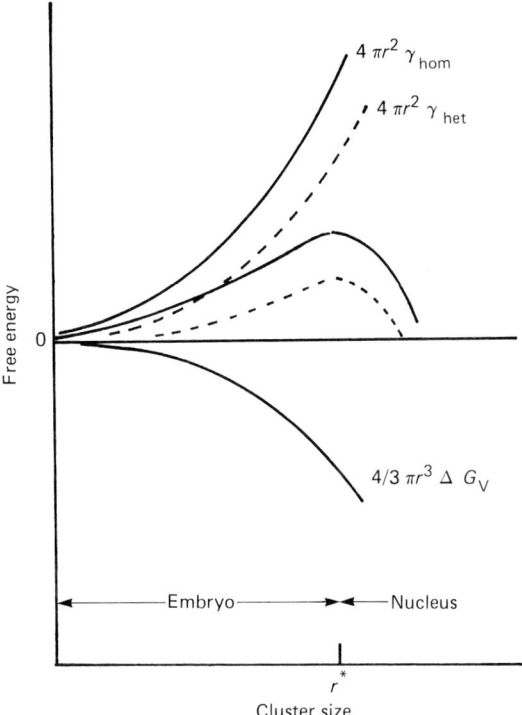

Figure 3.2 Schematic plot of the free energy barrier to nucleation at constant temperature for ——, homogeneous nucleation; – – –, heterogeneous nucleation

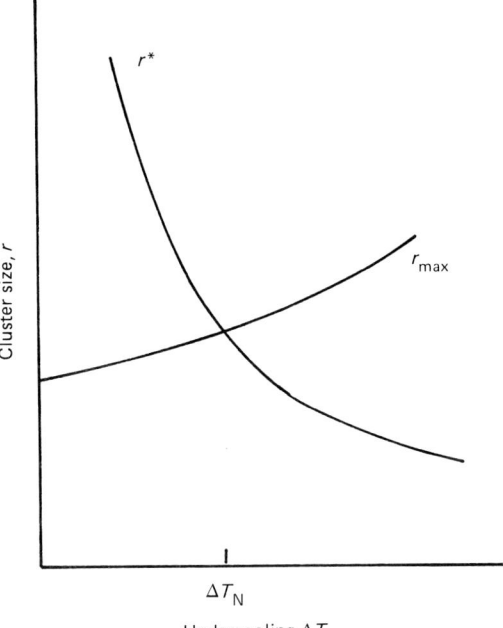

Figure 3.3 The relationship between critical nucleus size, r^*, and the maximum cluster size in the liquid, r_{max}, indicating the undercooling, ΔT_N, at which nucleation is possible

where

$$\Delta T = T_E - T.$$

A liquid is in a state of dynamic equilibrium. Although atomic positions appear to be completely random over a period of time, an instantaneous picture would reveal small close-packed atomic clusters exhibiting the same crystallographic order as in the solid. These embryonic clusters span a range of sizes. The distribution within this range is such that at temperatures just below T_E there are many clusters of small radius. Only a few clusters approach the critical r^* value. As the temperature falls, the number of larger clusters increases and the number of smaller clusters decreases. The number of spherical clusters of radius r is given by:

$$n_r = n_0 \exp - (\Delta G/kT)$$

where n_0 is the total number of atoms in the system and ΔG is given by Equation (3.2).

Equation (3.2) shows that effectively there is a maximum cluster size in the liquid and that it increases as the temperature decreases as shown in *Figure 3.3*. The maximum cluster size exceeds r^* at an undercooling ΔT_N.

Nucleation is possible when the liquid cools to this temperature, but the rate of nucleation, I, depends on the rate at which further atoms join a cluster of critical size. Thus:

$$I_{hom} = K \exp - \left[\frac{\Delta G^*_{hom} + \Delta G_D}{kT}\right]$$

Figure 3.4 shows that the rate of nucleation increases abruptly as the liquid undercooling exceeds ΔT_N which is approximately $0.33\, T_E$ for homogenous nucleation.

Heterogeneous nucleation occurs at much lower undercooling. It is achieved by lowering the free energy barrier, usually by effectively reducing the interfacial energy. This can be achieved by elements adsorbing on the interface (S in cast iron) or with the aid of a suitable substrate such as the mould wall or deliberately added particles, for example sulphides added as a result of inoculation in cast iron. A model is shown in *Figure 3.5*. The change in free energy accompanying the formation of an embryo with this configuration is given by:

$$\Delta G = -V_S \Delta G_V + A_{SL}\gamma_{SL} + A_{SX}\gamma_{SX} - A_{SX}\gamma_{XL}$$
$$= f(\theta)(-\tfrac{4}{3}\pi r^3 \Delta G_V + 4\pi r^2 \gamma_{SL})$$

where

$$f(\theta) = \frac{(2 + \cos\theta)(1 - \cos\theta)^2}{4}$$

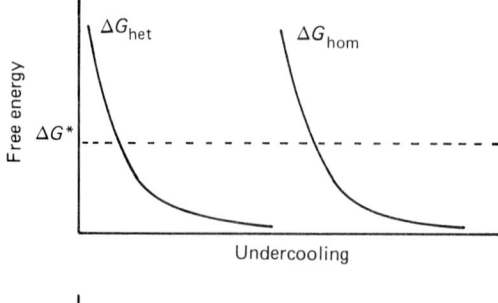

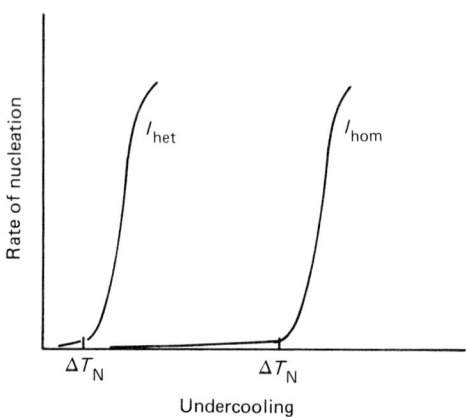

Figure 3.4 The relationship between the critical free energy barrier, ΔG^*, with undercooling and rate of nucleation for homogeneous and heterogeneous nucleation

and

$$\cos \theta = (\gamma_{XL} - \gamma_{SX})\gamma_{SL} \qquad (3.3)$$

Consequently, $\Delta G^*_{het} = f(\theta)\Delta G^*_{hom}$, but $r^*_{het} = r^*_{hom}$ as illustrated in *Figure 3.2*. The rate of heterogeneous nucleation is given by:

$$I_{het} = K \exp - \left[\frac{f(\theta)\Delta G^*_{hom} + \Delta G_D}{kT} \right]$$

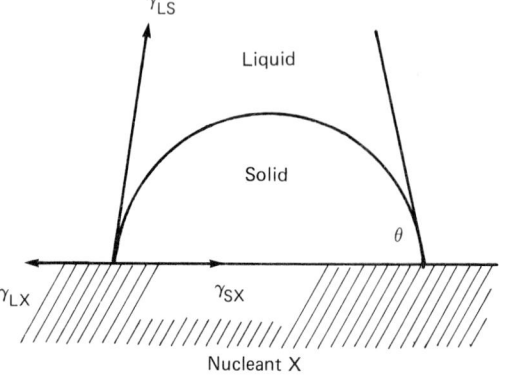

Figure 3.5 Schematic representation of a cap-shaped nucleus

Figure 3.4 shows that the free energy barrier to nucleation is overcome at a lower undercooling by heterogeneous nucleation. This model suggests that the criterion for an effective nucleant is a low θ value. Unfortunately a contact angle approach to the selection of an effective nucleant is of limited value because, in addition to doubts concerning the shape of the nucleus and the precise meaning of a contact angle when applied to a nucleation event involving only a hundred or so atoms, θ is difficult to measure. The paucity of interfacial free energy values makes its calculation from Equation (3.3) difficult.

As a result, attempts have been made to correlate nucleant effectiveness with factors that contribute to γ_{XL} and γ_{SX} in Equation (3.3). A complex crystal structure is associated with a high γ_{XL} value. A low γ_{SX} will occur when there is a small lattice disregistry between solid and nucleant, X.

However, satisfaction of this criterion does not always ensure effective nucleation. For example, Hellawell[2] has shown that a high nucleation rate can be precluded if insufficient undercooling develops prior to recalescence. High superheat and long holding times destroy potential nucleants. A rough substrate surface with crevices increases nucleation potency and the number of nucleation centres can be increased during solidification by dendrite remelting or fragmentation.

The nucleation events of importance in cast irons are those of austenite, graphite and carbide. Several studies have identified additions that promote austenite nucleation in grey[3,4] and spheroidal irons[5,6]. *Figure 3.1* shows the reduced undercoolings associated with the heterogeneous nucleation of austenite in Ca–Si inoculated iron and oxidized iron compared with an uninoculated iron of the same composition[7]. Ca–Si is added primarily to nucleate eutectic graphite but, as with several inoculants, it promotes austenite nucleation. An oxidized iron contains numerous SiO_2 particles which have a small lattice disregistry with austenite[8].

One of the most important liquid iron treatments is inoculation. This is the provision of heterogeneous nucleation centres for eutectic graphite nucleation. Effective treatment prevents excessive undercooling of the liquid during solidification. This avoids the formation of undesirable graphite morphologies and chill formation which will occur if the liquid cools below the carbide eutectic temperature when carbide nucleation occurs readily.

The growth process

The second stage in the solidification process is growth of the nucleus by atom transfer from the liquid to the solid at the solid–liquid interface. Major factors influencing this process, in which the structural features of the casting are developed, are the structure of the solid–liquid interface and the thermal and solute gradients present in the liquid at the interface.

Solid–liquid interface structures are of two types, atomically smooth or atomically rough[9,10]. Atomically smooth interfaces occur in metals (alloys) with a high entropy of fusion (solution). They offer a limited number of sites for atomic additions during growth.

In the absence of structural defects the initiation of a new growth layer requires the formation of a stable interface atom grouping to provide a growth step. This 'nucleation' event must be repeated for each new growth layer. The driving force required for this event determines the interface undercooling for growth. In the presence of structural defects such as screw dislocations, a self-perpetuating growth step is created on the interface and the undercooling required for growth is reduced.

High entropy of fusion materials display considerable growth rate anisotropy resulting in faceted structures. Graphite is a high entropy of fusion material. On the other hand, atomically rough interfaces occur in metals (alloys) with a low entropy of fusion (solution). Many common metals are in this category and they display non-faceted growth forms. Both the undercooling required for growth and the growth rate anisotropy are much smaller. Transitions from faceted to non-faceted growth behaviour have been observed with decreasing growth temperature. Impurities adsorbing on the growth interface can change the interface structure.

The undercooling in the liquid required for solidification can be generated in several ways. *Figure 3.6* illustrates four situations that can develop during solidification. The liquidus temperature is the temperature below which solidification is possible. The actual temperature, which is controlled by the cooling conditions, can increase (positive gradient) or decrease (negative gradient) into the liquid. If the actual temperature is below the liquidus temperature, the liquid is undercooling. The shaded areas in *Figure 3.6* identify the extent of the undercooling. The liquidus temperature in *Figures 3.6a* and *3.6b* is independent of distance into the liquid. This occurs with a pure metal and the liquidus temperature will remain constant during solidification. It will also occur with an alloy when solute mixing in the liquid is sufficient to ensure a liquid of uniform solute concentration at all stages of solidification. This concentration will increase progressively during solidification (C_0, C_1 etc. in *Figure 3.6e*) resulting in a continuous decrease in the liquidus temperature.

Figures 3.6a and *3.6b* represent conditions of thermal undercooling in the liquid. The liquidus temperature increases with distance into the liquid in *Figures 3.6c* and *3.6d*. This situation can only arise in alloys and when mixing in the liquid is insufficient to ensure a uniform solute concentration in the liquid. In that case solute accumulates progressively in the

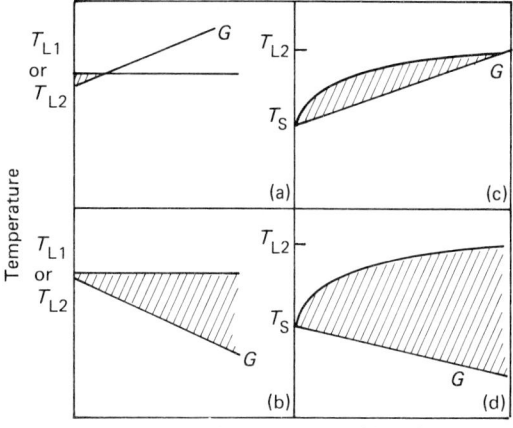

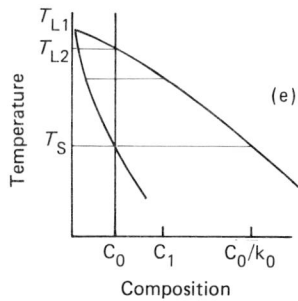

Figure 3.6 Four situations in which undercooling is generated ahead of the solid–liquid interface during solidification (a) thermal undercooling with a positive temperature gradient in the liquid; (b) thermal undercooling with a negative temperature gradient in the liquid; (c) constitutional undercooling with a positive temperature gradient in the liquid; (d) constitutional undercooling with a negative temperature gradient in the liquid; (e) phase diagram

liquid at the interface during solidification. This gives rise to the variation in liquidus temperatures shown. The liquidus variation shown is for the most severe conditions of no mixing in the liquid when the solute concentration in the liquid at the interface quickly attains the value C_0/k_0. This establishes a steady-state condition that can exist for most of the solidification process. The liquidus temperature under these conditions falls to the solidus temperature.

Undercooling generated in this way is termed constitutional undercooling. Solutes most effective in generating constitutional undercooling have an equilibrium distribution coefficient, k_0, that deviates appreciably from unity. Several solutes in cast iron including P, S, Bi, Mn and Mo fall into this category. The condition for the onset of constitutional undercooling is:

$$G/v < mC_0(1 - k_0)/Dk_0 \qquad (3.4)$$

where G is the temperature gradient in the liquid, v is the growth velocity, C_0 is the alloy composition and m

is the slope of the liquidus and D is the diffusion coefficient.

The undercooling generated in *Figures 3.6b, 3.6c* and *3.6d* can be sufficient to alter the course of the solidification process. It can lead to the instability of a planar solid–liquid interface, resulting in an alternative growth morphology, or it may be sufficient for further nucleation to occur. An example of further nucleation are the eutectic cell nucleation conditions created by the 'impurity restricted growth of eutectic cells'.

Austenite primary phase formation

Austenite is the first phase to form in hypoeutectic irons. Once nucleated, austenite primary phase develops into a network by dendritic growth as the liquid cools. Austenite is a non-faceting phase and the dendrite morphology results from instability of a planar growth front. The essential stages of dendritic pattern formation have been established from observations and analysis of the free dendritic growth of pure substances and, in particular, succinonitrile, a transparent body centred cubic material, and pivalic acid, a transparent face centred cubic material. Both solidify in a non-faceted manner similar to that of low entropy of fusion metals[11-23].

The unconstrained growth of these thermal dendrites is controlled by heat flow. It requires a negative temperature gradient in the liquid ahead of the interface to dissipate the latent heat evolved (see *Figure 3.6b*). These observations show that the dendrite tip very quickly attains a parabolic shape and characteristic speed, irrespective of the shape of the nucleus and the nature of the instability that initiates dendritic growth.

Figure 3.7 illustrates the time-independent parameters that characterize the steady-state growth of the dendrite tip. They are the tip radius, R, the primary space λ_1 and the initial secondary arm space λ_2'. The tip radius scales inversely with the undercooling. The relationship between R and growth velocity shows a maximum. The operative growth point on this curve has been defined using a maximum growth criterion. More recently, a marginal stability criterion, which considers the tip to sharpen until it reaches its slowest stable growth rate has been used. The slowest stable growth rate is the growth velocity at which the tip is just marginally stable to perturbations. Secondary arm perturbations initiate at a distance behind the tip. The ratio λ_2'/R is constant over several magnitudes of growth velocity. Coarsening of the secondary arms commences a fixed distance (8 spacings in succino-nitrile) behind the tip, irrespective of the growth velocity. This means that coarsening will always occur and will initiate sooner the greater the cooling rate or undercooling.

We are concerned with the formation of alloy

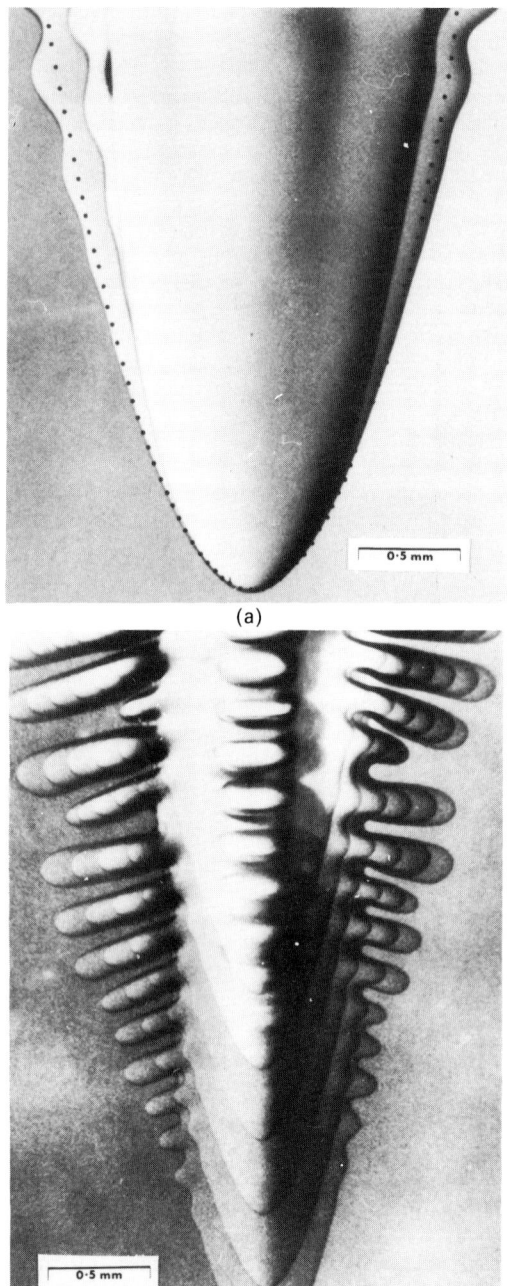

(a)

(b)

Figure 3.7 (a) The tip of a growing succinonitrile dendrite showing the fit to a common parabolic curve (black dots); (b) Superimposition of time-lapsed photographs of a growing succinonitrile dendrite viewed along a [100] direction and showing the development of secondary arms (Courtesy of *Acta Met.*)

dendrites in cast irons. This is a more complex situation to analyse because it requires the simultaneous solution of the thermal and solutal transport equations. Free alloy dendritic growth (*Figure 3.6d*) can occur during equiaxed grain formation. The presence of solute results in the growth velocity showing a maximum at a certain alloy concentration. The tip radius exhibits a minimum value at a somewhat higher solute concentration[24-26].

A more common situation in castings is heat extraction from the liquid through the solid (*Figures 3.6a* and *3.6c*). This heat flow pattern is promoted by high pouring temperature. The induced directional mode of freezing also assists in promoting feeding. The columnar structure generated grows with a positive temperature gradient in the liquid ahead of the interface. If the liquid is pure, the solid–liquid interface is planar. However, in alloys, solute accumulation at the interface creates constitutional undercooling, which destabilizes the planar interface. This promotes a cellular interface when the temperature gradient is high, for example, in welds or a cellular–dendritic interface, or when the temperature gradient is lower, as in castings[27]. This mode of growth is referred to as constrained dendritic growth.

The steady state features of constrained dendritic growth have been analysed with the aid of studies on transparent systems[27,31]. Many of these characteristics are similar to those of pure dendrites. The primary spacing passes through a maximum with increasing growth velocity that correlates with the cell–dendritic transition. At higher growth velocities $\lambda_1 \propto v^{-1/4}$. The primary spacing increases as the alloy concentration increases and varies inversely with $G^{1/2}$. The initial secondary arm spacing is independent of temperature gradient and varies inversely with the growth velocity and solute concentration.

However, the properties of a casting are influenced more by the time-dependent features that develop behind the dendrite tips, independently of their steady-state growth, and whether the mode of solidification is columnar or equiaxed. The formation of equiaxed grains is favoured by low pouring temperature, numerous and efficient heterogeneous nuclei, convection, low temperature gradient and increased solute content[33]. Equiaxed grains usually grow more slowly than columnar grains and develop rod-like dendrites. In contrast, columnar grains exhibit a plate-type structure.

The important phenomena which occur behind the tips include arm coarsening and interdendritic segregation. Coarsening occurs during growth as a result of diffusion processes driven by surface tension. Several mechanisms, including the disappearance of side arms, can operate at different stages during solidification. The arm spacing in the solidified alloy depends on the time, t_f, from the onset of solidification, with a small independent effect of the cooling conditions. Measurements have shown that the final

secondary arm spacing is given by $\lambda'_2 \propto t_f^n$ where t_f is the local solidification time and n is approximately $\frac{1}{3}$ (refs. 33 and 34).

The cooling curve in *Figure 3.1* for the oxidized iron shows that austenite formation occurred over a longer period than in the other irons. This produces a greater volume fraction of dendrites and allows longer for coarsening. Quantitative models of microsegregation based on the Scheil equation for solute segregation modified for the effect of solute diffusion in the solid have been developed by Flemings and co-workers[1,35]. Dendrite arm coarsening influences microsegregation. In general, it reduces it[36].

Microsegregation can be extensive. For example, Bi, Pb and Sn have distribution coefficients approaching 0.1 in cast iron. Their concentration can increase eight times in the last 10% of the liquid to solidify. Initially, interdendritic segregation creates the liquid condition necessary for eutectic solidification. In the later stages, minor elements may segregate sufficiently to change the graphite morphology or to create sufficient constitutional undercooling to promote further eutectic nucleation or even result in the formation of intercellular carbides.

Several studies have reported the influence of solidification conditions and minor elements on proeutectic austenite dendrite formation and on iron properties[37-40]. The observations can be explained using the concepts described above. For example, Ti additions increase dendrite length and interaction area, thus increasing the iron strength. This is attributed to TiN or Ti(CN) acting as a heterogenous nucleant for austenite and Ti promoting constitutional undercooling. When excess Ti is present, it reacts with S instead of Mn and the resulting TiS particles are not as effective as MnS in nucleating graphite. Consequently, delayed eutectic graphite nucleation allows more primary phase formation.

Ti refines the secondary arm spacing in higher C.E.V. irons by enhancing nucleation but increases the spacing in lower C.E.V. irons where the longer freezing period allows extensive coarsening. Small amounts of Ti (0.005–0.01%), through their influence on austenite dendrite formation, can be used to reduce the susceptibility to spiking and associated casting defects[41].

Spiking is the phenomenon of the formation of coarse, aligned austenite dendrites at a temperature just below the austenite liquidus temperature. It was first associated with the observation of black zones on the fracture surface of feeder contacts. The presence of large coarse dendrites hinders the flow of liquid iron from feeder to potential shrinkage locations. Shrinkage then comes to the surface and the interdendritic voids are oxidized and give rise to the dark coloration on the fracture surface. Conditions found to favour spiking include a high liquid iron temperature and excessive holding time, oxidizing melting conditions, high steel content in the charge, slow

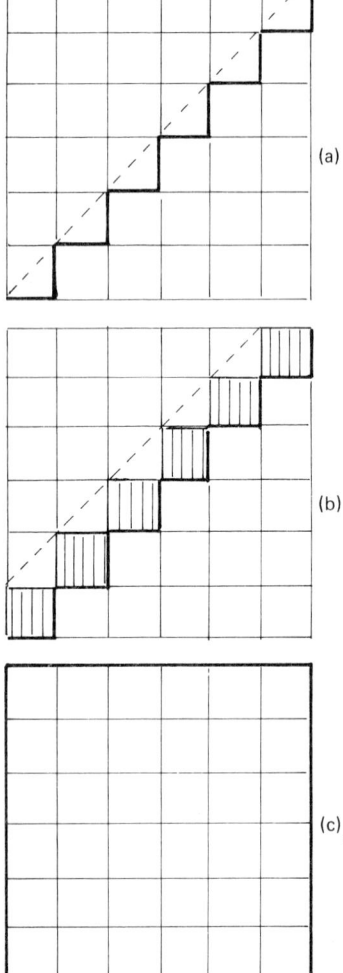

Figure 3.8 (a) Stepped structure on the (11) face of a two dimensional crystal; (b) Advance of the (11) face by addition of atoms to the step; (c) All the high index faces have grown out of the crystal which is bound by low index faces

cooling rate and low C.E.V.. Suggested remedies include Bi and Ti additions, inoculation and effective control over oxidizing conditions in the liquid[42]. Spiking manifests itself internally in spheroidal irons in the form of ferritic bands which contain aligned spheroids and oxide inclusions. This reduces the ductility of the iron[43].

Graphite primary phase formation

Graphite is the first phase to form in high C.E.V. hypereutectic irons. It has a high entropy of fusion and differs from austenite in its solidification behaviour. Whereas austenite grows at a rate controlled by diffusion with all atoms finding a place on its atomically rough solid–liquid interface, graphite growth is controlled by the rate at which different solid–liquid interface structures can incorporate atoms. If a multi-faced particle nucleates in the liquid, the high-index stepped interfaces will grow relatively quickly and disappear, leaving the particle bound by low-index faces. This is illustrated for a simple two dimensional crystal in *Figure 3.8*.

In the case of graphite, the resulting crystal is bound by six prism faces and two close packed basal faces as shown in *Figure 3.9*. Continued growth of the particle under solidification conditions that do not cause interface instability depends on the structure of the growth interfaces. If the faces are structurally perfect, growth requires a nucleation event before the interface can advance one atomic layer by the passage of a step across the interface. However, both surfaces contain defects which provide steps and make growth easier. Growth in the 'c' direction occurs from the step of screw dislocations intersecting the interface[44]. The step of a screw dislocation becomes a spiral during growth. Atoms attach at the step of the spiral and the resulting rotation produces growth of one step height per revolution. A second step defect is a twist or rotational boundary[45]. Growth in the 'a' direction

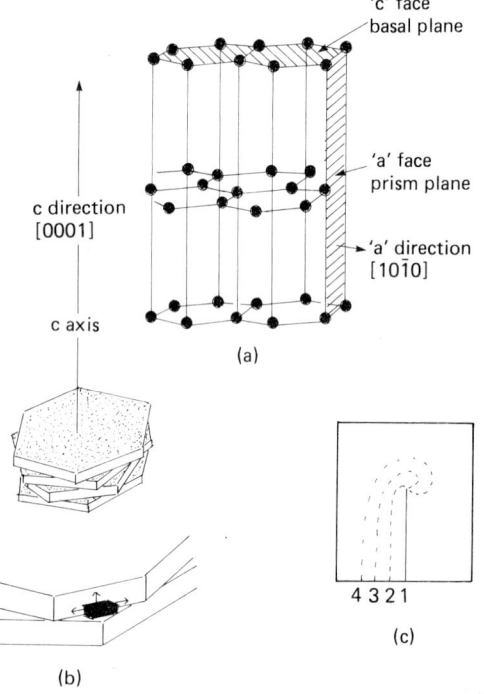

Figure 3.9 (a) Graphite crystal structure; (b) Growth in the [10$\bar{1}$0] by growth of (10$\bar{1}$0) faces as a result of nucleation on steps at rotational faults; (c) Growth in the [0001] by growth at a dislocation step. The successive positions of the step of a screw dislocation show the development of a growth spiral

from prism faces occurs with the aid of this step defect.

Growth rate anisotropy still exists and determines the shape of the graphite phase such that the primary phase in commercial flake irons occurs as flat hexagonal plates known as Kish graphite. Primary graphite formation can be troublesome in the production of high C irons for applications such as ingot moulds. The phenomenon called 'graphite flotation' results in a layer of high C content on the cope surface of a casting. This occurs primarily in heavy, high-C iron castings with extended solidification times. Recent studies have shown that the primary phase forms at a free liquid surface as a consequence of supersaturation that is controlled by composition and temperature[46-48] and not in the melt followed by flotation. These findings throw doubt on the effectiveness of the established practice of holding the melt to dissolve graphite particles. This practice anticipated separation of graphite particles by flotation and attempted homogenization by heating.

Graphite flotation occurs in compacted and spheroidal irons when the primary phase structure is one of several forms of exploded graphite. These graphite morphologies develop when the solid–liquid interface is unstable. Growth instabilities in low entropy of solution alloys that display isotropic surface energy and interface kinetics can be defined by perturbation analysis or constitutional undercooling theory, as outlined in the previous section. Solute has a similar destabilizing effect on a faceted interface but the surface energy and interface kinetics are not isotropic. Consequently, the behaviour of such a surface must be examined in terms of its structure, step source behaviour and growth kinetics. It is found that the resulting growth morphologies are more varied than the dendrite morphology observed in metals.

Indeed, interface kinetics can lead to an increase in the interface stability. Consider a cube growing in a liquid undercooled by a constant amount. Part of this undercooling is used for atomic attachment to the solid and part for solute and heat diffusion, the latter in particular. Heat flow is easier from the corner than from the centre of a facet. Hence, the driving force available for layer growth is greater at the corners. Steps nucleate in these areas and propagate across the interface at a decreasing rate, v, determined by the undercooling available for atomic attachment. This decreases towards the centre of the facet. This leads to closer spacings, d, of the steps on the interface at the centre of the facet. The facet can advance with the same velocity everywhere and remains stable provided v/d is constant across the interface. This renders the faceted interface relatively stable compared to a non-faceted interface, which would develop perturbations. Eventually, the diffusional undercooling at the centre of the facet becomes so large that the undercooling available for atomic attachment becomes very small.

Steps cease to propagate in this area and the planar interface breaks down. This is predicted to occur when:

$$vr/D \; > \; \Delta T/mC_0(1 \, - \, k)$$

where ΔT is the undercooling, D the diffusion coefficient, v the growth velocity and r the approximate radius of the faceted crystal. Once a critical crystal size is exceeded, the corner effect can predominate, leading to faceted dendritic or hopper crystal growth[35,49].

Chernov[50] has analysed the instability of faceted surfaces in a similar way and related the changes in orientation to changes in the kinetic coefficient. Chernov also considers the formation of hillocks on unstable facets, showing that the angle of the hillock increases with the degree of supersaturation. Minkoff[51] has examined the growth features of graphite in several alloys including cast irons. A central feature of his explanation of cast iron structures is the instabilities that occur on primary and eutectic graphite during growth in contact with the liquid. These instabilities include:

1. primary graphite branching dendritically from $(10\bar{1}0)$ faces;
2. eutectic crystals branching out of the edge of flakes;
3. steps on graphite becoming unstable and forming ledges (see *Figure 6.21* in Chapter 6);
4. these ledges becoming unstable and leaving the surface to grow into the liquid (see *Figure 6.26* in Chapter 6);
5. ledges growing round crystals;
6. pyramidal projections growing on (0001) faces (see *Figure 6.12* in Chapter 6) and
7. elongated surface pyramidal crystals.

Eutectic solidification

Eutectic solidification is a nucleation and growth process. Nucleation occurs at temperature TES in *Figure 3.1*. Inoculants present in the liquid and the solidification conditions determine the nucleation temperature and the eutectic cell count. They also influence the solidification conditions when growth commences. The main structural features are determined during growth. Directional solidification has been used to study eutectic growth without interference from nucleation. Alloys have been divided into two groups on the basis of these studies[52].

1. normal eutectics, in which both phases of a binary alloy show non-faceting growth behaviour and
2. anomalous eutectics, in which one phase has a high entropy of solution and is capable of faceting.

Figure 3.10 shows a section through a three dimensional model showing the main types of eutectic structure observed in pure binary alloys. The section

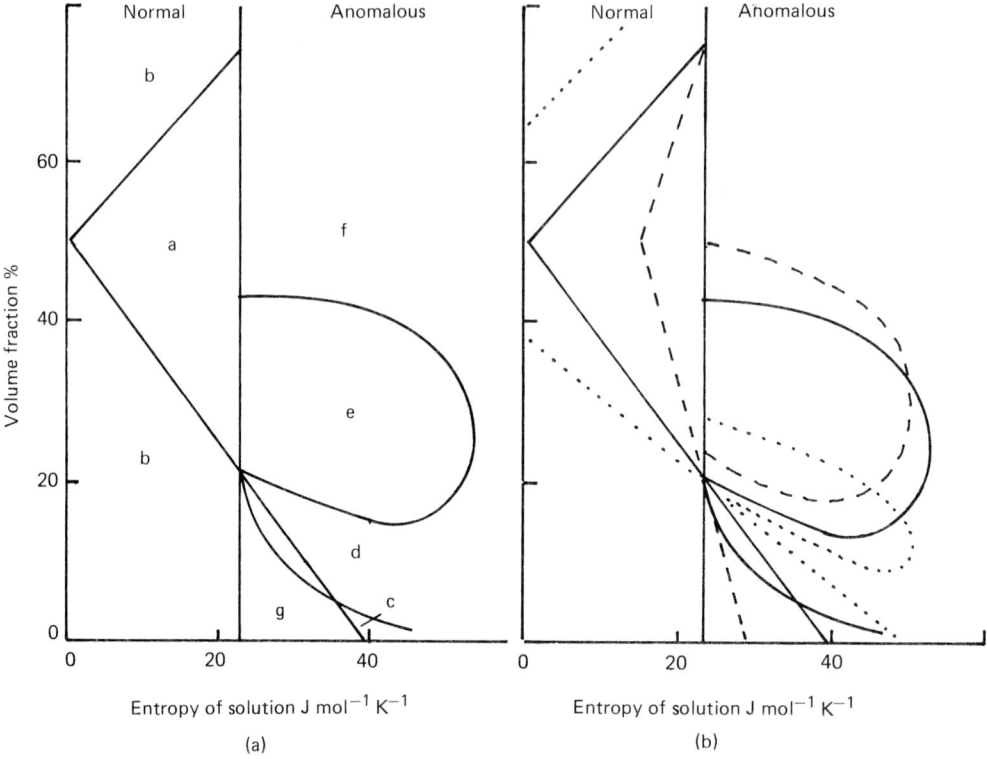

Figure 3.10 Classification of eutectic microstructures in terms of entropy of solution and volume fraction of the second phase (a) at a growth velocity of $5\,\mu m\,s^{-1}$ a, normal lamellar; b, normal rod; c, anomalous broken lamellar; d, anomalous irregular flake; e, anomalous complex regular; f, anomalous quasi-regular; g, anomalous fibrous structure (b) location of the boundaries at different growth velocities; — —, $0.5\,\mu m\,s^{-1}$; ———, $5.0\,\mu m\,s^{-1}$; – – –, $50\,\mu m\,s^{-1}$ (after ref. 49)

shows the influence of entropy of solution and volume fraction of the second phase on structure type and the influence of growth velocity on the position of several boundaries dividing the structural types. The vertical line at $\Delta S = 23\,\text{J mole}^{-1}\,\text{K}^{-1}$ divides normal and anomalous types.

Normal structures which form between two non-faceting phases display lamellar or rod morphologies. The boundary dividing these two structures falls to lower volume fraction values with increasing ΔS because the interphase boundary energy anisotropy which stabilizes the lamellar structure increases as the tendency towards faceting increases. The two phases in a normal structure have similar growth characteristics. Hence growth occurs in a coupled manner at a relatively isothermal interface which is undercooled a few degrees.

This growth process has been analysed on several occasions[1], in particular by Jackson and Hunt[53,54]. They presented a steady-state solution for the solute diffusion equation for lamellar growth at a planar interface. The interface must undercool for growth and this undercooling was attributed to:

1. ΔT_d–the undercooling required for solute diffusion in the liquid; this allows the two phases to be in contact with different compositions in the liquid;
2. ΔT_c–the undercooling required for the formation of phase boundaries and
3. ΔT_k–the undercooling required for atomic attachment at the interface; this contribution is negligible compared to the other two undercoolings for normal eutectics.

Their analysis relates total interface undercooling, ΔT, with growth velocity and interlamellar spacing, λ, by a 'growth' equation:

$$\Delta T/m = v\lambda Q + a/\lambda \qquad (3.5)$$

where m, Q and a are constants.

Equation (3.5) is represented in *Figure 3.11* for a constant growth velocity. This growth curve does not provide the unique solution that experiment suggests exists between v, ΔT and λ because the analysis assumes, a priori, a steady state of uniform spacing and does not contain information about the states of transient evolution leading to the steady state. The

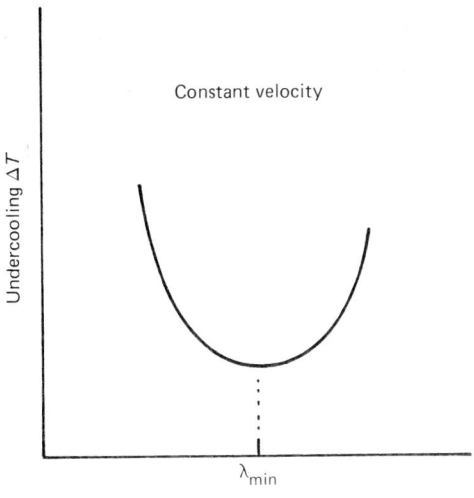

Figure 3.11 Growth curve relating undercooling and spacing at constant velocity

operative growth point on the curve has been discussed extensively[1] and the concensus is that growth occurs close to, but to the right of, the minimum. For the minimum point:

$$\lambda^2 v = a/Q$$

$$\Delta T^2/v = 4m^2 aQ$$

and

$$\Delta T\lambda = 2ma.$$

This form of relationship has been observed in several lamellar systems, but uncertainty in the values of m, Q and a makes assessment of the agreement between theory and experiment difficult. There must be a structural mechanism available for rapid spacing changes if growth is to be maintained close to the minimum under non-steady state conditions. The interlamellar spacing adjusts readily to the value appropriate for a changed growth velocity by the rotation of lamellar faults[55].

Temperature gradient in the liquid has little effect on the growth of lamellar structures in pure binary alloys. However, it plays a role in the development of constitutional undercooling in the presence of impurities. This leads to interface instability, as in single phase alloys, and a cellular or colony structure. The use of the term 'cell' in this context should not be confused with its long established use to describe a eutectic grain in cast irons.

Many casting alloys, including cast irons, display anomalous eutectic structures. The different growth behaviour of the two phases produces a different growth pattern than in normal eutectics. Growth is less coupled at a non-isothermal interface. This results in branched and irregular morphologies. The

structural features depend on the growth mechanism of the faceting phase, the temperature gradient in the liquid and the difference in response of the two phases to growth rate fluctuations.

When the eutectic contains a small volume fraction of faceting phase, overgrowth of this phase by the non-faceting phase in response to a growth rate fluctuation is restricted by local branching or twinning of the faceting phase. The broken lamellar structure predominates when $V_f < 10\%$ (region 'c' in *Figure 3.10*) and local branching or splitting of the faceting phase occurs within the lamellar plane. When the volume fraction is greater, the faceting phase branches outside the lamellar plane. This produces the flake structure (region 'd'). The frequency of overgrowth decreases as the volume fraction of faceting phase increases. When V_f exceeds about 20%, cell formation at the interface becomes the preferred mechanism for overcoming growth restrictions. Region 'e' contains an increasing amount of complex regular structure as V_f increases[56,57]. When V_f exceeds about 40%, the quasi-regular structure containing regular dispersions of faceting phase is found in region 'f'. Eventually, as V_f exceeds 50%, the high entropy phase becomes the matrix[58,59]. Curvature effects at the solid–liquid interface can prevent the high entropy phase from faceting, and a regular structure is formed.

The Fe–Fe$_3$C eutectic falls into this category. Region 'g' defines a fibrous structure often found at higher growth velocities in systems that otherwise display broken lamellar, irregular flake or complex regular structures. Regions 'c' and 'd' tend to shrink and region 'g' to expand as the growth velocity increases. This accounts for growth-velocity-dependent transitions such as quench modification (flake–fibre transition) in Al–Si alloys.

The flake structure is the most studied anomalous structure because of its commercial significance. Experiment has shown that both the interface undercooling and the interparticle spacing are greater than in lamellar growth. They decrease as the temperature gradient in the liquid increases[60,61].

Attempts to account for these differences have been discussed by Elliott[1]. The first analyses attributed the differences to the kinetic control of the growth of the faceted phase which was neglected in the lamellar growth analysis[1,62]. These analyses, based on a planar interface, were only partially successful and did not account for the temperature gradient dependence. More recent analyses[63,64] consider the diffusional and curvature contributions to the undercooling to be the significant ones. They also discuss the modification required to the Jackson and Hunt analysis in order to explain the larger spacing and undercooling.

Lamellar growth analysis indicates that interface non-planarity can have an appreciable effect on solute diffusion and its importance increases with increasing asymmetry of the eutectic point in the phase diagram.

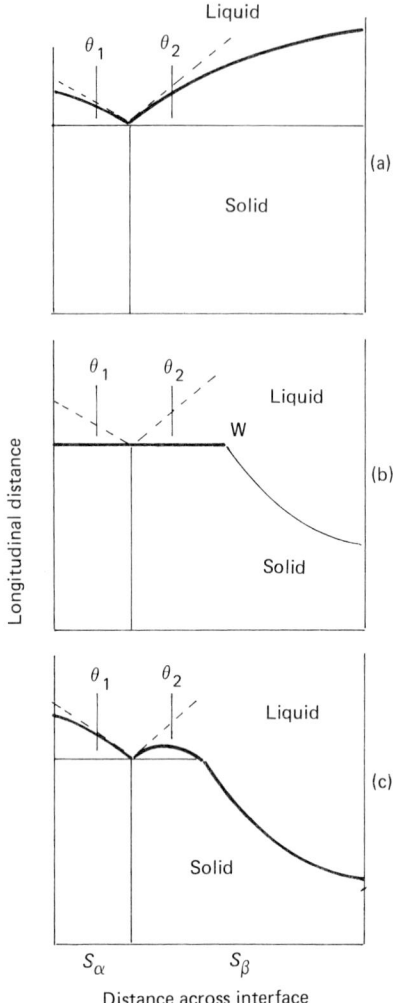

Figure 3.12 Eutectic interface geometries analysed by (a) Jackson and Hunt[53]; (b) Sato and Sayama[63]; (c) Fisher and Kurz[64]

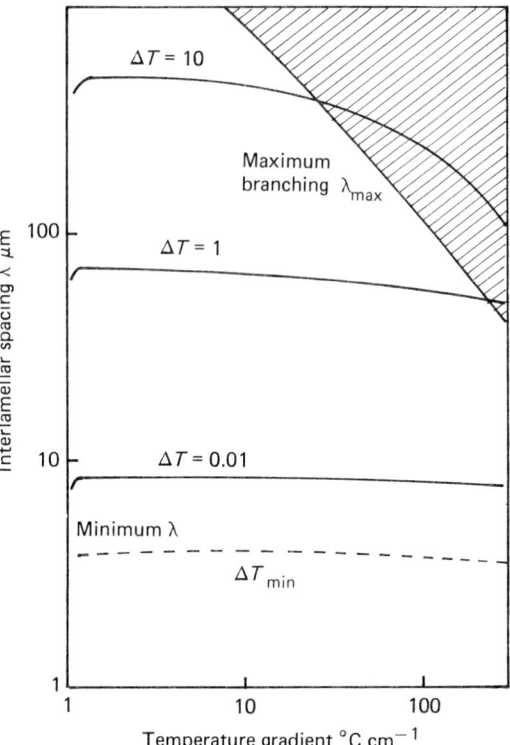

Figure 3.13 The variation of undercooling with spacing and temperature gradient for a constant growth velocity of $10 \, \mu m \, s^{-1}$ calculated by Fisher and Kurz[64]. The shaded area is inaccessible because branching occurs at the edge of the area (after ref. 64)

Sato and Sayami[63] consider partially coupled growth by assuming that part of the major phase falls away from the isothermal interface (*Figure 3.12*) and only that part of the major phase that remains isothermal (up to point 'w') with the minor phase needs to be considered. Calculation of the growth equation follows the Jackson and Hunt model, but the solute and curvature undercoolings for the major phase are modified. The growth equation is of the same form as Equation (3.5), but with different constants. This model leads to larger spacings and undercoolings for partially coupled growth. The predicted value for grey iron is smaller than the observed value. This discrepancy and the influence of S on the $\lambda^2 v$ parameter were discussed[65] in terms of spacing changes that occur as a result of pinching off or branching

phenomena within the groove that forms the major phase. This behaviour depends on the G/v ratio.

Fisher and Kurz[64] used a similar model, but argue that if the geometry on the undepressed part of the major phase is fixed, it must exert an influence over the shape of the depressed part through surface tension effects.

These arguments are developed to include the temperature gradient in the liquid in the growth equation. An example of the solution to this equation is given in *Figure 3.13*. It is necessary to decide which spacing the eutectic will choose when the growth velocity and temperature gradient are fixed. This chosen spacing must be related to a spacing change mechanism as in lamellar growth. Observations made on organic analogues suggest that a spacing that is too small is easily adjusted by a mechanism in which the flake ceases to grow. This mechanism is similar to that observed in normal eutectics. Consequently, the minimum local spacing corresponds to the extreme value on the growth curve. If the spacing is locally too large, the anisotropic growth of the faceted phase prevents easy decrease in the spacing in contrast to lamellar growth. The flakes then continue to grow in a divergent manner until the undercooling at the flake interface is sufficient to cause interface instability and branching of the flake.

A simple Mullins and Sekerka stability criterion was used to define the spacing at which branching occurs. This defines the shaded area in *Figure 3.13* at the boundary of which branching occurs. The observed spacing is the average of λ_{min} and λ_{branch}. This analysis predicts much larger spacings and undercoolings than the Sato and Sayama analysis. A decrease in spacing and undercooling is predicted as the temperature gradient increases. The effect of temperature gradient is explained in terms of its effect on the overall extent of the solid–liquid interface in the growth direction. The model suggests that the divergence of two flakes leads to a depression in the non-faceted matrix due to the accumulation of solute. This results in a depletion of solute at the flake interface and decreases the tendency to branch. Increasing the temperature gradient flattens the interface. This reduces the efficiency of solute rejection from the faceted interface. It also increases the constitutional undercooling at the interface, resulting in a greater instability and a decrease in the maximum spacing thereby producing the decrease in the average spacing shown in *Figure 3.13*.

The interface depression that develops on the major phase as the flake spacing increases may breakdown to give a dendritic matrix structure depending on the degree of interface undercooling. A structure that is

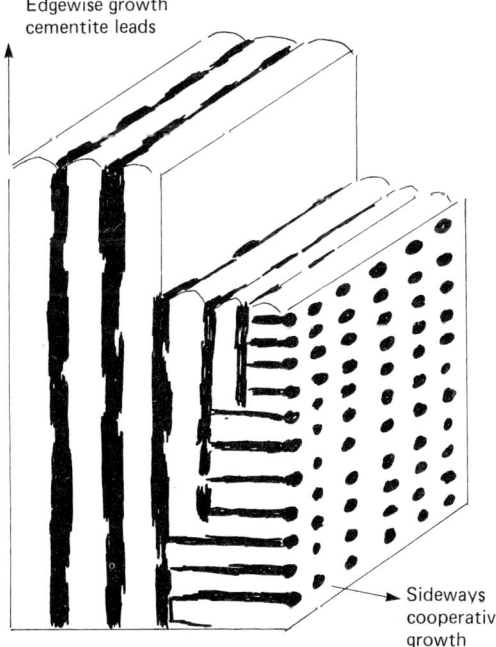

Figure 3.14 The solidification of white cast iron. Fe_3C is the white phase and austenite is the black phase (after ref. 66)

observed occasionally in grey irons consists of two distinct flake spacings; the finer-spaced flakes appear between those of coarse spacing. This structure has been referred to as undercooled because the finer structure is similar to that observed at higher growth velocities and grows behind the tips of the coarse flakes.

Fisher and Kurz show that the 'undercooled' eutectic forms in the depression on the major phase as an alternative means of relieving the undercooling that occurs as a result of the inability of the coarse flakes to adjust their spacing. The Fisher and Kurz analysis is not based simply on a larger kinetic undercooling associated with a faceted interface. Indeed, the analysis uses the basic premise that a large growth rate anisotropy, which is characteristic of high entropy of solution phases, prevents easy decrease in flake spacing.

Cast iron eutectic growth morphology

The characteristic structures of pure eutectic alloys have been described in the previous section. The growth morphologies exhibited by cast irons are described and related to the classification in this section.

White irons

The metastable unalloyed $Fe–Fe_3C$ ledeburite eutectic, illustrated in *Figure 3.14*, is classified as quasi-regular. Hillert and Subba Rao[66] have shown that once Fe_3C has nucleated, edgewise growth of Fe_3C plates occurs rapidly. They lead austenite at the interface and an orientation relationship develops;

$$(104)\,Fe_3C \,/\!/\, (101)\,\gamma$$

$$(010)\,Fe_3C \,/\!/\, (\bar{3}\bar{1}0)\,\gamma$$

Fe_3C and austenite also grow in a co-operative manner and form a rod structure perpendicular to and between Fe_3C plates. These two modes of growth form the quasi-regular structure but the edgewise growth is more rapid than the sidewise mode and dominates the structure (see *Figure 6.1* in Chapter 6). Directional solidification experiments[67] have shown that the operative point on the growth curve for the quasi-regular structure is close to the extremum point. Powell[68] has shown that the quasi-regular structure can be quench modified, but impurity modification has not been explored.

The improved toughness of alloyed white irons is associated with the different eutectic morphology of the $Fe–M_7C_3$ eutectic. This is an anomalous structure with a broken lamellar structure. The structure is a mixture of blades and hollow faceted rods. The proportion of rods in the structure increases as the cooling rate increases.

Grey flake irons

Grey cast irons do not fit as easily into the classification system. The characteristic structure of pure Fe–C–Si eutectic alloys is spheroidal[69]. Although the flake structure is often described as characteristic, it only forms in the presence of impurities. The significance of minor elements has been emphasized already. It is important to gain an understanding of their role in controlling graphite morphology in order to produce the family of structures that range from flake through intermediate forms such as coral and compacted to spheroidal, and to avoid the formation of undesirable degenerate spheroidal structures.

The growth of the flake structure is well defined. Once graphite has nucleated, the eutectic cell grows in an approximately radial manner within the constraints imposed by surrounding austenite dendrites. The flakes bend, twist and branch as depicted in *Figure 3.15*. Microstructures of irons quenched during cell growth show that the graphite leads at the interface. Each flake is wetted by austenite up to the growing edge. The flakes can only grow by the extension of the flake in the close packed, strong bonding 'a' direction. Significant growth in the 'c' direction is precluded by the adjacent austenite.

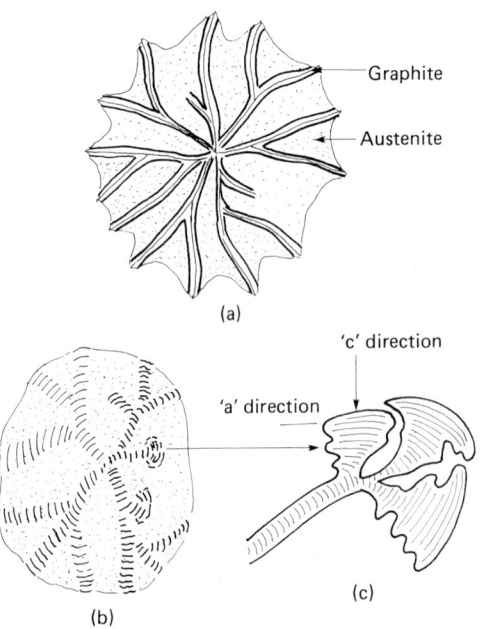

Figure 3.15 (a) Schematic representation of the growth of a flake graphite eutectic cell; (b) Eutectic cell growth in a compacted iron showing graphite layer growth along the c axis; (c) Growth of a spheroid at the tip of compacted graphite after losing contact with the liquid

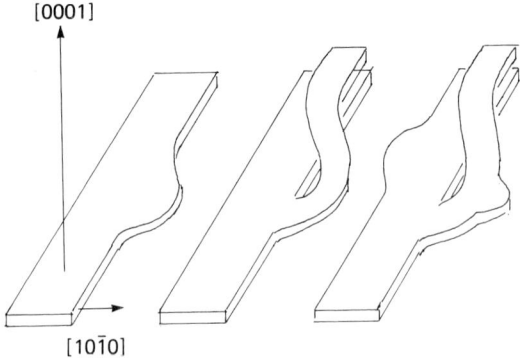

Figure 3.16 Branching mode of graphite in a eutectic cell (after ref. 70)

The graphite is continuous through the cell, but is not a single crystal. Defects are introduced as a result of growth constraints imposed by the surrounding austenite, the fourfold increase in specific volume of graphite relative to the liquid from which it forms, and by impurities. The presence of layered rotational faults[45] provides a means for flakes to branch in many directions within their own plane. Twin/tilt boundaries allow branching outside the plane.

Flakes branch during growth by lateral division and rarely by splitting along the plane of the sheets. A model of branching from the $(10\bar{1}0)$ plane is illustrated in *Figure 3.16*[70]. This is an example of an anomalous flake structure and the branching frequency is considered to be proportional to the interface undercooling. Undercooled graphite is a fine flake graphite grown at a higher undercooling.

Different explanations have been given for the role of impurities, such as O and S, in promoting the flake morphology. On the one hand, it has been suggested that O and S in graphite impede slip and prevent curved graphite growth that is considered an essential feature of spheroidal growth. In addition, S lowers the austenite–liquid interfacial free energy, thus creating a mushy zone in which austenite dendrites are separated by continuous liquid films (*Figure 3.17*). This configuration is not conducive to spheroidal growth. An alternative view is that small concentrations of impurities in solution can have a considerable effect on interface mobility and the driving force for growth.

The effect on mobility has been studied by Gilmer[71] and quantified using computer simulation of the dynamics of crystal growth. In keeping with these predictions, it is suggested that several impurities, including O and S, adsorb on the growth interface. This makes the prism plane non-faceted and lowers the kinetic undercooling required for growth in the 'a' direction. The role of impurities that bond strongly with the host lattice (ordering impurities such as O) can be distinguished from those that bond weakly (clustering impurities such as S). The influence of

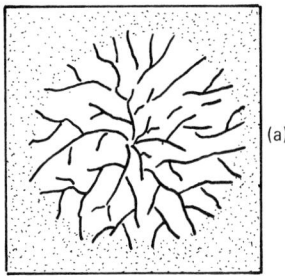

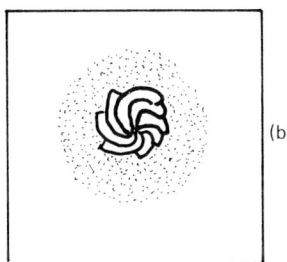

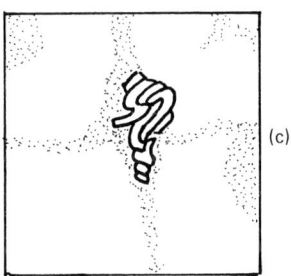

Figure 3.17 Schematic representation of the growth of eutectic cells of (a) flake graphite; (b) spheroidal graphite; (c) degenerate graphite and their associated mushy zone configurations; ■ graphite; □ austenite; ▨ liquid

minor elements on the driving force is through the generation of constitutional undercooling.

Nieswaag and Zuithoff[65] have shown that the flake spacing first increases and then decreases with increasing S content in irons grown at a constant velocity. The increase in spacing is attributed to S adsorption, which lowers the kinetic undercooling and hence, the interface undercooling and the frequency of branching. Higher S concentrations lead to a second influence. S builds up in the liquid and promotes constitutional undercooling, causing an increase in the frequency of branching.

Spheroidal irons

This is a divorced eutectic. Graphite nucleates in liquid iron pockets and grows into a spheroid independently of the austenite phase. Growth proceeds

from the centre of the spheroid outwards until enveloped by austenite. Growth after envelopment must be limited and incidental to the formation of the spheroid. The 'c' axis is approximately parallel to the radial direction. These are established features of spheroidal growth.

There is no unanimity on growth mechanism. Some authors[69,72] have proposed a circumferential growth model involving growth in the 'a' direction as indicated in *Figure 3.18* (see also *Figure 6.24* in Chapter 6). This model is based on the known ability of graphite for curved growth in pure alloys. The observation of serrated 'herring-bone' patterns

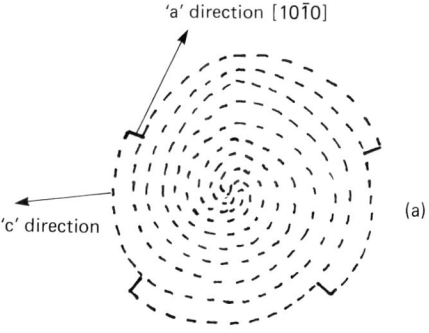

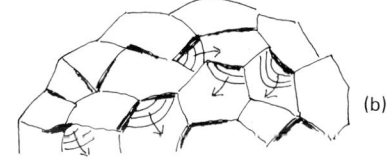

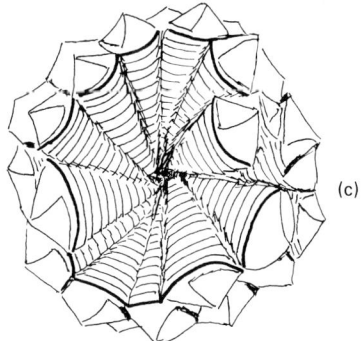

Figure 3.18 (a) Schematic representation of the circumferential growth of a spheroid; (b) Schematic representation of the surface of a spheroid showing a variety of tilt–twist boundaries between individual segments which provide numerous locations for layer growth; (c) Model for the initial stages of spheroid growth (after ref. 73)

associated with distinct curvature of the basal planes in the core of spheroids has led to the suggestion[73] that these irregularities are not incompatible with a variety of sections through conical helices. It has also been suggested that the initial stages of growth may consist of a loose combination of conical helices as shown in *Figure 3.18*.

Slight misorientations between numerous radial filaments will produce a very uneven multi-stepped surface with myriads of tilt and twist boundaries. These defects will allow one subgrain to grow over another. This creates many sites of active growth and quickly renders single dislocation step sources relatively unimportant. The growth form has been likened to a cabbage leaf pattern. The role of spheroidizing agents is that of scavengers removing elements such as S from solution, thus permitting curved growth. Through its influence on the austenite–liquid, the spheroidizing agent interfacial energy creates mushy zone configurations conducive to the unconstrained growth of spheroids (see *Figure 3.17*).

Johnson and Smartt[74] have proposed a slightly different model based on Auger microprobe measurements. These measurements show that O and S are adsorbed at interfaces in flake irons, but that interfaces in spheroidal irons were free of minor elements. Spheroids are considered to grow in liquid iron pockets in the 'c' direction with the aid of screw dislocations with a negative temperature gradient in the liquid. When the spheroids are small, growth perturbations are damped by capillary forces. Once the diameter exceeds 300–500 Å, perturbations develop, and a star-shaped spheroid is produced. Cavities between perturbations eventually solidify, thus producing a macroscopically smooth surface. A high density of screw dislocations generated as a result of stress introduced by the incorporation of non-C atoms in the lattice is suggested to promote rapid growth. Splitting and branching of the graphite combined with its high flexibility allows the relative orientation of the basal planes to change during growth.

A similar model was presented by Oldfield *et al.*[75] based on observations made on transparent glycerol-water solutions. Interface breakdown is a central feature of this model. The starting point is a polyhedral crystal growing with a stable faceted interface. Once a critical size is exceeded the facets become unstable. If a protuberance formed on the face projects far enough into the undercooled liquid it will grow at the same velocity as the corner in *Figure 3.19*. It will always lag behind the corners by a distance, *d*, which separated it from the spheroid envelope at initiation. However, as growth proceeds, the assembly increasingly approximates to a spherical form. The initial crystal in spheroidal graphite is considered to be bound by faceting planes and the growth instability to be dendritic. Growth occurs by

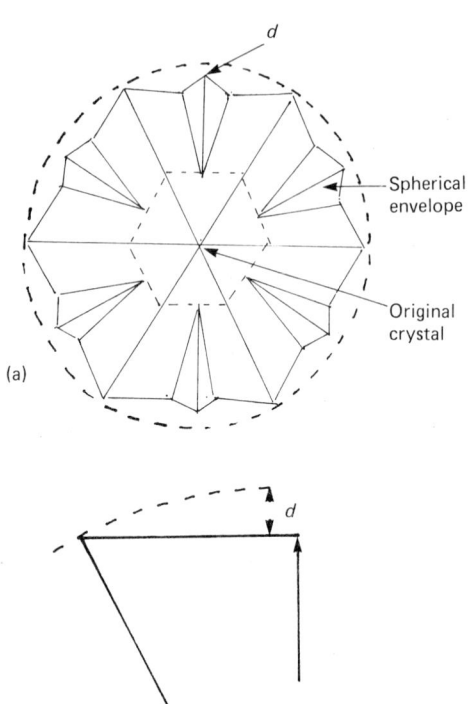

Figure 3.19 (a) Growth of a spheroid by interface breakdown; (b) Crystallite develops at centre of the face when the crystal is of size *r*. The spherical envelope is a distance *d* ahead of the growing crystallite (after ref. 75)

a screw dislocation mechanism. Each screw dislocation has an edge component which results in the graphite twisting in space. This ensures that the dendrites have sufficient flexibility to develop into a spheroid.

More recently, Minkoff and co-workers[51] have studied interface instability in several eutectic systems which contain graphite as a faceting phase. Their interpretation of structural observations places emphasis on the role of interface undercooling and its dependence on the minor element concentration. Reactive elements, including Mg, Ce and rare earths, are considered to adsorb on the graphite–liquid interface. In addition to neutralizing growth from the prism plane by rendering it faceted, they are thought to poison growth on the basal plane. When insufficient reactive element is adsorbed for complete coverage of the (0001) interface, growth is impeded only at dislocation steps where adsorption has occurred. This localized effect leads to the formation

that have been overtreated with rare earths or irons of hypereutectic composition that exhibit graphite flotation. The growth characteristics of different forms of exploded graphite are represented schematically in *Figure 3.24*.

Degenerate flake graphite forms result in a considerable loss in mechanical strength. Widmanstätten graphite is a degenerate flake form in which fine needles or sheets of graphite are arranged in a regular pattern to the side and at the ends of otherwise normal flake graphite (see *Figures 6.15–6.18* in Chapter 6). This growth form is found in irons containing Pb and H (ref. 108). A small concentration of Al which promotes absorption of H by reaction with moisture in the mould or environment also produces Widmanstätten graphite in irons containing Pb. More extensive Widmanstätten graphite formation occurs in alloy irons. Mesh graphite is another undesirable flake form that occurs in the presence of Te and can be associated with the excessive use of Te mould or core wash[109].

The coupled zone concept

Unexpected features of grey iron solidification include the formation of primary austenite dendrites in irons of eutectic and slightly hypereutectic composition and the formation of an austenite halo around primary Kish graphite. These structures can be rationalized in terms of the coupled zone. This zone defines the solidification conditions under which two eutectic phases can grow at a common interface at a rate that exceeds that of either of the component phases separately. The shape of the coupled zone is determined by the growth characteristics of the two phases and the solidification conditions.

Figure 3.25 shows typical zone shapes for isothermal growth ($G = 0$) and illustrates the influence of growth behaviour on zone shape. A symmetrical zone forms when both phases grow in a non-faceted manner. An asymmetrical zone forms when one phase shows non-faceting and the other faceting behaviour. *Figure 3.26* shows how the zone shape changes when growth occurs with a positive temperature gradient in the liquid. Temperature–growth velocity relationships are given in the figures and can be used to locate the zone boundaries. The procedure is to define the temperature–growth velocity relationship for α and β phases and for the α–β eutectic. The preferred growth form at any velocity is the one with the highest growth temperature. For the alloy composition C_0 it is given by the solid lines on the temperature–growth velocity curves. This analysis is repeated for different compositions when the coupled zone boundaries can be defined.

Growth temperature measurements suggest that the eutectic phase follows a relationship of the form:

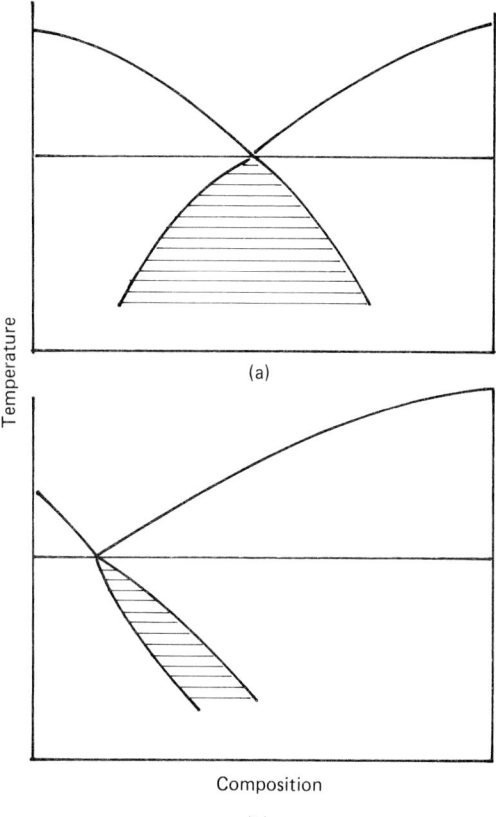

Figure 3.25 Typical coupled zone shapes for (a) normal eutectic alloys; (b) anomalous eutectic alloys; for isothermal growth conditions ($G = 0$)

$$\Delta T_{\text{eut}} = K_1 v^{1/2} = T_{\text{E}} - T_{\text{i}} \qquad (3.6)$$

and is independent of G for normal eutectics but not for anomalous eutectics. The relationship for non-faceting, dendritic primary phase growth is of the form:

$$\Delta T_{\text{den}} = \frac{GD}{v} + K_2 v^{1/2} = T_{\text{l}} - T_{\text{i}} \qquad (3.7)$$

The eutectic and primary phase interface temperatures, T_{i}, are equal at the coupled zone boundary, consequently:

$$T_{\text{l}} - T_{\text{E}} = \frac{GD}{v} + (K_2 - K_1) v^{1/2}$$

However from the phase diagram:

$$T_{\text{l}} - T_{\text{E}} = m\Delta C$$

where m is the liquidus slope and ΔC measures the half width of the coupled zone. Therefore:

$$\Delta C = \frac{1}{m}\left[\frac{GD}{v} + (K_2 - K_1)v^{1/2}\right] \qquad (3.8)$$

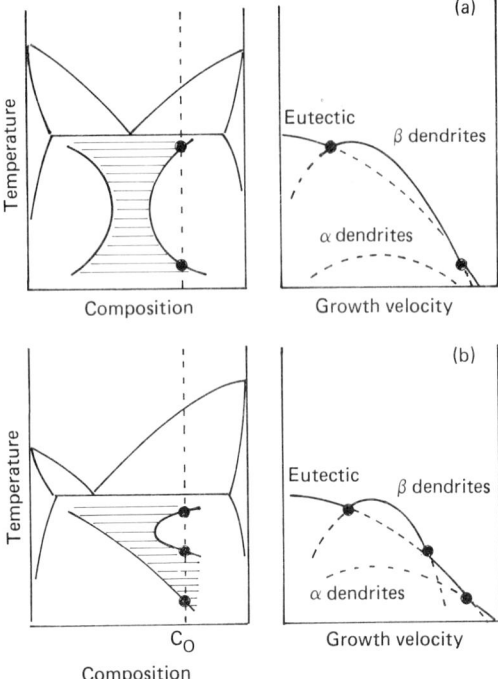

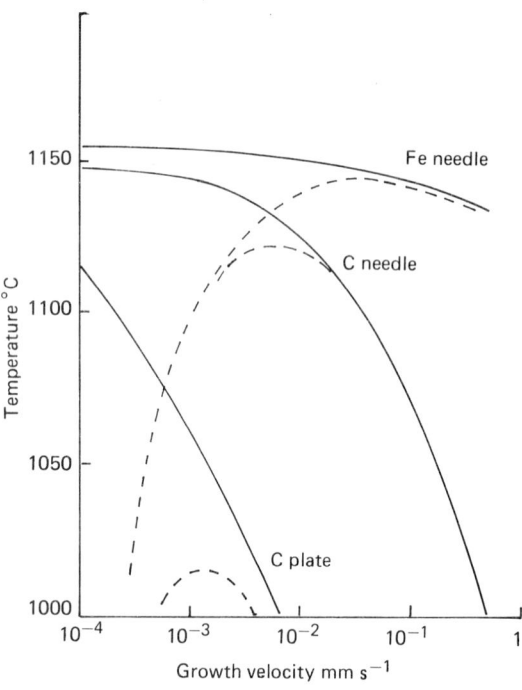

Figure 3.26 The coupled zone and temperature–growth velocity relationship for (a) normal eutectic alloys; (b) anomalous eutectic alloys for a positive temperature gradient in the liquid

Figure 3.27 Interface growth temperature as a function of growth velocity for different primary phases and morphologies in a Fe–4.27 weight% C alloy; ——, $G = 0$; ---, $G = 100\,°\text{C/cm}$ (after ref. 110)

Cast irons belong to the anomalous group of eutectics. The α phase half width is negative for an asymmetrical coupled zone. This occurs when K_1 in Equation (3.8) is large. Experimental measurements have shown that the undercooling of anomalous eutectics can be several orders of magnitude greater than that of a normal alloy. A second requirement is a large β phase half width. This requires a large value of K_2 for the β phase. This is the faceting phase in an anomalous system. It may adopt different morphologies, depending on the solidification conditions and melt purity. Consequently, it is not surprising that very few growth temperature–velocity relationships have been measured for faceting primary phases.

In an attempt to provide a quantitative analysis for cast irons Kurz and Fisher[110] have considered the faceting phase to grow as a plate and have derived growth curves for needle (dendritic) and plate forms. They conclude that the undercooling for both morphologies can be described by the equation:

$$\Delta T = \frac{GD}{v} + K_3 v^w \qquad (3.9)$$

and suggest that for austenite needles $K_3 = 17.6\,\text{K s}^{0.46}\,\text{mm}^{-0.46}$ and $w = 0.46$. For graphite plates $K_3 = 932\,\text{K s}^{0.32}\,\text{mm}^{-0.32}$ and $w = 0.32$.

Figure 3.27 shows the growth curves calculated for isothermal growth and for $G = 100\,\text{K/cm}$. Recently, Jones and Kurz[11] have measured dendrite tip and eutectic interface temperatures in an Fe–C alloy (4.8 wt % C) and used this data in Equations (3.6) and (3.9) to predict the coupled zones for grey and white iron solidification for $G = 70\,\text{K/cm}$ as shown in *Figure 3.28*. The value of K_1 in Equation (3.6) for the flake eutectic was $102\,\text{K s}^{1/2}\,\text{mm}^{-1/2}$. An asymmetrical coupled zone is predicted for the graphite system and a symmetrical zone for the carbide eutectic. This is because this alloy displays a more regular eutectic structure and the K_3 values for austenite needles and cementite plates both exceed the K_1 value for the Fe–Fe$_3$C eutectic.

The solidification of grey flake irons of various compositions can be related to the coupled zone using *Figure 3.29*. A eutectic alloy undercools to point 1 when nucleation occurs. The location of point 1 depends on the solidification conditions and the heterogeneous nucleants present in the liquid. Austenite is more easily nucleated than graphite and at location 1, outside the coupled zone, the austenite growth rate exceeds that of the eutectic and that of graphite. Consequently, austenite dendrites will develop in the liquid. They will enrich the interdendritic areas in C until graphite nucleation occurs at

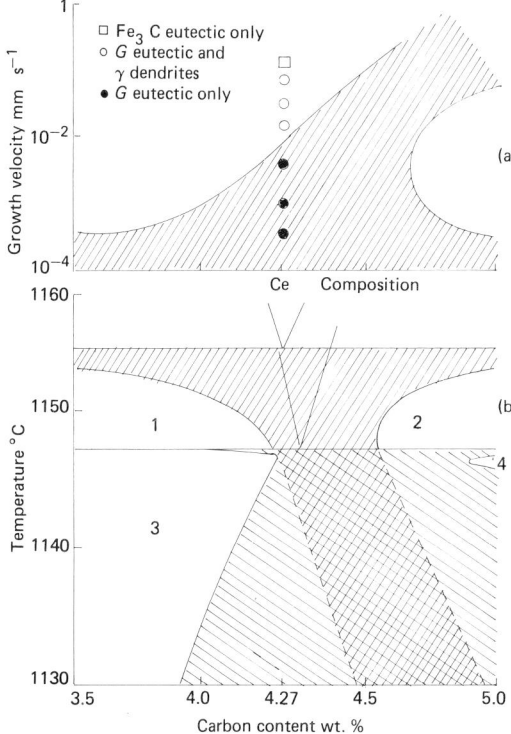

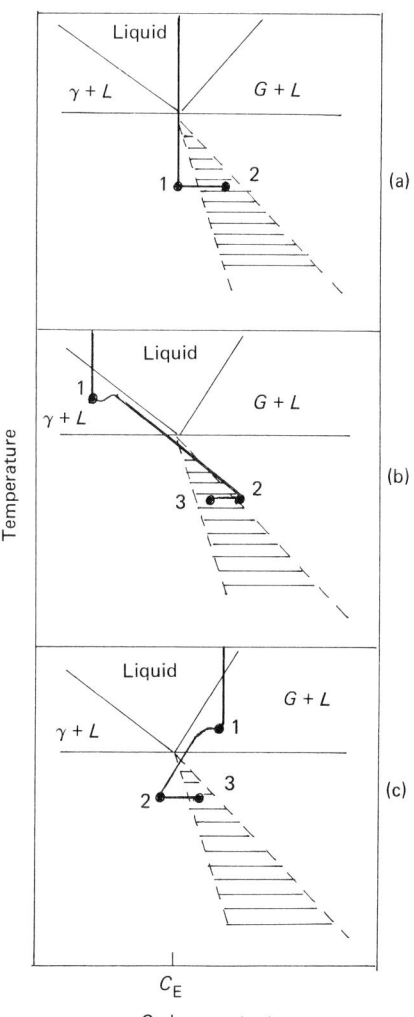

Figure 3.28 The coupled zone in directionally solidified Fe–C alloys with $G = 70\,°C/cm$. (a) Velocity-composition plot showing calculated boundaries and experimental points; (b) Temperature-composition plot with superimposed phase diagram; ▨, Fe–G coupled zone; ▧, Fe–Fe$_3$C coupled zone; Region 1 shows γ austenite dendrites and G eutectic; Region 2 shows G plates and G eutectic; Region 3 shows γ austenite dendrites and Fe$_3$C eutectic; Region 4 shows Fe$_3$C plates and Fe$_3$C eutectic (after ref. 111)

Figure 3.29 The cooling of irons with different C equivalents using the coupled zone concept (a) an iron of eutectic composition; (b) a hypoeutectic iron; (c) a hypereutectic iron

point 2. Because this point lies inside the coupled zone, the austenite graphite eutectic forms.

The solidification of a hypoeutectic alloy commences with the nucleation of austenite at point 1 in *Figure 3.29*. As the austenite dendrites develop during cooling, the interdendritic liquid is enriched in C until graphite nucleates at point 2. This event moves the liquid composition into the coupled zone where the eutectic will form. With a hypereutectic iron, primary graphite solidifies at point 1. Growth of the graphite phase during cooling causes liquid close to the primary phase to become depleted in C. This localized depletion results in liquid of less than eutectic composition. In this liquid austenite nucleates at point 2 and grows rapidly as a halo surrounding the primary graphite.

The growth of the austenite halo along with the accompanying segregation of C returns the composition into the coupled zone. Eutectic solidification follows in the same manner as in a liquid of eutectic composition. *Figure 3.28* shows that the coupled zone for white irons is symmetrical. Therefore a completely eutectic structure forms over a wider range of solidification conditions. Spheroidal irons form a divorced eutectic with the graphite and austenite phases solidifying independently. The spheroidal iron eutectic grows with a larger undercooling than flake iron. The absence of a coupled zone is attributed to the eutectic growth curve not cutting the austenite growth curve.

Primary cast iron solidification structures

An overall picture of the primary solidification structure of a cast iron can only be obtained when the collective effect of changes in the individual events, described in the previous section, is considered. A cooling curve has the ability of recording and identifying all the events that occur in the solidification process and offers the opportunity of evaluating the effect of changes made by the foundryman on the primary solidification structure.

The major element content (C, Si, Mn, S etc.), nature of the original melting charge, melting method and liquid metal treatment produce a liquid iron of a certain graphitization potential and with a base nucleation level[112,113]. The solidification of this iron without inoculation or spheroidizing treatment may be considered in relation to the cooling curve defined in *Figure 3.1*. The first solidification event is the nucleation and growth of austenite dendrites commencing at temperature TAL. The volume fraction of dendrites is determined mainly by the C.E.V. although other factors also influence the dendrite network characteristics. The second significant event is the nucleation of eutectic at TES. The location of TES and that of TEU depends on the nucleation level in the liquid. It has been shown that if the melt is uninoculated the most effective nuclean for graphite is MnS particles[114,115]. Thus, the Mn:S ratio is of significance.

Table 3.1 The influence of inoculation on nucleation and undercooling in a flake iron (after ref. 89)

Inoculant % CaSi	0	0.05	0.1	0.2
Eutectic undercooling °C	24	15	4	2
Cell count/cm²	55	108	160	215

Table 3.2 The influence of cooling rate on the nucleation and undercooling in a flake iron (after ref. 89)

Cooling rate °C/min	60	120	200	375
Eutectic undercooling °C	12	14	18	22
Cell count/cm²	57	75	94	113

The effects of the inoculation of liquid iron are evident from the cooling curves in *Figure 3.30*. There are few MnS particles when the Mn:S ratio is high; the Mn and S are unbalanced and the iron is poorly inoculated. Both TEU and TER are low and the graphite flakes are Type D. Decreasing the Mn:S ratio increases the number of nuclei (cell count). TEU and TER also increase and the flake morphology gradually changes to Type A. However, the lowest Mn:S ratio leaves free S in the liquid, which promotes undercooling and, in this iron, white iron formation. Inoculation raises the nucleation level and increases all the eutectic temperatures defined in *Figure 3.1* to an extent depending on the amount and type of inoculant. It also promotes Type A graphite structures[116]. This effect is illustrated in *Table 3.1*.

On the other hand, as indicated in *Table 3.2*, increased cooling rate raises the nucleation level but lowers the cooling curve temperatures, as shown in *Figure 3.31*. The graphite morphology in a flake iron

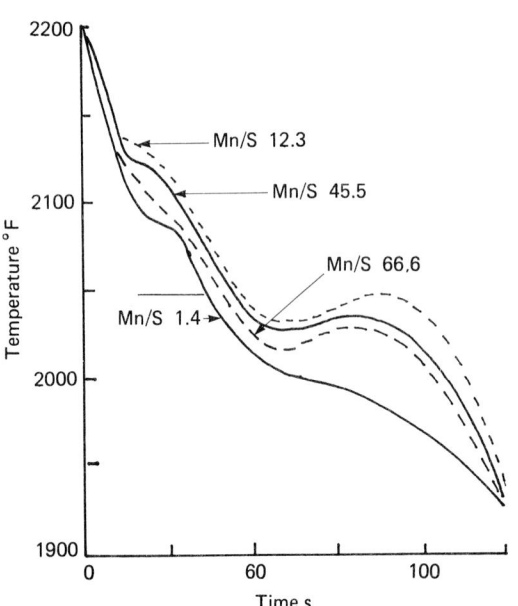

Figure 3.30 Effect of S and Mn on the cooling curve of an uninoculated grey iron of composition 3.4% C, 1.9% Si, 0.07% P (after ref. 115)

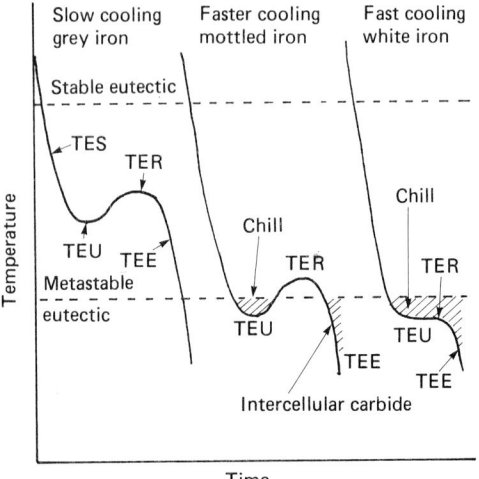

Figure 3.31 Schematic representation of the effect of cooling rate on the cooling curve and structure of cast irons

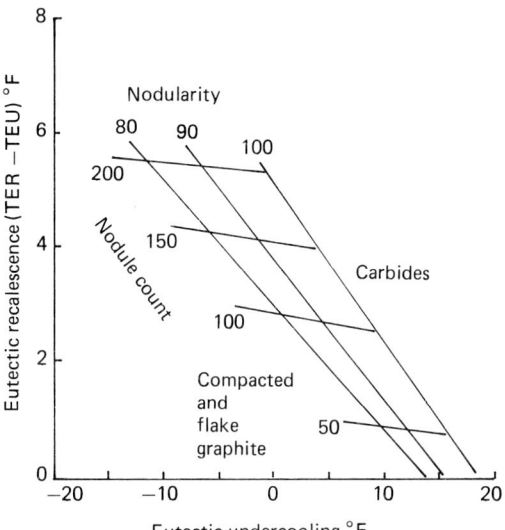

Figure 3.39 Prediction chart using the temperature parameters calculated from a cooling curve (after ref. 141)

TAL = 1594.4 − 102.2 C.E.L.

for melting under oxidizing conditions   (3.12)

TAL = 1540 − 92.06 C.E.L.

for melting under deoxidized argon at temperatures > 1454 °C   (3.13)

The relationship due to Moore[134] and Equations (3.11) to (3.13) are shown in *Figure 3.37*. Although there is overlap and some agreement at intermediate C.E.L. values, there is disagreement between relationships 1 and 4 in *Figure 3.37* over the C.E.L. corresponding to malleable and grey and spheroidal irons. The deviation of the measured TAL from that given by Equation (3.11), used as a reference, has proved a valuable tool in measuring the degree of oxidation of malleable, spheroidal and grey liquid irons[139].

Typical measurements for an induction melted spheroidal iron over a period of relatively high oxidation and one of more acceptable levels are shown in *Figure 3.38*[152]. These differences represent differences in charges and melting practice, for example, changes in the per cent returns charged, melt temperature, preheat temperature, type and quality of additives and of the scrap used. The deviations are measured either as the degree of change in TAL, that is, the difference between the measured TAL and that calculated using the chemically analysed C.E.L. and Equation (3.11), or as ΔC.E.L., defined as the difference between the chemically determined C.E.L. and the effective C.E.L. as determined from the liquidus measurement.

Changes in TAL reflect changes in the activity of C and Si in the liquid iron. Dissolved oxygen, which results from higher superheating temperatures and holding times, melting in air and rust steel, reduces C and Si activity and with Bi additions makes C.E.L. more negative. Al, Ti and C additions have the opposite effect.

Movements in ΔC.E.L. are indicators of potential casting defects. Frequently large positive values of ΔC.E.L. indicate mottle in malleable iron[138]. Large negative deviations correlate with increasing chill in spheroidal or grey irons. Increased austenite formation occurs with increasingly negative ΔC.E.L. at constant C.E.V.. This promotes spiking which leads to feeding difficulties[137].

Tests based on the shape of the cooling curve

Several tests have been suggested for predicting iron type and quality based on the shape of the grey iron cooling curve. A plain tectip is used for measurement and conditions must be standardized for reproducible results[141–144]. *Figure 3.39* is a predictive diagram relating undercooling and recalescence with microstructure. It enables spheroid count and quality to be assessed. The diagram shows that for a given eutectic recalescence the spheroid quality (nodularity) increases as the eutectic undercooling increases. This is followed by a sharp transition to a carbide structure. The spheroid count is strongly dependent on the eutectic recalescence. A similar prediction chart can be drawn up on the basis of time rather than temperature parameters from the cooling curve.

A more simple analysis[145] has shown that iron type

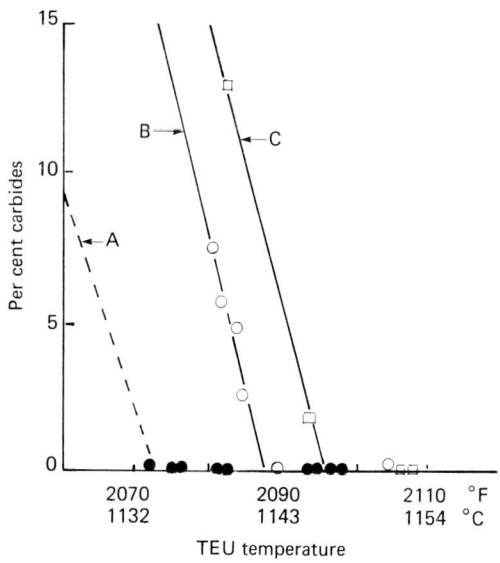

Figure 3.40 Percentage carbide in castings as a function of the TEU value determined from the cooling curve. A, 0.025 in. thick, 1.75 lb. weight; B, 0.25 in. thick, 1.16 lb. weight; C, 0.19 in. thick, 0.70 lb. weight (after ref. 140)

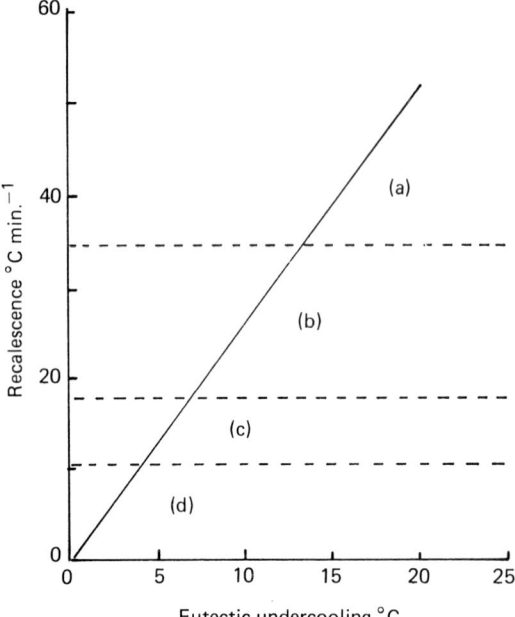

Figure 3.41 The relationship between recalescence and relative eutectic undercooling for different graphite morphologies (a) > 85% compacted graphite; (b) 50–85% compacted graphite; (c) > 85% spheroidal graphite; (d) flake graphite (after ref. 147)

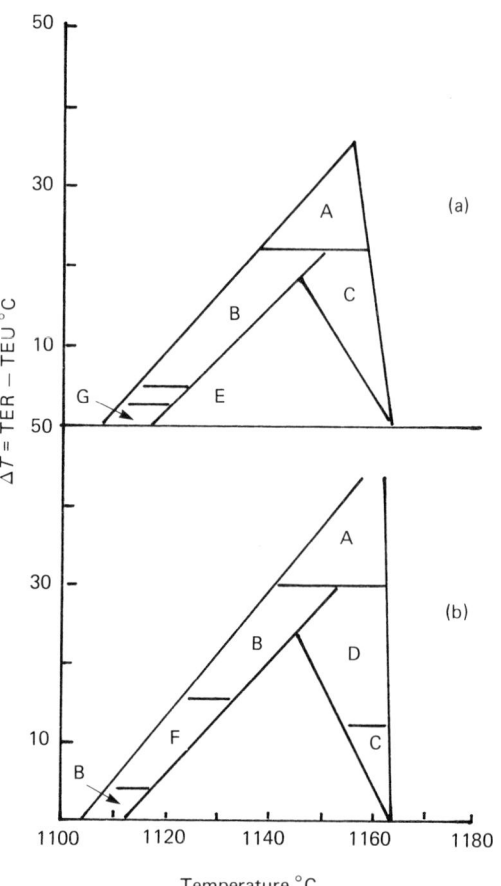

Figure 3.42 Correlation between eutectic recalescence temperature TER and $\Delta T = $ TER $-$ TEU; (a) La- and Ce-treated irons; (b) rare earth silicide-treated irons; A, $\geqslant 85\%$ compacted graphite; $< 9\%$ carbides; B, $< 85\%$ compacted graphite; irregular graphite; 9–27% carbides; C, flake graphite; D, compacted graphite; $\geqslant 20\%$ spheroidal graphite; 0–21% carbides; E, irregular graphite; 9–27% carbides; F, irregular graphite; 27–38% carbides; G, very irregular graphite; $> 38\%$ carbides (after ref. 148)

can be predicted with slightly over 80% confidence, based on the eutectic growth temperature, TER. Growth temperatures in the range 1130–1135 °C were associated with good spheroidal structures in Mg-treated irons. A method that compares selected regions of the cooling curve with a superimposed visual rating chart, so avoiding the measurement of temperatures and times, has been proposed[146]. *Figure 3.40* shows measurements taken in practice[140]. The minimum TEU value permissible in the tectip before carbides formed in spheroidal iron castings of different weights and thickness is shown. Thermal analysis measurement made just before a carbide-sensitive casting is poured can be used with the chart to determine whether carbide formation will be a problem and, if so, allows time for corrective liquid metal treatment. For example, castings A and B will run free of carbide if TEU is 1143 °C but casting C will run with about 5% carbides unless the liquid is treated to raise TEU to > 1145 °C.

Backerad, Nilsson and Steen[147] have shown that flake, spheroidal and compacted morphologies can be distinguished using a plot of rate of recalescence against undercooling (TER–TEU) derived from tectip cooling curves. This is shown in *Figure 3.41*. Compacted structures relate to points higher on the curve. Flake and spheroidal structures relate to points lower on the curve. Flake and spheroidal structures

can be distinguished because there is a 15 to 20 °C difference in the absolute growth temperature of these structures. An independent study[148] has shown that graphite morphology is related to the eutectic recalescence (TER–TEU) in alloys of eutectic and slightly hypoeutectic composition. Recalescence values greater than a critical value indicate compacted graphite. Lower values indicate flake or spheroidal graphite as shown in *Figure 3.42*. Some exceptions were found to this classification which were not explained.

The recent introduction of computers into the foundry has made possible more detailed analysis of the cooling curve[149–154]. One procedure has already been illustrated in Chapter 1, *Figure 1.26*. Chen and

Stefanescu[149] have reviewed the use of thermal analysis as an analytical and process control tool and demonstrated that computer-aided differential thermal analysis (CA–DTA) is a viable method. The neutral body in classical differential thermal analysis is simulated with a computer data acquisition system which also records the cooling and its first and second derivatives. Eutectic type (white or grey) and graphite shape (spheroidal, compacted, flake) can be predicted relatively easily from the shape of the CA–DTA curves. The first derivative curve can be used to calculate the latent heat of solidification and hence, the amount of austenite and eutectic formation.

Results obtained suggest that the graphite shape and the C.E.V. influence the latent heat of solidification. Predicting nodularity in spheroidal irons is more complicated. Good correlation was identified between critical points on the CA–DTA curves and nodularity, but difficulties were experienced in predicting nodularity in the region (80%) of practical interest due to data scatter. The scatter is caused by variations in C.E.V. and pouring temperature. The error margin can be narrowed by statistical analysis of the data.

References

1. ELLIOTT, R., *Eutectic solidification processing*, Butterworths, London (1983)
2. HELLAWELL, A., Heterogeneous nucleation and grain refinement in aluminium castings, in *Solidification and Casting of Metals*, Metals Society, p. 161 (1979)
3. GHORESHY, M., ZEHTAB-JAHEDI, M. and KONDIC, V., Primary austenite dendrites in grey cast iron, *Brit. Foundryman*, **73**, 277 (1980)
4. QUINTERO, M. and KONDIC, V., Technology of structural control of flake graphite cast iron, in *Proc. Solidification Processing*, Sheffield (1987)
5. PARKS, T.P. and LOPER, C.R. JR., A study of conditions promoting dendritic growth in ductile iron, *A.F.S. Trans.*, **77**, 90 (1969)
6. NARO, R.L. and WALLACE, J.F., Minor elements in grey iron, *A.F.S. Trans.*, **78**, 229 (1970)
7. KAYAMA, N., MURAI, K., SUZUKI, K. and KATAOKA, Y., Effects of inoculation and oxidation on the shape of austenite dendrites in cast iron, *Rept. Castings Research Lab. Waseda University*, **33**, 33 (1982)
8. KAYAMA, N., NASHIMOTO, K. and SUSUKI, K., Effect of SiO_2 on primary austenite nucleation in hypoeutectic cast iron, *Rept. Castings Research Lab. Waseda University*, **35**, 15 (1984)
9. JACKSON, K.A., *Liquid metals and solidification*, A.S.M., Cleveland, Ohio, p. 174 (1958)
10. JACKSON, K.A., Crystal growth kinetics, *Materials Sci. and Eng.*, **65**, 7 (1984)
11. HUANG, S.C. and GLICKSMAN, M.E., Fundamentals of dendritic solidification: I. Steady state tip growth, *Acta. Met.*, **29**, 701 (1981)
12. HUANG, S.C. and GLICKSMAN, M.E., Fundamentals of dendritic solidification. II. Development of side branch structure, *Acta. Met.*, **29**, 716 (1981)
13. LANGER, J.S., Dynamics of dendritic pattern formation, *Mater. Sci. and Eng.*, **65**, 37 (1984)
14. LANGER, J.S., Pattern selection in solidification, *Trans. Met. A.*, **15A**, 961 (1984)
15. OLDFIELD, W., Computer model studies of dendritic growth, *Mater. Sci. and Eng.*, **11**, 201 (1973)
16. TRIVEDI, R. and TILLER, W.A., Interface morphology during crystallization: I. Single filament unconstrained growth from a pure melt, *Acta. Met.*, **26**, 671 (1974)
17. LANGER, J.S. and MULLER-KRUMBHAAR, H., Stability effects in dendritic crystal growth, *J. Crystal Growth*, **42**, 11 (1977)
18. LANGER, J.S. and MULLER-KRUMBHAAR, H., Theory of dendritic growth: I. Elements of a stability analysis, *Acta. Met.*, **26**, 1681 (1978)
19. LANGER, J.S. and MULLER-KRUMBHAAR, H., Theory of dendritic growth: II. Instabilities in the limit of vanishing surface tension, *Acta. Met.*, **26**, 1689 (1978)
20. LANGER, J.S. and MULLER-KRUMBHAAR, H., Theory of dendritic growth: III. Effects of surface tension, *Acta. Met.*, **26**, 1697 (1978)
21. MULLER-KRUMBHAAR, H. and LANGER, J.S., Sidebranching instabilities in two dimensional model of dendritic solidification, *Acta. Met.*, **29**, 145 (1981)
22. LAXMANAN, V., Dendritic solidification: I. Analysis of current theories and models, *Acta. Met.*, **33**, 1023 (1985)
23. TEWARI, S.N. and LAXMANAN, V., A critical examination of the dendrite growth models: comparison of theory with experimental data, *Acta. Met.*, **35**, 175 (1987)
24. TRIVEDI, R. and TILLER, W.A., Interface morphology during solidification: II. Single filament unconstrained growth from a binary alloy melt, *Acta. Met.*, **26**, 679 (1979)
25. GLICKSMAN, M.E., Free dendritic growth, *Mater. Sci. and Eng.*, **65**, 37 (1984)
26. LIPTON, J., GLICKSMAN, M.E. and KURZ, W., Dendritic growth into undercooled alloy melts, *Mater. Sci. and Eng.*, **65**, 37 (1984)
27. McCARTNEY, D.G. and HUNT, J.D., A numerical finite difference model of steady state cellular and dendritic growth, *Met. Trans. A.*, **15A**, 983 (1984)
28. LAXMANAN, V., Dendritic solidification in a binary alloy melt; comparison of theory and experiment, *J. Crystal Growth*, **83**, 391 (1987)
29. ESKA, H. and KURZ, W., Columnar dendritic growth: experiments on tip growth, *J. Crystal Growth*, **72**, 578 (1985)
30. TRIVEDI, R. and SOMBOONSUK, K., Constrained dendritic growth and spacing, *Mater. Sci. and Eng.*, **65**, 65 (1984)
31. TRIVEDI, R., Interdendritic spacing: Part II. A comparison of theory and experiment, *Trans. Met. A.*, **15A**, 977 (1984)
32. HUNT, J.D., Steady state columnar and equiaxed growth of dendrites and eutectic, *Mater. Sci. and Eng.*, **65**, 75 (1984)
33. KIRKWOOD, D.H., A simple model for dendrite arm

coarsening during solidification, *Mater. Sci. and Eng.*, **73**, L1 (1985)

34. GLICKSMAN, M.E. and VOORHEES, P.W., Ostwald ripening and relaxation in dendritic structures, *Trans. Met. A.*, **15A**, 995 (1984)

35. FLEMINGS, M.C., *Solidification processing*, McGraw-Hill Inc., New York (1974)

36. KIRKWOOD, D.H., Microsegregation, *Mater. Sci. and Eng.*, **65**, 101 (1984)

37. ZEHTAB-JAHEDI, M. and KONDIC, V., The effect of cast structure on the tensile strength of flake graphite cast irons, in *Proc. of solidification technology in the foundry and casthouse*, Metals Society, London (1980)

38. GHORESHY, M. and KONDIC, V., Structure, mechanical and casting properties of Fe–C–Al cast iron, in *Proc. of solidification technology in the foundry and casthouse*, Metals Society, London (1980)

39. HEINE, R.W. and LOPER, C.R. JR., On dendrites and eutectic cells in grey iron, *A.F.S. Trans.*, **77**, 185 (1969)

40. BASUTKAR, P.B., YEW, S.A. and LOPER, C.R. JR., Effect of carbon additions to the melt on the as-cast dendritic microstructure of grey cast iron, *A.F.S. Trans.*, **77**, 321 (1969)

41. RUFF, G.F. and WALLACE, J.F., Control of graphite structure and its effect on mechanical properties of gray iron, *A.F.S. Trans.*, **84**, 705 (1976)

42. LOPER, C.R. JR. and HEINE, R.W., Dendritic structure and spiking in ductile iron, *A.F.S. Trans.*, **76**, 547 (1968)

43. GAGNE, M. and GOLLER, R., Plate fracture in ductile iron castings, *A.F.S. Trans.*, **91**, 37 (1983)

44. MINKOFF, I., Crystal growth of graphite from the melt, in *Proc. of solidification of metals*, I.S.I., P110, p. 253 (1968)

45. MINKOFF, I. and LUX, B., Graphite growth from metallic solution, in *Metallurgy of cast irons*, Georgi Publ. Co., St Saphorin, Switzerland, p. 473 (1975)

46. TAKHARCHENKO, E.V., AKIMOV, E.P. and LOPER, C.R. JR., Kish graphite in gray cast iron, *A.F.S. Trans.*, **87**, 471 (1979)

47. SUN, G.K. and LOPER, C.R. JR., Graphite flotation in cast iron, *A.F.S. Trans.*, **91**, 841 (1983)

48. LOPER, C.R. JR. and ZAKHARCHENKO, E.V., Investigation of the development of Kish graphite in hypereutectic grey irons, *A.F.S. Trans.*, **92**, 281 (1984)

49. CAHN, J.W., On the morphological stability of growing crystals, *J. Crystal Growth*, 681 (1966)

50. CHERNOV, A.H., Stability of faceted shapes, *J. Crystal Growth*, **24/25**, 11 (1974)

51. MINKOFF, I., Physical metallurgy of cast iron, J. Wiley and Sons, London (1983)

52. CROKER, M.N., FIDLER, R.S. and SMITH, R.W., The characterisation of eutectic structures, *Proc. Roy. Soc.*, **335A**, 15 (1973)

53. JACKSON, K.A. and HUNT, J.D., Lamellar and rod growth, *Trans. Met. Soc. AIME*, **236** 1129 (1966)

54. SERIES, R.W., HUNT, J.D. and JACKSON, K.A., The use of an electric analogue to solve the lamellar eutectic diffusion problem, *J. Crystal Growth*, **40**, 221 (1977)

55. DOUBLE, D.D., Imperfections in lamellar eutectic crystals, *Mater. Sci. and Eng.*, **11**, 325 (1973)

56. BARAGER, D., SAHOO, M. and SMITH, R.W., The structural modification of the complex regular eutectics of Bi–Pb, Bi–Sn and Bi–Ti, *J. Crystal Growth*, **41**, 278 (1977)

57. BARAGER, D., SAHOO, M. and SMITH, R.W., Complex regular growth in the Bi–Pb eutectic, Solidification and Casting of Metals, Metals Society, Conference Proceedings, p. 88 (1979)

58. SAVAS, M.A. and SMITH, R.W., Quasi regular eutectic growth; a study of the solidification of some high volume fraction faceted phase anomalous eutectics, *J. Crystal Growth*, **71**, 66 (1985)

59. SAVAS, M.A. and SMITH, R.W., The modification of binary eutectics by a third element, in *Proc. Solidification Processing*, Sheffield, (1987)

60. TOLOUI, B. and HELLAWELL, A., Phase separation and undercooling in Al–Si eutectic alloy – the influence of freezing rate and temperature gradient, *Acta. Met.*, **24**, 565 (1976)

61. ELLIOTT, R. and GLENISTER, S.M.D., The growth temperature and interparticle spacing in aluminium silicon eutectic alloys, *Acta. Met.*, **28**, 1489 (1980)

62. LESOULT, G. and TURPIN, M., Coupled growth of graphite and austenite in grey cast iron, in *Metallurgy of cast irons*, Georgi Publ. Co., St Saphorin, Switzerland, p. 255 (1975)

63. SATO, T. and SAYAMA, Y., Completely and partially co-operative growth of eutectics, *J. Crystal Growth*, **22**, 259 (1974)

64. FISHER, J.D. and KURZ, W., A theory of branching limited growth of irregular eutectics, *Acta. Met.*, **28**, 777 (1980)

65. NIEWSWAAG, H. and ZUITHOFF, A.J., The effect of S, P, Si and Al on the morphology and graphite structure of directionally solidified cast iron, in *Metallurgy of cast irons*, Georgi Publ. Co., St Saphorin, Switzerland, p. 327 (1975)

66. HILLERT, M. and SUBBA RAO, V.V., Grey and white solidification of cast iron, in *Proceedings of Solidification of Metals*, I.S.I., P110, p. 204 (1968)

67. JONES, H. and KURZ, W., Relation of interphase spacing and growth temperature to growth velocity in Fe–C and Fe–Fe$_3$C eutectic alloys, *Zeit. Metallkunde*, **72**, 792 (1981)

68. POWELL, G.L.F., Morphology of eutectic M$_3$C and M$_7$C$_3$ in white iron castings, *Metals Forum*, **3**, 37 (1980)

69. SADOCHA, J.P. and GRUZLESKY, J.E., The mechanism of graphite spheroid formation in pure Fe–C–Si alloys, in *Metallurgy of cast irons*, Georgi Publ. Co., St Saphorin, Switzerland, p. 443 (1975)

70. LUX, B., MINKOFF, I., MOLLARD, F. and THURY, E., Branching of graphite crystals growing from metallic solution, in *Metallurgy of cast irons*, Georgi Publ. Co., St Saphorin, Switzerland, p. 495 (1975)

71. GILMER, G.H., in *Proceedings of symposium on modelling of casting and welding processes;* eds H.D. Brody and D. Apelion, AIME, p. 385 (1981)

72. DOUBLE, D.D. and HELLAWELL, A., Growth structure of

various forms of graphite, in *Metallurgy of cast irons,* Georgi Publ. Co., St Saphorin, Switzerland, p. 509 (1975)

73. DOUBLE, D.D. and HELLAWELL, A., Conical helix growth forms in graphite, *Acta. Met.,* **22,** 481 (1974)

74. JOHNSON, W.C. and SMARTT, H.B., Role of interphase boundary adsorption in the formation of spheroidal graphite in cast iron, in *Solidification and Casting of Metals,* Metals Society, p. 125 (1979)

75. OLDFIELD, W., GEERING, G.T. and TILLER, W.A., Solidification of spheroidal and flake graphite cast iron, in *Proc. of solidification of metals,* I.S.I., P110, p. 256 (1968)

76. MINKOFF, I. and NIXON, W.C. Scanning electron microscopy of graphite growth in iron and nickel alloys, *J. Applied Physics,* **37,** 4848 (1966)

77. MUNITZ, A. and MINKOFF, I., The relationship between graphite form and undercooling, *45th International Foundry Congress, Budapest,* (1978)

78. MINKOFF, I., The mechanism of spherulite growth in cast iron, *Third International Symposium on the Physical Metallurgy of Cast Iron, Stockholm,* (1984)

79. MINKOFF, I. and LUX, B., Step growth on graphite crystals, *Micron,* **2,** 282 (1971)

80. MINKOFF, I. and LUX, B., Branching during graphite growth, *Practical Metallography,* **8,** 69 (1971)

81. UDOMON, U.H. and LOPER, C.R. JR., Comments concerning the interaction of rare earths with subversive elements in cast iron, *A.F.S. Trans.,* **93,** 519 (1985)

82. LIU, P.C. and LOPER, C.R. JR., Segregation of certain elements in cast irons, *A.F.S. Trans.,* **92,** 289 (1984)

83. MUNITZ, A. and MINKOFF, I., Instability of graphite crystal growth in metallic systems, in *Solidification and Casting of Metals,* Metals Society, p. 70, (1979)

84. MINKOFF, I., 'Delta Tee' method for structure control of cast iron, *Third International Symposium on the Physical Metallurgy of Cast Iron, Stockholm,* (1984)

85. RIPOSAN, I., CHISAMERA, M., SOFRONI, L. and BRABIE, V., Contribution to the study of the solidification mechanism and of the influence of structure on the properties of compacted/vermicular graphite cast irons, *Third International Symposium on the Physical Metallurgy of cast, Iron, Stockholm,* (1984)

86. XIJUN, D., PEIYUE, Z. and QIFU, L., Structure and formation of vermicular graphite, *Third International Symposium on the Physical Metallurgy of Cast Iron, Stockholm,* (1984)

87. STEFANESCU, D.M., MARTINEZ, F. and CHEN, I.G., Solidification behaviour of hypoeutectic and eutectic compacted graphite cast irons: Chilling tendency and eutectic cells, *A.F.S. Trans.,* **91,** 205 (1983)

88. SU, J.Y., CHOW, C.T. and WALLACE, J.F., Solidification behaviour of compacted graphite, *A.F.S. Trans.,* **90,** 565 (1982)

89. BASDOGAN, M.F., KONDIC, V. and BENNETT, G.H.J., Graphite morphologies in cast iron, *A.F.S. Trans.,* **90,** 263 (1982)

90. GAN, Y. and LOPER, C.R. JR., Observations on the formation of graphite in compacted and spheroidal graphite cast irons, *A.F.S. Trans.,* **91,** 781 (1983)

91. PAN, E.N., OGI, K. and LOPER, C.R. JR., Analysis of the solidification process of compacted/vermicular graphite cast iron, *A.F.S. Trans.,* **90,** 509 (1982)

92. SUBRAMANIAN, S.V., ZHONG, W. and PURDY, G.R., Compacted graphite structure evolution in a hypoeutectic cast iron, in *Proc. Solidification Processing, Sheffield,* (1987)

93. RIPOSAN, I., CHISAMERA, M. and SOFRONI, L., Contribution to the study of some technological and applicational properties of compacted graphite cast iron, *A.F.S. Trans.,* **93,** 35 (1985)

94. CHAO, C.G., LU, W.H., MERCER, J.L. and WALLACE, J.F., Effect of treatment alloys and section size on the compacted graphite structures produced by the in-mould process, *A.F.S. Trans.,* **93,** 651 (1985)

95. PAN, E.N. and LOPER, C.R. JR., A study of the production of compacted/vermicular graphite cast irons in thin sections, *A.F.S. Trans.,* **93,** 523 (1985)

96. CORNELL, H.H. and LOPER, C.R. JR., Variables involved in the production of compacted graphite cast iron using rare earth containing alloys, *A.F.S. Trans.,* **93,** 435 (1985)

97. LIU, J., DING, N.X., MERCER, J.L. and WALLACE, J.F., Effect of type and amount of treatment alloy on compacted graphite produced by the Flotret Process, *A.F.S. Trans.,* **93,** 675 (1985)

98. ITOFUJI, H., KAWANO, Y., YAMAMOTO, S., INOYANA, N., YOSHIDA, H. and CHANG, B., Comparison of substructure of compacted/vermicular graphite with other types of graphite, *A.F.S. Trans.,* **91,** 313 (1983)

99. BUHR, B.K., Vermicular graphite formation in heavy section nodular iron castings, *A.F.S. Trans.,* **76,** 497 (1968)

100. SUBRAMANIAN, S.V., KAY, D.A.R. and PURDY, G.R., Compacted graphite morphology control, *A.F.S. Trans.,* **90,** 589 (1982)

101. SUBRAMANIAN, S.V., KAY, D.A.R. and PURDY, G.R., Graphite morphology control in cast iron, in *Third International Symposium on the Physical Metallurgy of Cast Iron, Stockholm,* (1984)

102. HRUSOVSKY, J.P and WALLACE, J.F., Effect of composition on solidification of compacted graphite iron, *A.F.S. Trans.,* **93,** 55 (1985)

103. LIU, P.C. and LOPER, C.R. JR., S.E.M. study of the graphite morphology in cast iron, *Scanning Electron Microscopy,* **1,** 407 (1980)

104. LOPER, C.R. JR., Graphite morphology in cast irons, *Third International Symposium on the Physical Metallurgy of Cast Iron, Stockholm,* (1984)

105. LIU, P.C., LI, C.L., WU, D.H. and LOPER, C.R. JR., S.E.M. study of chunky graphite in heavy section ductile iron, *A.F.S. Trans.,* **91,** 119 (1983)

106. TANG CHONG, X.I., FARGUES, J. and HECHT, M., Formation and prevention of chunky graphite in slowly cooled nodular irons, *Third International Symposium on the Physical Metallurgy of Cast Iron, Stockholm,* (1984)

107. LIU, P.C., LOPER, C.R. JR., KIMURA, T. and PAN, E.N., Observations on the graphite morphology of compacted graphite cast iron, *A.F.S. Trans.*, **91**, 341 (1983)

108. GREENHILL, J.M., Some practical observations on lead contamination of cast iron, *Brit. Foundryman*, **77**, 370 (1984)

109. VOROS, A. and FARAGO VOROS, E., Influence of Tellurium on the solidification and properties of cast iron, *A.F.S. International Cast Metals Journal*, **4**, 41 (1979)

110. KURZ, W. and FISHER, D.J., Dendritic growth in eutectic alloys; the coupled zone, *International Met. Review*, **No. 244**, (1977)

111. JONES, H. and KURZ, W., Growth temperature and the limits of coupled growth in unidirectional solidification of Fe–C eutectic alloy. *Trans. Metall. Soc. AIME*, **11A**, 1265 (1980)

112. SWINDEN, D.J. and WILFORD, C.F., Review of factors influencing formation of graphite eutectic cells in grey cast irons, *Foundry Trade Journal*, **140** (1976)

113. SWINDEN, D.J. and WILFORD, C.F., The nucleation of graphite from liquid iron: a phenomenological approach, *Brit. Foundryman*, **69**, 118 (1976)

114. GUOGING, X., ZONGSEN, Y. and MOBLEY, C.E., Solidification and structures in rare earth-inoculated grey irons, *A.F.S. Trans.*, **90**, 943 (1982)

115. WALLACE, J.F., Effects of minor elements on the structure of cast irons, *A.F.S. Trans.*, **83**, 363 (1975)

116. QUINTERO, M. and KONDIC, V., Measurement of inoculation effects in flake cast iron, *Brit. Foundryman*, **77**, 220 (1984)

117. WILFORD, K.B. and WILSON, F.G., The effect of titanium on the microstructure and properties of heavy section iron castings: Part I. Ferritic cast irons, *Brit. Foundryman*, **78**, 301 (1985)

118. WILFORD, K.B. and WILSON, F.G., The effect of titanium on the microstructure and properties of heavy section iron castings, *Brit. Foundryman*, **78**, 364 (1985)

119. MAGNIN, P. and KURZ, W., Competitive growth of stable and metastable Fe–C eutectic, *Third International Symposium on the Physical Metallurgy of Cast Iron, Stockholm*, (1984)

120. WOLF, G., FLENDER, E. and SAHM, P.R., Solidification behaviour of technical metastable near eutectic iron carbon alloys, *Third International Symposium on the Physical Metallurgy of Cast Iron, Stockholm*, (1984)

121. FREDRICKSSON, H. and SVENSSON, I., Computer simulation of the structure formed during solidification of grey cast iron, *Third International Symposium on the Physical Metallurgy of Cast Iron, Stockholm*, (1984)

122. EVANS, W.J., CARTER, S.F. JR. and WALLACE, J.F., Factors influencing the occurrence of carbides in thin sections of ductile iron, *A.F.S. Trans.*, **89**, 293 (1981)

123. LIU, P.C. and LOPER, C.R. JR., Electron microprobe study of the intercellular compounds in heavy section ductile iron, *A.F.S. Trans.*, **89**, 131 (1981)

124. GUNDLACH, R.B., JANOWAK, J.F. and RÖHRIG, K., On the problems with carbide formation in grey cast iron, *Third International Symposium on the Physical*

Metallurgy of Cast Iron, Stockholm, (1984)

125. STEFANESCU, D.M., LOPER, C.R. JR., VOIGT, R.C. and CHEN, I.G., Cooling curve-structure analysis of compacted/ vermicular graphite cast irons produced by different melt treatments, *A.F.S. Trans.*, **90**, 333 (1982)

126. KANETKAR, C.S. and STEFANESCU, D.M., Macromicroscopic simulation of solidification of eutectic and off eutectic alloys, in *Proc. Solidification Processing, Sheffield*, (1987)

127. SMICKLEY, R.J. and RUNDMAN, K.B., The effect of aluminium on the structure and properties of grey cast iron, *A.F.S. Trans.*, **89**, 205 (1981)

128. MOORE, A., Rapid carbon determination using BCIRA carbon calculator, *BCIRA report no. 1126* (1973)

129. DONALD, W. and MOORE, A., Significance of carbon equivalent formulae and their applications in the foundry, *BCIRA report no. 1128* (1973)

130. BREEDEN, C.R., Determining C.E.L. values for malleable irons – the effect of melting process variables, *BCIRA report no. 1499* (1982)

131. HEINE, R.W., Carbon, silicon, carbon equivalent, solidification and thermal analysis relationships in gray and ductile cast irons, *A.F.S. Trans.*, **81**, 462 (1973)

132. HEINE, R.W., Liquidus and eutectic temperature solidification of white cast irons, *A.F.S. Trans.*, **85**, 537 (1977)

133. HEINE, R.W., The carbon equivalent Fe–C–Si diagram and its application to cast iron, *A.F.S. Cast Metals Research Journal*, **7**, 49 (1971)

134. BOOTH, M., Thermal analysis for composition determination of grey cast iron, *Brit. Foundryman*, **76**, 35 (1983)

135. EKPOOM, U. and HEINE, R.W., Thermal analysis by differential heat analysis DHA of cast irons – a modified differential thermal analysis technique, *A.F.S. Trans.*, **89**, 27 (1981)

136. EKPOOM, U. and HEINE, R.W., Metallurgical processing variables affecting the solidification of malleable base white cast iron, *A.F.S. Trans.*, **89**, 1 (1981)

137. LEON, C., EKPOOM, U. and HEINE, R.W., Relationship of casting defects to solidification of malleable base white cast iron, *A.F.S. Trans.*, **89**, 323 (1981)

138. JACOBS, F.W., Practical application of liquidus control for malleable iron melting, *A.F.S. Trans.*, **89**, 261 (1981)

139. ALAGARSAMY, A., JACOBS, F.W., STRONG, G.R. and HEINE, R.W., Carbon equivalent vs. austenite liquidus. What is the correct relationship for cast irons?, *A.F.S. Trans.*, **92**, 871 (1984)

140. STRONG, G.R., Thermal analysis as a ductile iron molten metal processing evaluation tool, *A.F.S. Trans.*, **91**, 151 (1983)

141. CHAUDHARI, M.D., HEINE, R.W. and LOPER, C.R. JR., Potential applications of cooling curves in ductile iron process control, *A.F.S. Trans.*, **82**, 379 (1974)

142. HEINE, R.W., LOPER, C.R. JR. and CHAUDHARI, M.D., Characterisation and interpretation of ductile iron cooling curves, *A.F.S. Trans.*, **79**, 399 (1971)

143. CHAUDHARI, M.D., HEINE, R.W. and LOPER, C.R. JR.,

Principles involved in the use of cooling curves in ductile iron process control, *A.F.S. Trans.*, **82**, 431 (1974)

144. HRIBOVSEK, B.T. and MARINCEK, B., Thermal analysis of bulk solidified cast iron melts, in *Metallurgy of cast irons*, Georgi Publ. Co., St Saphorin, Switzerland, p. 659 (1975)

145. RYNTZ, E.F. JR., JANOWAK, J.F., HOCHSTEIN, A.W. and WARGEL, C.A., Prediction of nodular iron microstructure using thermal analysis, *A.F.S. Trans.*, **79**, 161 (1971)

146. MONROE, R. and BATES, C.E., Thermal analysis of ductile iron samples for graphite shape prediction, *A.F.S. Trans.*, **90**, 307 (1982)

147. BACKERUD, L., NILSSON, K. and STEEN, N., Study of nucleation and growth of graphite in magnesium treated cast iron by means of thermal analysis, in *Metallurgy of cast irons*, Georgi Publ. Co., St Saphorin, Switzerland, p. 625 (1975)

148. STEFANESCU, D.M., LOPER, C.R. JR., VOIGT, R.C. and CHEN, I.G., Cooling curve structure analysis of compacted/vermicular graphite cast irons produced by different melt treatments, *A.F.S. Trans.*, **90**, 333 (1982)

149. CHEN, I.G. and STEFANESCU, D.M., Computer aided differential thermal analysis of spheroidal and compacted graphite cast irons, *A.F.S. Trans.*, **92**, 947 (1984)

150. CORBETT, C.F., Computer aided thermal analysis and solidification simulation, paper presented at *84th IBF Annual Conference*, Buxton, June 1987

151. ABLEIDINGER, K. and RAEBUS, K., Thermoanalytical computer control and correction of S.G. iron melts with constant feeding condition, *Brit. Foundryman*, **79**, 320 (1986)

152. KNOTHE, W., Practical application of thermal analysis as an aid to strict metallurgical control, in *S.G. iron – the next 40 years.*, BCIRA Conference, Warwick, April 1987

153. SMILEY, L.E., Solidification modelling on a P.C. A foundryman's approach, *Modern Casting*, **77**, 32 (1987)

154. MARINCEK, B., The cooling curve method for controlling structural quality of flake and spheroidal graphite cast irons, in *Proc. Solidification Processing*, Sheffield, (1987)

Chapter 4

Solid state transformations

Introduction

The strong structure property relationships that are the basis of cast iron technology have been emphasized and illustrated in previous chapters. Both solidification structures and solid state transformation structures contribute to this relationship. Various forms of structural diagrams have been proposed for grey and malleable irons for use as a guideline for predicting the solidification structure[1]. An example is shown in *Figure 4.1*. The important structural change at low Si concentration is the liquid–solid transformation. The feature of interest is the relative nucleation and growth rates of graphite and carbide phases. The solid state transformation of austenite is important at higher Si concentrations. The relative proportions of ferrite and pearlite in cast structures are particularly important. Cooling rate influences both transformations. Increased solidification rate favours the metastable eutectic and increased cooling rate favours the pearlite reaction in solid state transformations. Structural diagrams have been produced showing the influence of elements in solution.

Although economical advantages are to be gained if the iron structure can be achieved during the casting process, heat treatment is an integral part of processes such as malleabilizing and offers advantages in others. Heat treatment is remedial. It can be used to remove undesirable carbides and to relieve residual stresses. It also confers machinability and makes the realization of the full range of mechanical properties possible (see *Figure 1.7* in Chapter 1). Heat treatment offers a guarantee of quality by compensating for small variations in metal composition, metal treatment and pouring operations. In addition the matrix structure is more consistent in castings of varying section size after heat treatment. Indeed, some foundries cast spheroidal irons to a single composition and use heat

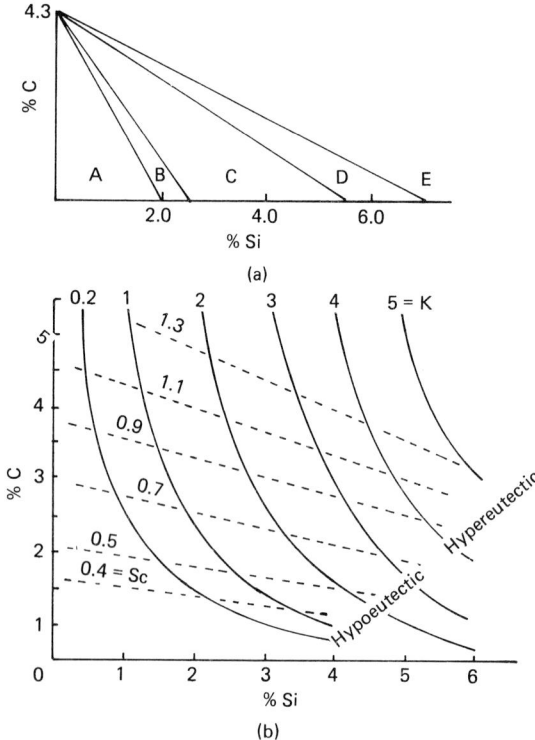

Figure 4.1 (a) Maurer structural diagram; A, white structure; B, mottled structure; C, pearlitic structure; D, pearlitic and ferritic; E, ferritic.(b) Laplanche structural diagram. Cast iron structures are determined by the graphitizing coefficient K and the degree of saturation coefficient S_c. For a cylindrical bar, 30 mm diameter, $K = 4Si/3(1 - 5/[3C + Si])$. $S_c = $ total C/(4.3 − 0.3Si). S_c is selected by the desired position of the alloy in relation to (4.3 − 0.3Si) and K by the desired structure. The intersection of the two gives the desired C and Si contents

treatment to produce the range of grades in the speci-fication. Heat treatment practices are described in this chapter.

The fundamentals of heat treatment

There are three main allotropic forms of Fe, δ and α ferrite which are b.c.c. and γ (austenite) which is f.c.c. The temperature range over which these phases are stable in a plain C steel are shown in *Figure 4.2*. The γ phase which formed during solidification or on subsequent heat treatment changes its composition according to the solubility line ES in *Figure 4.2* as the temperature falls. On further cooling below the eutectoid temperature transformation to pearlite, bainite or martensite is possible as indicated in *Figure 4.3*.

Pearlite formation is favoured by slow cooling. The transformation begins with the nucleation of either ferrite or cementite heterogeneously on the proeutec-toid phase or on the austenite grain boundary. When cementite nucleates on the boundary, the energy barrier opposing nucleation is reduced by the formation of an orientation relationship with one of the austenite grains (γ_A) (see *Figure 4.4*).

A semi-coherent, low mobility interface is formed with γ_A. An incoherent mobile interface is formed with γ_B. This nucleation event depletes the surround-ing austenite of C which makes ferrite nucleation easier. The nucleation occurs adjacent to the cementite with an orientation relationship with γ_A. Repetition of these events expands the pearlite colony

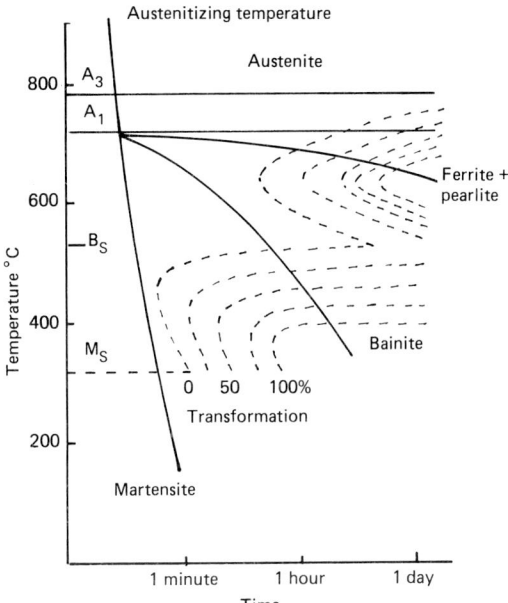

Figure 4.3 Isothermal transformation diagram

sideways along the grain boundary. The main colony growth occurs edgewise by movement of the incoherent, mobile interfaces into γ_B. This growth process is similar to eutectic lamellar growth and is controlled by the diffusion of C in the austenite phase.

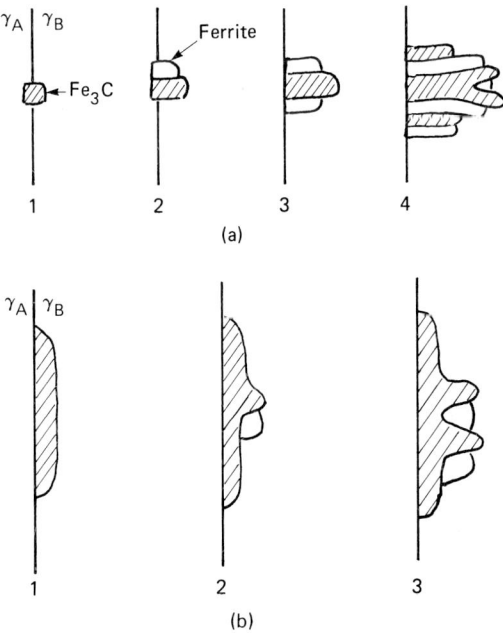

Figure 4.4 Successive stages (1, 2, etc.) in the development of a pearlite colony when (a) nucleation of cementite occurs on an austenite grain boundary; (b) nucleation of pearlite occurs on proeutectoid cementite

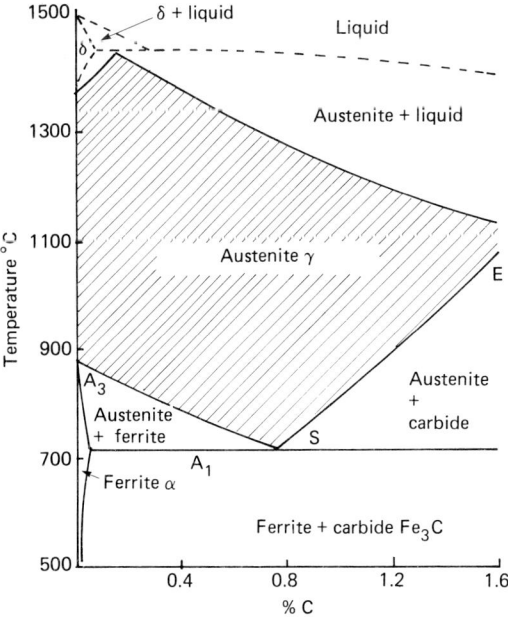

Figure 4.2 The iron carbon phase diagram for steel

The pearlite spacing varies inversely with under-cooling and the growth velocity varies as the square of the undercooling[2]. The kinetics of the pearlite transformation can be represented on a temperature, time, transformation (TTT) curve as shown in *Figure 4.3*. This details the progress of the isothermal transformation by plotting the fraction transformed as a function of time at various temperatures. The C-shaped curve is characteristic of a nucleation and growth process. The driving force for the transformation is small just below the eutectoid temperature where the undercooling is low. As the undercooling increases so does the driving force and the transformation occurs more rapidly. It reaches a maximum rate at the nose of the curve. Although the driving force increases at temperatures below the nose, the reaction is slowed by the lack of diffusivity of the rate-controlling element.

In common with the eutectic reaction, the coupled zone concept has been used to define the conditions under which the proeutectoid phase forms, as shown in *Figure 4.5*. The majority of alloying elements displace the TTT curve to the right. This slows the pearlite reaction and increases the hardenability. Redistribution of the alloying element at the austenite–pearlite interface is a thermodynamic requirement for growth at low undercooling when ferrite and cementite with compositions close to equilibrium must be formed. Carbide-forming elements Mn, Cr and Mo partition to cementite. Ni, Si and Co partition to the ferrite. However, below a composition-dependent temperature, the no-partition temperature, pearlite growth can occur without partitioning at the interface. It has been suggested that steels alloyed with Mn and Ni, which lower the eutectoid temperature, show a no-partitioning tem-

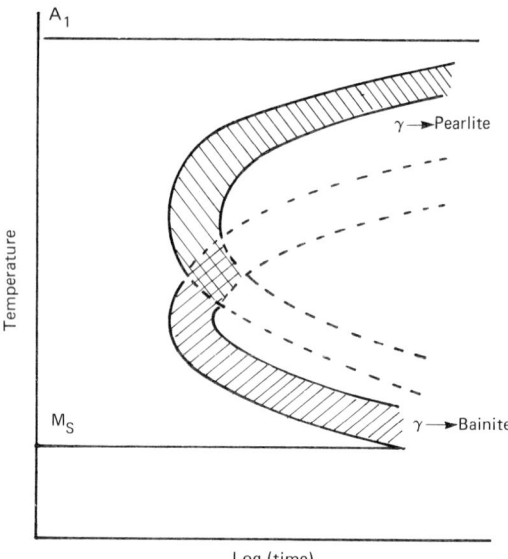

Figure 4.6 Schematic diagram showing the relative positions of the transformation curves for pearlite and bainite in plain C eutectoid steel

perature. In contrast additions such as Cr, No, Si and Co, which raise the temperature, are forced by thermodynamics to partition over a wide range of temperatures.

Bainite is the second transformation product and forms between 550 and 250 °C. These temperatures are intermediate between those for pearlite and martensite formation. A temperature, B_s, exists above which bainite does not form. A temperature range exists below B_s in which the bainite transformation does not go to completion. However, the retained austenite will often transform during cooling to room temperature. The bainite and pearlite transformation curves often overlap on the TTT diagram as shown in *Figure 4.6*. Complete transformation to bainite is only possible, and the features on the transformation curve are only revealed, in alloy steels, for example, in B-containing low C-Mo steel in which Mo promotes the bainite reaction and B retards ferrite formation. Solute elements, C in particular, depress the B_s temperature. Upper bainite forms in the temperature range 550 to 400 °C and grows competitively with pearlite at higher temperatures.

Although the two structures form by different reactions they are difficult to distinguish microscopically. However, crystallography distinguishes the two structures. Upper bainite forms when ferrite nucleated at several sites on the austenite boundary grows in lath form into the austenite grain, γ_B (see *Figure 4.7*). A Kurdjumov-Sachs orientation relationship exists between γ_B and the ferrite. As growth proceeds C partitioning increases the C concentration of the austenite until cementite nucleates and grows.

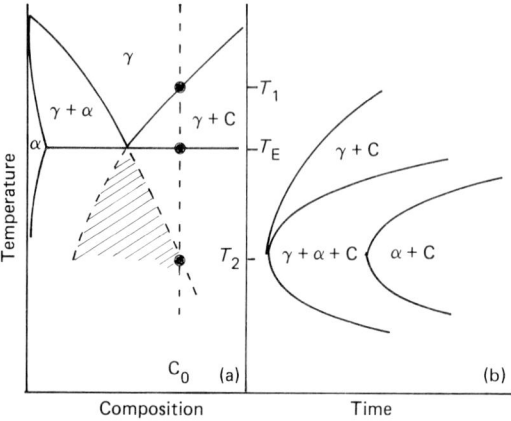

Figure 4.5 (a) Range of eutectoid transformation in a Fe-C alloy. The shaded area is the coupled zone. (b) An alloy of composition C_0 will precipitate C before pearlite in the temperature interval $T_1 - T_2$. The transformation is directly to pearlite at T_2

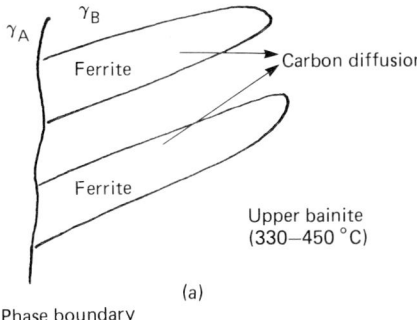

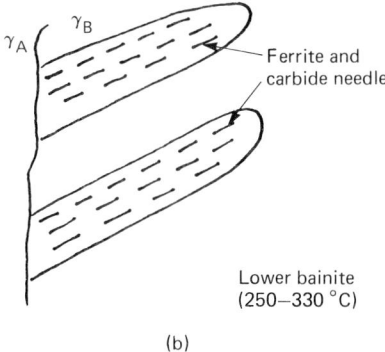

Figure 4.7 Schematic representation of the mechanism of formation of upper and lower bainite

Opinions still differ concerning the mode of growth. Although there is some evidence of surface relief effects, thermionic electron microscopy has shown that thickening of the laths occurs by the movement of small ledges as in the growth of Widmanstätten ferrite.

When the transformation temperature falls below 350 °C, the bainite microstructure changes from lath to plate ferrite with a much finer carbide dispersion. This lower bainite structure resembles tempered martensite. The ferrite nucleates at austenite boundaries and within the grains. The habit plane of the plates depends on the temperature and the C content. A distinguishing feature of lower bainite is the growth of carbide rods within the ferrite plates. The carbide can be cementite or ε-iron, depending on the temperature and alloy content.

An orientation relationship develops between the carbide and ferrite which is identical to that in tempered martensite. However, the bainitic carbide exhibits only one variant of the relationship. As a result the carbide forms in parallel arrays inclined at about 60° to the ferrite plate axis. These particles nucleate at the γ/α interface and grow as the interface advances. This suggests that the lower bainite reaction is interface-controlled and that the precipitation of carbide decreases the C content of the austenite. This increases the driving force for further transformation.

Although differences of opinion also exist concerning this growth mode, it appears that the lower bainite transformation is partly martensitic in character. However, it displays several characteristics of a diffusion controlled process. The difference in modes of transformation of upper and lower bainite produce different kinetics and separate C curves on the TTT diagram.

Rapid cooling rates suppress the pearlite and bainite transformations and promote martensite formation as shown in *Figure 4.3*. Martensite normally forms athermally rather than isothermally. Formation begins at temperature M_s, which can vary from 500 °C to below room temperature depending on alloy content, and finishes at the M_f temperature. Occasionally all the austenite has not transformed on reaching M_f. Larger amounts of austenite are retained in steels with low M_f temperatures. Martensite forms from austenite by a sudden shear process without a change in chemical composition. The C which is in solid solution in γ remains in solution in the new body-centred tetragonal structure, a distorted form of body-centred cubic Fe.

Martensite is easily recognized by its distinctive microstructural features. Each austenite grain transforms by the sudden appearance of martensite laths. The laths have a well defined habit plane which changes as the C content increases. The laths occur on several variants of this plane within each grain[3]. An orientation relationship exists between the new martensite lath and the austenite.

Martensite is a strong but brittle phase. It is customary to temper quenched structures. Four distinct but often overlapping stages occur during tempering:

1. precipitation of ε-iron carbide and partial loss of martensite tetragonality up to 250 °C;
2. retained austenite decomposes between 200 and 300 °C;
3. ε-iron carbide is replaced by cementite between 200 and 300 °C and
4. cementite coarsens and spheroidizes and ferrite recrystallizes above 350 °C.

The kinetics of the basic reactions and the products of the reactions can change in the presence of alloying elements. For example, the phenomenon of secondary hardening is important in producing high strength steels.

Practical heat treatment involves transformations during continuous cooling. Consequently, TTT diagrams which describe isothermal transformations are of limited use. Continuous cooling transformations (CCT) diagrams describe the progress of a transformation with decreasing temperature for a series of cooling rates. A diagram for a low alloy grey iron is shown in *Figure 4.8*.

CCT diagrams identify the A_1 and A_3 temperatures, the percentage transformation product and the

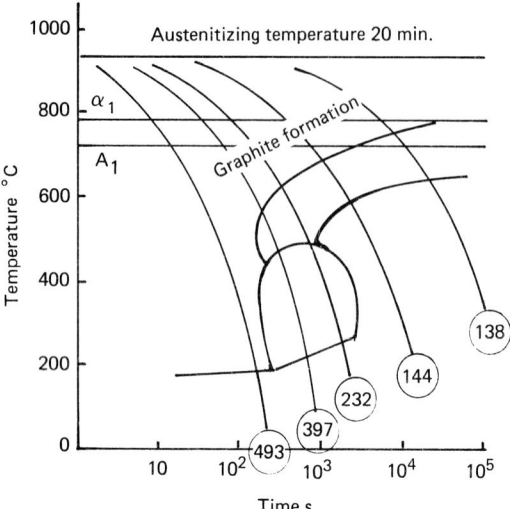

Figure 4.8 CCT diagram for a low alloy Mo-Cu grey cast iron of composition: 3.17% C, 2.0% Si, 0.75% Mn, 0.3% Mo, 0.6% Cu

hardness of the alloy (circled at the end of the cooling curve). It is necessary to establish the cooling rate for the section size of a casting before the diagram can be used. When the section size varies, a conservative estimate of structure and properties can be made using the cooling rate at the centre of a representative casting section. The structure of the iron in *Figure 4.8* consists of austenite and graphite immediately after solidification. Secondary graphite precipitation occurs in the temperature range close to A_1. The slowest cooling rates result in pearlite formation, whereas cooling into the nose of the pearlite transformation gives a mixed structure of pearlite and bainite.

The heat treatment of cast irons

Heat treatment considerations in cast iron differ from those in steel in several significant ways. Whereas C content is of prime importance in selecting the austenitizing temperature for a steel, the Si content is the significant factor for cast irons. The presence of Si introduces a three phase region in which ferrite, austenite and graphite coexist. This excludes cementite from the equilibrium phase diagram. The eutectoid temperature of a steel is expanded into a critical temperature range. The location and size of this range depends on the Si content as shown in *Figure 4.9* (ref. 4). The influence of the main elements in cast iron on the critical temperature range is indicated in *Table 4.1*.

The solid state transformations in a grey iron during cooling after solidification occur in the

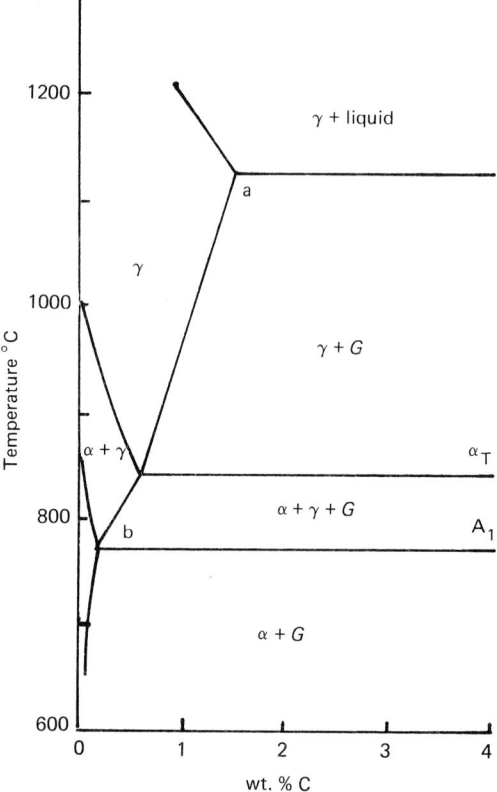

Figure 4.9 Section of the Fe-G-Si phase diagram at 2% Si (after ref. 4)

following way. Immediately after solidification the structure consists of austenite (%C, point a in *Figure 4.9*) and graphite in flake or spheroidal form. The solubility of C in austenite decreases on further cooling. The rejected C deposits on the graphite flakes or spheroids. On cooling below the α_T temperature at a rate that enables equilibrium to be attained, more C is deposited on the graphite. Ferrite nucleates and grows until at point b, all the austenite has transformed into ferrite and graphite.

Fully ferritic structures are more likely to occur in heavy section castings which cool slowly. The cooling rate in most castings leads to non-equilibrium cooling

Table 4.1 The influence of the main elements in cast iron on the critical temperature range

Element	%	Upper critical temperature effect per 1%	Lower critical temperature effect per 1%
Si	0.3–3.5	37 increase	29 increase
P	0–0.2	220 increase	220 increase
Mn	0–1.0	37 decrease	130 decrease
Ni	0–1.0	17 decrease	24 decrease

and some pearlite formation. The pearlite content of a grey iron is an important metallurgical parameter which controls hardness, yield and tensile strength, fatigue properties, wear characteristics and machinability. Pearlite forms during non-equilibrium cooling when some austenite remains on reaching the A_1 temperature and then transforms eutectoidally below this temperature. Austenite retention is influenced by the austenite C content.

Any factor inhibiting C diffusion during cooling will increase the propensity for austenite retention and pearlite formation. Such factors include rapid cooling and coarse graphite structures. Si promotes ferrite formation by decreasing the C solubility in austenite. This promotes diffusion by increasing the α_T–A_1 temperature interval. Alloying additions may be made to counter the effect of Si and promote pearlite formation.

The ferrite in as-cast mixed structures may appear as grain boundary or Widmanstätten ferrite or as silico-ferrite in irons containing more than 3% Si. (See *Figure 6.40* in Chapter 6.) Perhaps the most well known form is the bullseye structure in spheroidal iron. The ferrite ring surrounding the spheroid may form prior to eutectoid transformation or as a result of cementite decomposition[5].

Stress relief heat treatment

Castings are rarely free from residual stress, but its effect on casting performance and the need for removal depends on its magnitude and location with respect to service stresses. Stress develops in a body during cooling because the surface loses heat more rapidly than the interior. This causes differential cooling and differential contraction, which leads to stress. However, no residual stress should remain when the body has cooled to room temperature unless plastic deformation occurs during cooling as a result of the differential contraction stress.

Unfortunately, this situation occurs frequently for several reasons. Consider a casting of uniform section such as a plate. Once the surface regions have cooled below 650 °C they become sufficiently rigid to support and impose stress. At this stage heat is still lost more quickly from the surface and the hotter central areas can yield and relieve any imposed stress. In the later stages of cooling when this region is rigid it will contract more than the surface and impose compressive stresses in the surface and leave the centre in a state of tension. Both areas are unable to deform plastically and the stresses are permanent. They can cause buckling of the plate or cracks in the centre of the plate. For example, a flatness tolerance of 0.1 mm could not be maintained after machining as-cast clutch plates made from a low alloy flake iron. These tolerances can be achieved after stress relieving at 620 °C for 2 h.

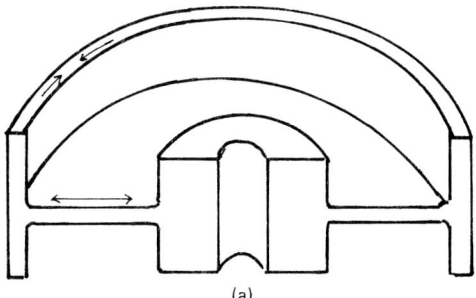

(a)

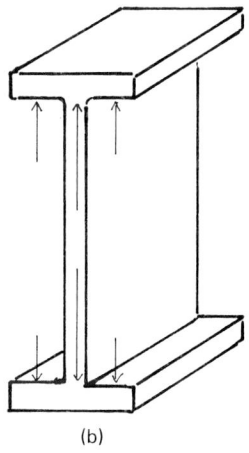

(b)

Figure 4.10 Residual stresses in castings due to (a) non-uniform casting sections; (b) mechanical mould restraint

Thermal gradients and residual stresses can be more pronounced in castings with non-uniform cross-sections, as shown in *Figure 4.10*. This figure also shows how restriction or restraint of the thermal contraction of sections of a casting by a rigid mould can create the conditions leading to residual stress even in castings of uniform thickness. In *Figure 4.10*, the horizontal sections in a rigid mould restrict the thermal contraction of the vertical section leading to the residual stresses shown. Phase transformations are accompanied by a volume change. If it occurs at low temperature, as in martensite, it can induce residual stress. This is a problem in high Cr-Mo irons[6].

Compressive residual stresses may be introduced in the surface by shot peening. Thermal processes, such as welding for fabrication or repair and surface or through hardening, introduce residual stress. Mechanical mishandling and heavy machining operations can cause stresses large enough to cause concern. Considerations at the design stage and careful casting practice can minimize residual stresses and avoid the need for costly stress relieving heat treatments. Castings which incorporate frameworks with non-uniform sections are particularly prone to residual stress.

Careful consideration of the form of ribbing and the design of joints can minimize residual stress. For example, gating through one member only of a grid of uniform thickness can create sufficient heat to delay the cooling of the gated member and produce residual stress. On the other hand, gating through the thinnest member of a grid of non-uniform thickness can reduce the difference in cooling rate between members and reduce residual stress.

The time and temperature of knock-out from the mould can be used to control residual stress development. Knock-out at a high temperature followed by forced cooling of selected thicker sections can be used to eliminate temperature gradients prior to cooling below ∼ 650 °C or, alternatively, may be used to control temperature gradients so that the residual stresses are not harmful in service. Otherwise, slow cooling in the mould or in soaking pits, as practised for many years in cast roll production, may be preferable.

The basic principle involved in stress relief is closely related to the concept of creep. For stress relief, a casting has to be raised to a temperature (480–650 °C) at which it can deform plastically at low stress. However, to maintain its strength there should not be any structural change. Several factors influence the extent to which residual stress is relieved:

1. the initial severity of the residual stress;
2. the stress relieving time at temperature;
3. the heating–cooling cycle and
4. the chemistry and metallurgical structure.

In general, the higher the initial level of residual stress, the more readily stress will be relieved. The time–temperature relationship for stress relief is significant as a control factor. However, the economics of furnace capacity and operating costs invariably favour stress relief at higher temperatures for shorter times.

The cooling rate following stress relief treatment is important. Rapid cooling of castings with non-uniform sections will induce thermal gradients and re-establish residual stresses. Therefore, it is highly desirable that uniform cooling occurs. This is usually achieved by furnace cooling to about 200 °C. Alloying elements such as Mn, Cr and Mo increase elevated temperature strength. Such irons require higher stress relieving temperatures.

Examples of stress relieving treatments are given in *Table 4.2*. Some of the factors described above are illustrated in the practice described in *Table 4.3*. This table concerns a large diesel engine cylinder block made from flake iron containing 3.25% C, 2.20% Si

Table 4.2 Typical stress relieving treatments for grey cast irons

Cast iron	Section thickness	Stress relief cycle
Unalloyed grey iron	up to 50 mm	2 h at 565–580 °C
	50 to 100 mm	1 h/20 mm at 565–580 °C
	> 100 mm	6 h at 565–580 °C
High strength low alloy grey iron	up to 50 mm	2 h at 565–595 °C
	50–100 mm	1 h/20 mm at 565–595 °C
Spheroidal iron	> 100 mm	6 h at 565–595 °C
High alloy irons	up to 50 mm	2 h at 595–650 °C
	50–100 mm	1 h/20 mm at 595–650 °C
	> 100 mm	6 h at 595–650 °C

Table 4.3 Influence of various treatments on the residual stress in cylinder blocks

Treatment	Closure cm/cm	Stress relief %
(1) Shake out after 6 h, cool to room temperature over a 10 h period	0.16	0 reference
(2) Same as (1) but cores removed at shake out	0.076	52
(3) Cooled in the mould for 16 h	0.064	60
(4) Same as (1) then stress relieved at 535 °C for 2 h, furnace cooled to 370 °C	0.06–0.076	52–62
(5) Same as (1) then stress relieved at 620 °C for 2 h, furnace cooled to 370 °C	0.012–0.015	92

and 0.3% Cr (ref. 7). A saw cut was made through the end of the block to the first cylinder bore and parallel to two vertical lines drawn on the end of the block. The residual stress was estimated by measuring the distance between the lines before and after sawing. The amount of close-in is related to the residual stress. The results illustrate the reduction in residual stress achieved by reducing the temperature gradients in the casting by slow cooling in the mould and by core removal at shake-out. The marked dependence on stress relief temperature is evident.

The most important features of a furnace suitable for stress relief are:

1. an ability to achieve controlled heating and cooling rates in the range 40 to 100 °C/h;
2. freedom from marked temperature gradients in the furnace and
3. freedom from direct flame or exhaust gas impingement on the casting.

Both continuous and batch type furnaces are used. They are electric, gas or oil-fired. Indirect heating with forced air circulation in the heated zone facilitates the achievement of the factors listed above.

The most serious consequences of incorrect stress relief treatment are distortion or cracking of castings. This may be caused by:

1. inadequate support of castings in a furnace;
2. overloading of castings during heat treatment due to stacking others on them;
3. temperature gradients generated in the casting due to flame impingement and
4. too rapid a heating and/or cooling rate.

Stress relief is usually carried out when the casting has been fettled and is ready for rough machining.

Annealing heat treatment

Annealing heat treatments are used to eliminate carbides and pearlite from as-cast structures thus producing graphite in a ferritic matrix. In this way:

1. grey and spheroidal irons are softened, ductility and machinability are increased and
2. white irons are converted to malleable irons.

Annealing of grey irons

A full annealing treatment involves two stages. The first is performed above the critical temperature range. It decomposes carbide and homogenizes the matrix. This disperses segregated elements that might otherwise lead to localized stabilization of carbides and pearlite. The second stage is performed at a temperature below the critical temperature range. It converts the matrix into ferrite by precipitating all the C in solution onto pre-existing graphite.

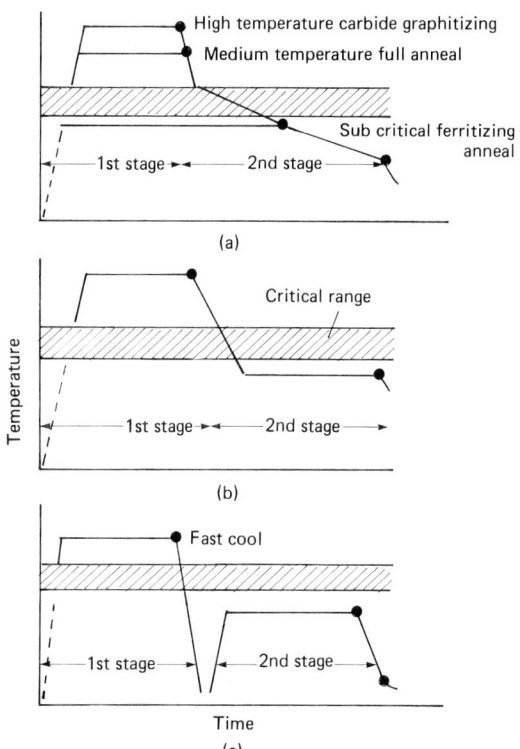

Figure 4.11 Typical full annealing processes: (a) various controlled slow cooling cycles; (b) interrupted cooling cycles; (c) air cool and reheat, two stage annealing

Various annealing cycles are used, as shown in *Figure 4.11*. The cycle used depends on the type and nature of the iron. Castings are soaked for up to 10 h in the temperature range 850 to 950 °C during first-stage annealing. Casting section size, iron composition, furnace loading and prior structure, if induction heating is used[8], determine the time and temperature. If carbides, particularly intercellular carbides are present, a prolonged annealing time at a higher temperature is necessary for their breakdown. In that case a high temperature carbide graphitizing cycle (900–950 °C) must be used. A medium temperature cycle (820–900 °C) can be used when only small quantities of dispersed carbide are present. First stage annealing can be eliminated in flake irons (sub-critical ferritizing anneal). It can be reduced considerably in spheroidal irons if the structure is carbide-free.

The use of a sub-critical anneal with spheroidal irons is not advised because it can lead to a reduction in mechanical properties due to substructure formation[9]. In addition, the rate of ferritization decreases rapidly below 650 °C and a medium temperature anneal can take less time and yield better properties. High Si content promotes carbide breakdown. The influence of minor elements on carbide and pearlite formation has been described

Table 4.4 The level of minor element concentration likely to lead to difficulties in full annealing operations

Carbide stabilizers	Pearlite stabilizers
Cr > 0.05%	As > 0.02%
V > 0.05%	Sn > 0.02%
S not balanced by Mn in flake irons	Cu > 0.05%
	Cr > 0.05%
N in flake irons	Mn unbalanced, particularly in heavy sections
B > 0.005%	Ni > 0.1%
	Unbalanced S in grey irons
	N in grey irons
	Mo > 0.05%

previously. Levels of minor elements at which difficulty might be expected during full annealing are indicated in *Table 4.4*.

Figure 4.11 shows that second stage annealing may be accomplished during slow furnace cooling by interrupted cooling or by a two stage treatment. A typical two stage process involves austenitizing at 900 °C, followed by transformation to pearlite during cooling to 675 °C, followed by ferritization of pearlite at 760 °C. This cycle can produce desired properties in shorter times and is utilized in the production annealing of spheroidal iron pressure pipes. Castings may be air cooled after second stage annealing unless susceptible to residual stresses when cooling should be at a rate between 50 and 100 °C/h to below 200 °C.

Malleablizing

Malleable irons have been described in Chapter 1. A full anneal, which constitutes malleabilizing, is an integral part of the production of these irons. However traditional treatment requires a long annealing time because the structure to be annealed is completely white. Consequently, the manufacturer must pay attention to factors that influence cycle time in addition to those controlling the annealed structure, particularly, the number and shape of the graphite clusters. If insufficient graphite nuclei are produced during first stage annealing, mechanical properties are below their optimum. Therefore, second stage annealing times are increased because of the long diffusion distances. Too many nucleation centres can lead to undesirable structures in which graphite clusters are aligned parallel to the boundaries of the original cementite. High cluster counts have been associated with low hardenability and non-uniform tempering in martensitic malleable irons. A cluster count of between 80 and 150/mm^2 is considered optimal.

The kinetics of first stage graphitization have been measured extensively in alloys of different compositions[10-13]. The reaction rate depends strongly on alloy composition. However, in all irons graphitiza-tion is a nucleation and growth process with overall kinetics described by:

$$y = 1 - \exp (t/k)^n$$

where y is the fraction of carbide decomposed in time t, k is a temperature dependent rate constant and n is a constant. The reaction commences when graphite is nucleated. This occurs predominantly at the interface between cementite and saturated austenite, but it also occurs on FeS particles in the presence of free S. Nucleation is promoted by high annealing temperature. However, too high a temperature distorts the casting. High C and Si contents promote graphitization and enhance nucleation, but the content of these elements must be limited to ensure that the casting solidifies white.

Graphitization continues with the growth of the nuclei. The nuclei depend on diffusion and carbide decomposition for their supply of C. The growth process is the rate controlling process. Various growth models have been proposed which suggest that growth is controlled by C diffusion[14], by the influence of solutes on C diffusion[15], by solute diffusion[16], by the dissolution of carbide[17] or by a combination of nucleus growth and coarsening with the balance between them controlled by the rate of carbide dissolution[18]. Minkoff has discussed various growth models[1].

A simple growth model based on C diffusion control identifies factors of importance in the malleabilizing process. C diffuses down the gradient $(C_2 - C_1)/x$ at temperature T in *Figure 4.12*. The solute flux J is approximately

$$J = D_s(C_2 - C_1)/x$$

This flux is maintained by the dissolution of a volume fraction V_f of Fe_3C of C concentration C. The transport of solute across an interface of area A of small volume Ax is:

$$JA = A \times \frac{d}{dt} CV_f$$

Combining the two equations for J and integrating gives factors that influence the time for complete graphitizing:

$$t \propto V_f \times \frac{C}{(C_2 - C_1)} \frac{x^2}{D_s}$$

Where x is the diffusion distance and D_s is the diffusion coefficient of C in Fe. Hence, malleabilizing time can be decreased by increasing temperature (D_s parameter) and minimizing x by using a fast cooling rate to produce a fine Fe-Fe_3C cast structure.

Si influences graphitization rate because it increases the separation of the limit of the metastable austenite solubility and the stable solubility limit. That is, it increases $(C_2 - C_1)$. In a similar manner Cr stabilizes the carbide phase, lowers its free energy, creates new

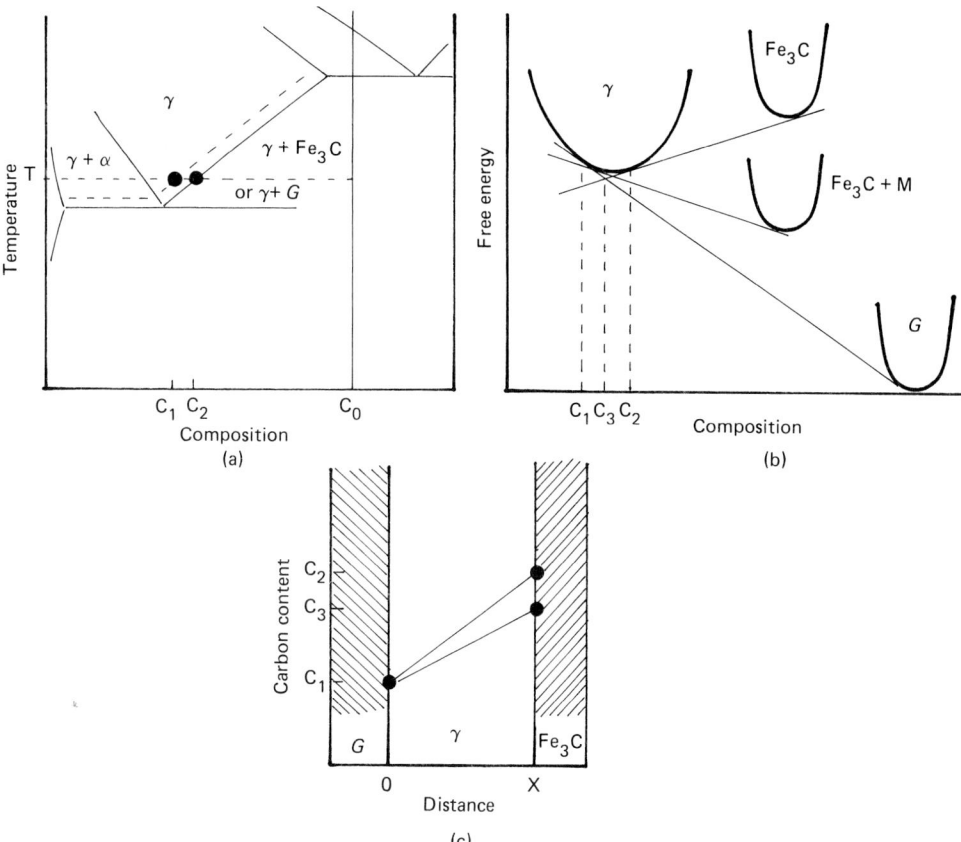

Figure 4.12 Malleabilizing of a white iron of composition C_0 at temperature T: (a) the Fe-G (– – – –) and Fe-Fe$_3$C (———) phase diagrams; (b) the free energy relationship between γ, G and Fe$_3$C defining $(C_2 - C_1)$ and showing how an alloying element such as Cr lowers $(C_2 - C_1)$ to $(C_3 - C_1)$; (c) the C concentration gradient $(C_2 - C_1)/x$ and $(C_3 - C_1)/x$ between the Fe$_3$C–γ interface and the G–γ interface

solubility relationships that decrease $(C_2 - C_1)$ and increases the graphitization time. On the other hand, Ni and Cu dissolve in austenite and lower its free energy. This changes the stability of the carbide phase and graphite in relation to austenite such that graphitization is enhanced.

The Si content is increased to 1.5% and the Cr limited to 0.1% in short cycle treatment, as indicated in *Table 1.8* in Chapter 1. The increased Si content renders the iron more susceptible to mottle formation, which can be countered by the addition of 0.01% Bi. This has no effect on annealability. The C content is limited to 2.35–2.45% to accommodate the increased Si content. The addition of 0.001% B increases annealability by accelerating carbide decomposition. However, the residual B should not exceed 0.0035% in order to avoid cluster alignment and carbide formation. Additions of 0.005% Al improve annealability without promoting mottle. Te is effective in suppressing mottle formation, but the residual concentration must be less than 0.003% otherwise annealing time is increased appreciably.

Recently the addition of 0.15% NaCl has been shown to be as effective as a combined B and Bi addition[19]. Mn and S contents must be balanced to ensure that all S is combined with Mn and only a safe, minimum quantity of excess Mn is present. An excess of S or Mn retards annealing[20]. The action of S has been attributed to interfacial segregation[21]. An excess of Mn increases carbide stability.

A typical short cycle anneal for a blackheart iron is shown in *Figure 4.13*. The casting is heated in a neutral atmosphere to a temperature in the range 940 to 960 °C and annealed for 3 to 10 h. Both temperature and time are adjusted to match the iron composition and casting section. The heat treatment is continued by rapid cooling to the upper temperature of the second stage annealing range (790 to 760 °C) and by controlled cooling through this range. Time must be allowed for C diffusion and its deposition on G nuclei for a fully ferritic matrix. This requires cooling rates in the range 3–10 °C/h depending on section size.

Graphite shape, as well as cluster count, influences

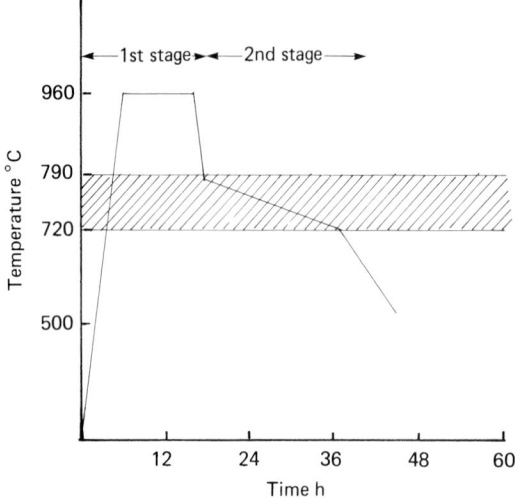

Figure 4.13 Typical heat treatment cycle for a short cycle blackheart iron

properties. Cluster shape varies from aggregate flake through compact clusters to spheroidal with corresponding changes in mechanical properties. Graphite growth in the solid state occurs from basal and prism planes and is influenced by instability effects[22]. S (contrary to its influence on growth from the liquid), H_2, Mg, Ce and rare earths promote spheroidal growth[23]. However the use of S for this purpose is limited because it reduces the rate of annealing.

Malleabilizing is performed in batch-type or continuous furnaces which are either electric or fuel fired. Whiteheart irons are annealed in a decarburizing atmosphere. Blackheart irons are annealed in a neutral atmosphere. Traditional practice in batch-type furnaces involved packing castings in cans containing live iron ore (whiteheart) or sealing in cans with spent ore (blackheart). The neutral atmosphere in the latter case is generated by CO_2 which is initially produced by reaction between entrapped air in the can and C at the metal surface. This CO_2 reacts with Fe_3C at the casting surface to yield CO. The CO_2 is then regenerated by the O in the ore. The steel cans in this operation limit heat transfer to the castings in the centre of the furnace. This leads to extended cycle times. The cycle temperature is limited because of problems of fusion of the casting with the ore.

Careful attention to furnace design, for example, effective sealing from the atmosphere, can eliminate the need for packing castings in cans. Careful design of heating elements can lead to energy savings and reduce cycle times[25]. Higher temperatures and shorter cycle times can be achieved in continuous furnaces. N_2 is used to maintain a neutral atmosphere. A decarburizing atmosphere for whiteheart irons is achieved by controlling the CO/CO_2 ratio in the atmosphere.

The proportion of gases present in the atmosphere is controlled by the equation:

$$H_2 + CO_2 \rightleftharpoons CO + H_2O$$

The equilibrium constant of this reaction at 1060 °C is 2.1. If the CO/CO_2 ratio, which is monitored during annealing, is maintained at 2.7, the H_2/H_2O ratio is 1.3 and the atmosphere composition will be 24.5% CO, 9.1% CO_2, 6.9% H_2O, 9.0% H_2 and 50.5% N_2.

The decarburization process is governed by the reactions:

$$C + CO_2 \rightleftharpoons 2CO + Fe(\gamma)$$

$$C + H_2O \rightleftharpoons CO + Fe(\gamma) + H_2$$

The austenite in equilibrium with the furnace atmosphere at 1060 °C contains less than 0.01% C with a CO/CO_2 ratio of 2.7. Hence, at equilibrium, surface decarburization should be complete. Surface oxidation scaling occurs according to:

$$Fe + CO_2 \rightleftharpoons FeO + CO$$

$$Fe + H_2O \rightleftharpoons FeO + H_2$$

A CO/CO_2 ratio of 2.7 is just sufficient to prevent oxidation scaling at 1060 °C.

Heat treatments to promote mechanical properties

The main use of heat treatment is to improve the mechanical properties of a casting compared to those displayed in the as-cast condition. The range of properties that can be attained in spheroidal irons are shown in Chapter 1 in *Figure 1.7*. The most common heat treatments are described below.

Normalizing

Normalizing is the cooling of an iron in air from a temperature above the critical range. This treatment is applied to obtain a higher hardness and strength than can be attained in either the as-cast or annealed conditions. Normalizing induces a fine pearlitic matrix structure, which combines good wear resistance with reasonable machinability. It also produces excellent response to induction or flame hardening, provided the cooling rate is fast enough and the hardenability sufficient for the section thickness.

Typical normalizing temperatures are 800–830 °C for malleable irons, 810–870 °C for high strength grey irons, 840–900 °C for lower strength grey irons and 820–900 °C for spheroidal irons. The heating rate must be controlled to avoid distortion or cracking. The time at temperature rarely exceeds that necessary to establish a uniform temperature distribution. The

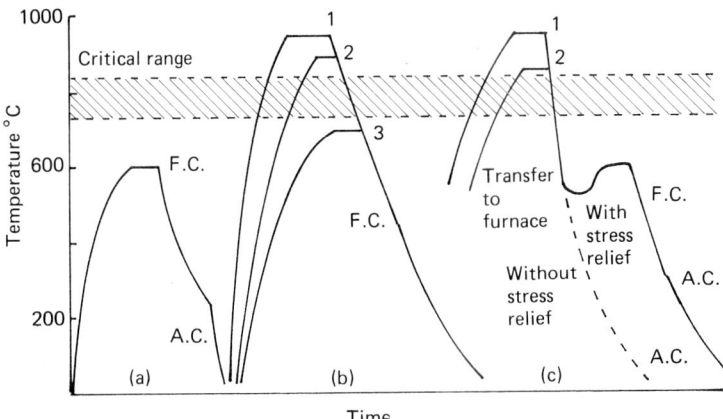

Figure 4.14 A comparison of the heat treatment cycles used for (a) stress relieving; (b) annealing, 1, high temperature carbide graphitizing; 2, medium temperature full anneal; 3, sub-critical ferritizing; (c) normalizing, 1, carbide graphitizing; 2, conventional normalizing; F.C. = furnace cool; A.C. = air cool

cooling rate can be varied from that of still air to that induced by large fans. The response to a normalizing treatment depends on section size and uniformity and alloy content. Alloying additions that increase hardenability allow larger sections to be normalized.

The hardening response of alloyed irons can be so great that tempering is required to obtain desired properties. Tempering after normalizing is usually performed in the temperature range 500 to 625 °C for 1 h per 2.5 cm of section thickness. Castings with non-uniform sections can require a stress relief anneal. These two operations can be combined by using a temperature that tempers to the desired level, followed by cooling in the furnace to about 300 °C, followed by air cooling. Tempering of a pearlitic structure produced by normalizing requires a temperature of up to 50 °C higher than that for tempering of a structure produced by through hardening to the same tempered hardness. Normalizing cycles are compared with stress relieving and annealing cycles in *Figure 4.14*.

Quenching and tempering treatments

Quenching and tempering treatments produce higher strengths than can be achieved by normalizing. They also produce good wear characteristics. In general, quenched and tempered structures are more machinable than pearlitic structures of the same hardness. A greater variety of irons can be furnace or salt-bath hardened than can be flame hardened or induction hardened. The relatively short time available for C solution in austenite in flame or induction hardening means that only irons with a relatively large amount of combined C can be hardened successfully. There is no such time limit in furnace or salt-bath processes and irons containing very little combined C can be hardened.

Austenitizing is performed up to 50 °C above the critical temperature range until the desired amount of C has been taken into solution in austenite. This controls the as-quenched hardness. Austenitizing time is important in irons with low combined C content. The amount of C in solution increases as the austenitizing temperature increases.

Si reduces the solubility of C in austenite. This necessitates the use of higher temperatures for maximum hardening. Higher temperatures can lead to more severe quenching and cracking. Consequently, it is found that high Si irons are less responsive to quenching and more prone to cracking.

Successful hardening is accomplished by quenching from the austenitizing temperature at a rate that exceeds, but is not greatly in excess of, the critical rate necessary to suppress pearlite formation. This rate can be determined from TTT or CCT diagrams or from hardenability data corrected for the influence of alloying additions and the severity of the quenching medium.

Hardenability is a measure of how rapidly the iron must be cooled to suppress its transformation. It is expressed as the diameter of a round bar that will just harden all the way through to 50% martensite at the end when quenched in water. The hardenability can be measured using the standardized Jominy end quench test. Alternatively, the critical diameter can be predicted for a known quenching medium from the iron composition[25].

This method uses experimentally determined relationships between hardenability and alloy element content. It is based on the observation that each element has a specific and independent influence on hardenability (see *Figure 4.15*). The starting point is the absolute hardenability, D_A, which is based on the iron C content. The critical diameter for water quenching is calculated by multiplying D_A by the

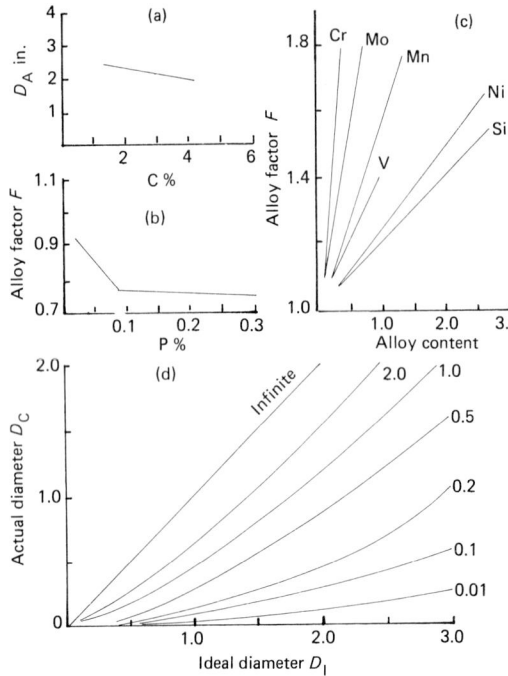

Figure 4.15 (a) Relationship between absolute hardenability D_A and C content; (b) and (c) Alloy factors (F) for various alloying elements; (d) Relationship between ideal diameter D_I and actual diameter, D_c for various quenching medium severity factors

Table 4.5 Quench severity factors for various media (after ref. 25)

Quenching medium	Quench factor	
	700–540°C	*200°C*
Aqueous solution 10% NaOH	2.06	1.36
Aqueous solution 10% H_2SO_4	1.22	1.49
Water at 0°C	1.06	1.02
Water at 18°C	1.00	1.00
Water at 25°C	0.72	1.11
Oil (rape seed)	0.30	0.055
Water at 50°C	0.17	0.95
Liquid air	0.039	0.033
Air	0.028	0.007

factor for each alloying element present as determined from *Figure 4.15*. The actual bar diameter that can be hardened is calculated from the ideal diameter:actual diameter relationship appropriate for the quenching medium used, as shown in *Figure 4.15*. Quench severity factors for various media are shown in *Table 4.5*.

Many iron castings cannot be hardened by quenching because of their size, complexity or the variation in section thickness. The cooling rate required necessitates water quenching. This distorts the casting or causes quench cracking. Water–polymer solutions have been used successfully on some castings to produce an efficient quench with less thermal stressing. However, irons are usually quenched in oil or even air-quenched, if heavily alloyed. Quenched irons are tempered to reduce residual stresses and the probability of cracking and to adjust mechanical properties. Tempering reduces hardness and improves tensile strength to an extent that depends on tempering temperature and iron type. Some typical cycles are given in *Table 4.6*.

Hardness is a convenient way of monitoring tempering but this control can be lost due to mechanical property impairment as a result of secondary graphitization. Secondary graphitization is the decomposition of a martensitic structure during tempering at higher temperatures (above 430°C) to form small graphite precipitates throughout the

Table 4.6 Examples of hardening and tempering cycles for flake and spheroidal irons

Purpose	Austenitizing temperature and time	Quench	Tempering cycle
Grey iron with maximum hardness	560°C preheat 870°C 1 h per inch section	Agitated oil quench to 120°C	200°C for 1 h Cool in still air
Grey iron with optimum strength and toughness	650°C preheat; 870°C 1 h per inch section	Agitated oil quench to 120°C	400°C for 1 h Cool in still air
To produce G 800/2 BS 2789 (1985)	900°C 1 h per inch section	Agitated oil (55–85°C)	480°C for 2 h; Furnace cool to 345°C; Air cool
To produce G 700/2 BS 2789 (1985)	900°C 1 h per inch section	Agitated oil	480°C for 2 h; Furnace cool to 345°C; Air cool
To produce G 700/2 BS 2789 (1985)	900°C 1 h per inch section	Air quench to 425°C	595°C for 2 h
Spheroidal iron preparation for flame hardening	900°C 1 h per inch section	Agitated oil	650°C for 2 h; Furnace cool to 345°C; Air cool

sound casting with a high count of good shaped spheroids in a desired matrix free from intercellular carbides. In addition, the iron must have sufficient, but not excessive, hardenability. The correct combination of alloying elements is necessary to produce a sufficient time interval between stages I and III to present the heat treater with a wide time processing window. The known influences of solutes on the solid state transformation and properties of ADI are discussed below.

Carbon. The C content of spheroidal irons is close to 3.6% in order to provide the desired solidification structure. Increasing the C content in the range 3 to 4% produces a progressive reduction in the tensile and proof strengths and little change in hardness and elongation to failure. This results from an increased graphite volume fraction which reduces the load bearing cross-section. The austenite C content depends on the alloying elements present, but can be varied by controlling the austenitizing temperature with the beneficial effect on mechanical properties shown in the previous section.

Silicon. Silicon influences hardenability and the equilibrium C content of the austenite phase. Several studies of the effect of Si on the mechanical properties have been reported[42,49,54,60,64,65]. The effect of austempering temperature on tensile properties is similar to that on unalloyed irons. However, increasing Si from 2.0 to 2.9% decreases the optimum austempering temperature for maximum strength from 325 to 275 °C and increases that for maximum ductility from 375 to 400 °C.

At any particular austempering temperature tensile and proof stress decrease and ductility increases as the Si content increases. This is due to increasing amounts of proeutectoid ferrite and pearlite in the structure. Other measurements have shown[54] that Si can partially neutralize the deleterious effect of Mn on the properties of austempered irons and vice versa. Si has a significant effect on the fracture toughness of irons which contain 0.6% Ni and 0.25% Mo and are austempered for 3 h at 325 °C. Limited measurements showed that the crack opening displacement (δ_c) and the equivalent plane strain fracture toughness K_{IC} increased by 23% and 12%, respectively as the Si content increased from 2.6 to 3.2%. This improvement is attributed to higher fractions of retained austenite. Luyendijk[64] has shown that Si increases impact strength and reduces the impact transition temperature. These changes suggest that Si might improve the fatigue strength.

Manganese. This element improves hardenability and severely retards the stage I reaction when present in amounts > 0.25% as shown in *Figure 4.29*. It also enhances stage III kinetics. It segregates strongly during solidification causing the formation of MISV with a loss of mechanical properties as illustrated in the previous section. Lo-Kan[66] has shown that

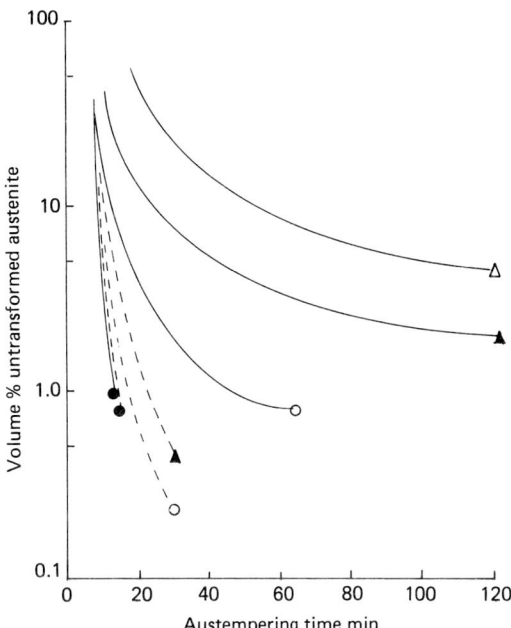

Figure 4.29 The variation of untransformed austenite volume with austempering time for $T_A = 315\,^{\circ}\text{C}$ and $T_\gamma = 927\,^{\circ}\text{C}$ for Mn- and Mo-alloyed spheroidal irons. (———●) 0.15% Mn; (———●) 0.25% Mn, 0.06% Mo; (———○) 0.35% Mn; (———○) 0.15% Mn, 0.40% Mo; (———▲) 0.57% Mn; (———▲) 0.15% Mn, 0.60% Mo; (———△) 0.97% Mn (after ref. 45)

increasing Mn from 0.07 to 0.74% in irons austempered at 280 °C reduced impact strength by 54%.

Molybdenum. This element increases hardenability, particularly in combination with Cu (ref. 67). It also retards stage I, but not as much as Mn (*Figure 4.29*). It segregates strongly during solidification and forms carbides. Its much reduced effect on stage I kinetics makes it a better addition than Mn for controlling pearlitic hardenability. However, additions must be limited to 0.3% because of strong segregation and carbide forming tendencies. *Figure 4.30* shows that tensile strength, hardness and elongation fall progressively as the Mo content increases[46]. It is likely that fatigue strength and fracture toughness will be reduced by Mo. More detailed studies[67] have shown tensile strengths in excess of 1000 N mm^{-2} and elongations in excess of 8% can be obtained by austempering a 1% Cu, 0.2% Mo iron at 355 °C. The time processing window extends from 0.75 to 24 h.

Copper. Additions of up to 1.5% only have a slight effect on the tensile strength and hardness of irons austempered between 300 and 400 °C, (see *Figure 4.31*). However, the elongation increased progressively with increasing Cu at austempering temperatures of 350 °C and below. Cu does not influence the amount of retained austenite. However it has been suggested

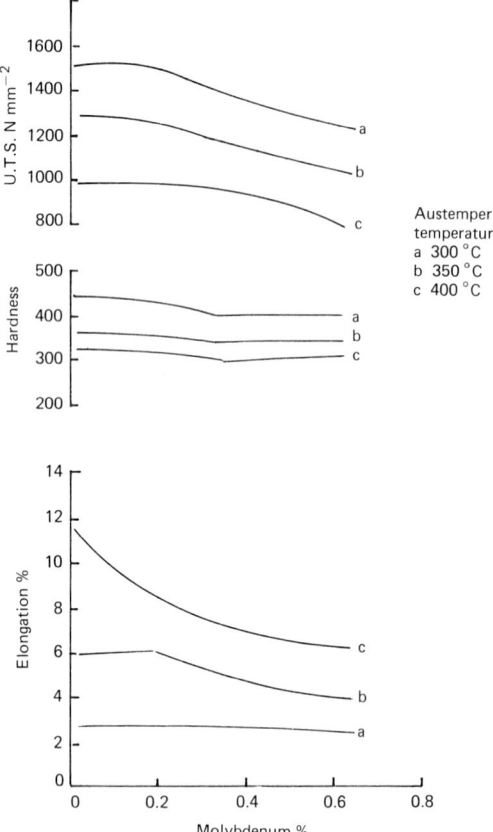

Figure 4.30 The effect of molybdenum content on the tensile properties of austempered spheroidal iron (after ref. 46)

that Cu suppresses carbide formation in lower bainite. This would account for the influence of Cu on ductility. This effect should lead to some improvement in the fatigue and fracture toughness properties.

Nickel. Figure 4.32 shows that with Ni contents in excess of 1%, tensile strength and hardness fall, particularly with austempering temperatures below 350 °C. The variation in ductility is greatest with austempering temperatures of 350 °C and less. It has been suggested that Ni acts in a similar way to Cu and suppresses carbide formation in lower bainitic structures. Segregation in irons containing more than 2% Ni may result in martensite formation during air cooling subsequent to the austempering treatment and may reduce ductility. The influence of Ni on other mechanical properties has not been reported to date.

The metallurgical characteristics of the transformation are well defined as indicated above. The structure property relationships are being established so rapidly that the time is approaching when it will be possible to select iron compositions and austempering conditions to satisfy a range of specifications. This has

been illustrated for a commercial 1.5% Ni–0.3% Mo spheroidal iron[52] as well as in several patented procedures.

There are several practical considerations to be satisfied before a component can be produced commercially, even after iron composition and austempering conditions have been defined. Die casting, permanent moulding, continuous casting and certain sand moulds (zircon) offer a selection of casting techniques for producing near net-shape castings. Factors discussed earlier in the chapter, for example, temperature distribution in the furnace, must be considered in austenitizing.

In addition, the high hardness of the heat-treated iron means that components should only need finish machining after austempering. Therefore, it is important to suppress decarburization and scaling. This can be achieved using fluidized beds, salt baths or controlled atmosphere furnaces using 'endothermic' gas or nitrogen. Proprietary stop-off compounds can be effective in preventing surface degradation when austenitizing is carried out in air.

Austempering involves quenching from the austenitizing furnace to the isothermal heat treatment furnace at a rate sufficient to avoid pearlite or ferrite

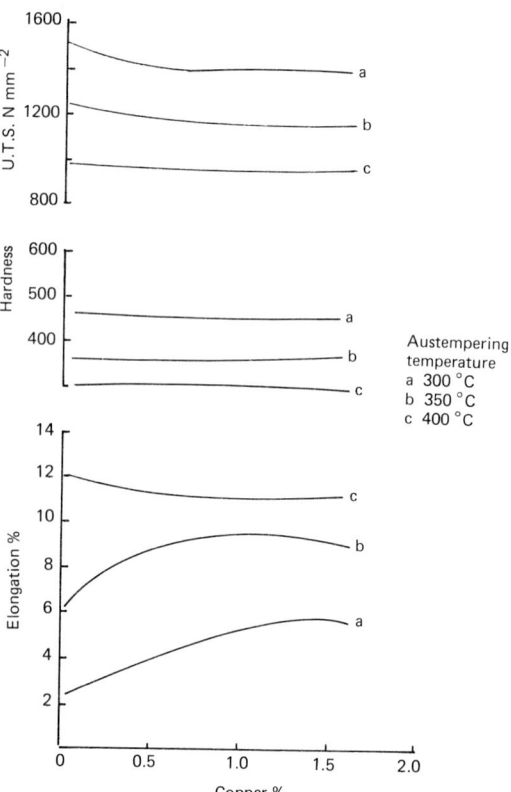

Figure 4.31 The effect of copper content on the tensile properties of austempered spheroidal iron (after ref. 46)

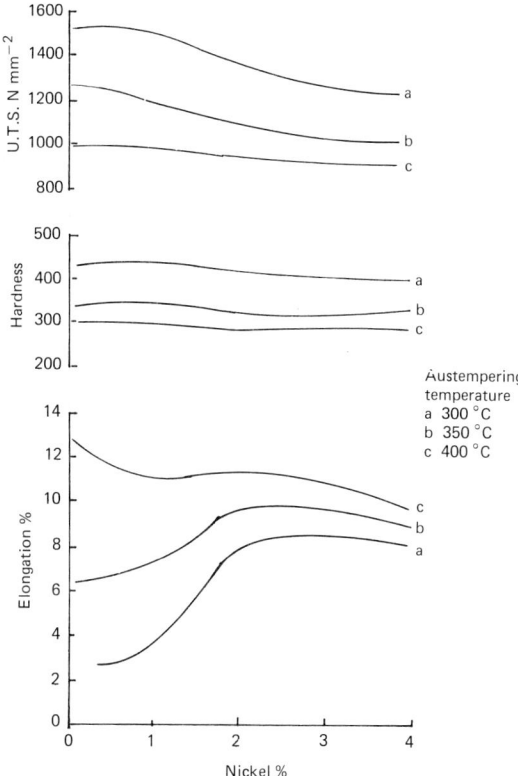

Figure 4.32 The effect of nickel content on the tensile properties of austempered spheroidal iron (after ref. 46)

formation. Austempering can be performed in oil or salt baths or in fluidized beds. Oil baths are limited to temperatures below 250 °C. Fluidized sand beds are not recommended for large castings because dead regions form easily above large flat areas. This creates uneven heat removal which leads to distortion.

The conventional heat treatment practice described above may not be the most economical, particularly if the mechanical property specification is not stringent. An alternative procedure is hot shakeout and controlled cooling. The casting is cooled in the mould to the austenitizing temperature, then shaken out of the mould and air cooled. When the desired austem-

pering temperature is reached, the component is transferred to an insulated hot box for the austempering treatment after which it is cooled to room temperature.

Alternatively, after stripping from the mould the casting is allowed to cool to room temperature. With the correct alloy content, the cast structure will be largely bainitic but with some martensite and large areas of retained austenite. When the casting is tempered in the bainitic region, this austenite will transform to bainite, thus producing an iron of high strength. This procedure may be used to advantage with a normalized iron.

Finally, a structure suitable for thermal applications can be produced by tempering at higher temperatures subsequent to producing a high strength bainitic structure. This treatment allows stage III to go to completion, thus producing a fully ferritic structure with the high strength of a bainitic iron. The high thermal conductivity and low thermal expansion of ferrite, the high strength of the fine pearlitic or bainitic structure and resistance to austenite formation above the A_1 temperature make the iron suitable for thermal cycling applications such as brake rotors. The properties obtained in an alloyed spheroidal iron using these different heat treatments are shown in *Table 4.7*.

The market for austempered irons is extremely large. Their attractive properties make them desirable not only for the manufacture of existing components with improved performance but for competing with other materials in new applications. Some applications relative to the idealized stages of the bainite reaction are shown in *Figure 4.33*. An already established, but still rapidly developing application, is for gears. Well established applications in Germany include timing gears, crown and pinion gears. ADI has been used by General Motors to replace case-hardened forged steel hypoid and ring gears.

Gear designing is based on preventing either bending or contact fatigue. Bending fatigue strength in ADI has been related directly to austempering temperature. Contact fatigue strength has been related inversely to austempering temperature. Fatigue properties obtained in various irons and steels are shown in *Figure 4.34*. Fatigue properties can be

Table 4.7 Mechanical properties of bainitic spheroidal iron produced by various methods. Iron composition: 3.5% C, 2.4% Si, 0.2% Mn, 1.5% Ni, 1.0% Cu, 0.5% Mo (after ref. 51)

Condition	U.T.S. $N\,mm^{-2}$	Yield $N\,mm^{-2}$	Elongation %	Hardness
Austempered 370 °C	972	455	8.1	309
Hot shake out interrupted cooling	876	503	7.9	269
Hot shake out and tempering 360 °C	1262	1014	1.0	461
Normalized and tempered 360 °C	1255	979	1.4	415
Austempered and tempered 480 °C	1048	814	3.1	300

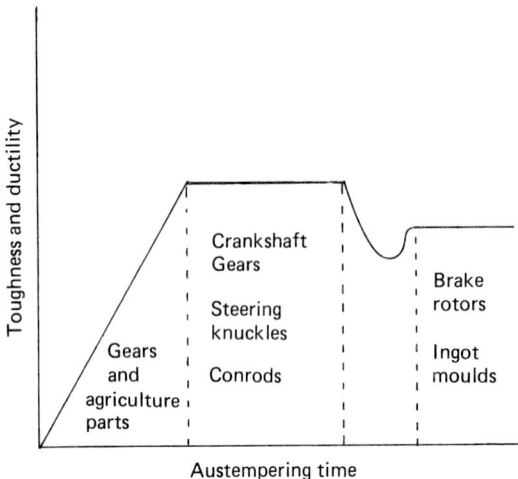

Figure 4.33 Potential applications for ADI in relation to the idealized toughness–austempering time curve

improved by inducing compressive stresses at the surface by shot peening or surface rolling.

An additional benefit in ADI is that cold working can cause retained austenite to transform to martensite, thereby increasing compressive stresses and improving wear resistance. The bending and

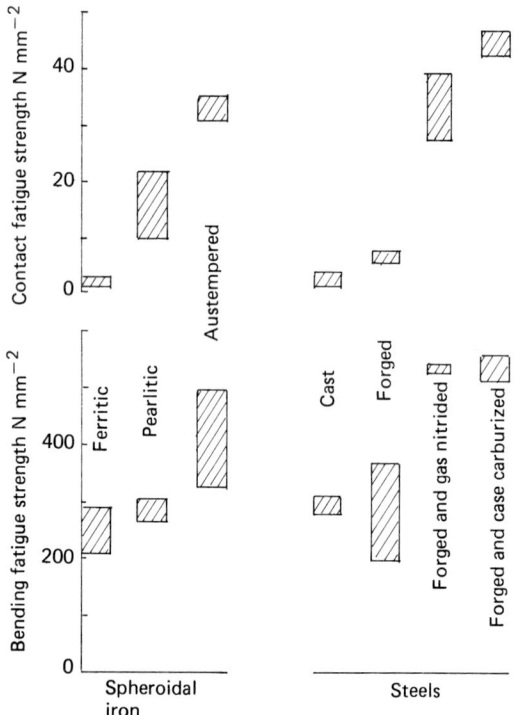

Figure 4.34 Typical ranges of measured values of bending fatigue strength and contact fatigue strength of gear teeth

contact fatigue strength can be improved by surface treatments such as nitriding prior to the austempering treatment. Other uses include crankshafts, camshafts, steering knuckles, suspension components, pump components, railway couplings and agricultural parts. Wear resistance approaching that of white iron, but with impact resistance combined with machineability before austempering, makes ADI suitable for sludge pumps, locomotive wheels, guides and blades for machinery.

In the absence of specifications for ADI, minimum property requirements have been suggested for several proposed grades as shown in *Table 4.8a*. The mechanical property measurements on which the BCIRA proposals are based have led to the more detailed, but as yet incomplete, engineering data sheet shown in *Table 4.8b*. Allowance is made in this table for a dependence of mechanical properties on section size. The light, medium and large section data was determined on bars 22 mm in diameter, blocks 25 × 45 mm and blocks 45 × 45 mm, respectively. The volume to surface area ratio of these sections is 5.50, 8.04 and 11.25, respectively. This corresponds approximately to thicknesses in large plate sections of 11, 16 and 23 mm.

Optimum compositions have not been established for the proposed grades to date. Alloy content must be increased as section size increases to ensure sufficient hardenability. The light section data in *Table 4.8b* was obtained using an unalloyed, 0.4% Mn iron. An iron containing 0.4% Mn and 0.6% Ni did not display properties fully representative of the light section data because the retained martensite content was high. On the other hand, decreasing the Mn content and adding a small amount of Ni may result in a better range of mechanical properties in small castings. An iron containing 0.3% Mn, 0.8% Ni and 0.1% Mo and one containing 0.3% Mn, 0.8% Ni and 0.25% Mo produced suitable structures and mechanical properties which satisfied those defined in *Table 4.8b* for medium and large sections, respectively.

Surface hardening

Surface hardening is an important aspect of the heat treatment of cast irons. It may take the form of transformation hardening or remelt hardening. It is an economical method of promoting wear resistance in selected areas. Strength is increased when highly stressed areas are treated. This is because the surface is left in compression, which reduces peak tensile stresses. Most irons, excepting white and highly alloyed irons, can be surface-hardened by flame, induction, electron beam, plasma and laser heating. Other techniques include diffusion hardening and chill casting.

Surface hardening without melting is achieved most

Table 4.8 (a) Proposed grades for austempered spheroidal iron; (b) Engineering data sheet proposed for BCIRA grades

Proposed USA Grades	Min. Tensile Strength $N\,mm^{-2}$	Min. 0.2% P.S. $N\,mm^{-2}$	Elongation %	Austempering Temperature °C	Characteristics
ADI 1	862	586	10 min	–	High ductility coupled with good strength
ADI 2	1034	689	7 min	–	Intermediate grade
ADI 3	1207	827	4 min	–	High strength with lower ductility
ADI 4	1379	965	2 min	–	Very high strength, low ductility
Georg Fischer Grades					
GF 90	850	600	5–12	380 ± 10	High ductility coupled with good strength
GF 120	1200	950	2–5	320 ± 10	High strength with lower ductility
GF 140	1400	1200	1–2	275 ± 10	Very high strength, low ductility
GF 100	1000	–	–	235 ± 10	Wear resistant, intermediate strength
BCIRA Proposed Grades					
ADI 950/6	950	697	6 min	375	High ductility coupled with good strength
ADI 1050/3	1050	792	3 min	350	Intermediate grade
ADI 1200/1	1200	947	1 min	325	High strength with lower ductility

Property	Minimum Mechanical Properties								
	ADI 950/6			*ADI 1050/3*			*ADI 1200/1*		
	Light section	Medium section	Large section	Light section	Medium section	Large section	Light section	Medium section	Large section
Tensile strength $N\,mm^{-2}$	950	950	925	1050	1050	1025	1200	1200	1175
Limit of proportionality $N\,mm^{-2}$	400	373	372	431	417	436	495	493	476
0.5% Proof stress $N\,mm^{-2}$	769	761	739	897	897	879	1087	1048	1018
Elongation %	6	6	6	3	3	3	1	1	1
Compressive strength									
Limit of proportionality $N\,mm^{-2}$	503	477	463	564	551	543	649	624	629
0.2% Proof stress $N\,mm^{-2}$	754	733	709	877	866	838	1060	1043	1003
Shear strength $N\,mm^{-2}$	855	855	833	945	945	923	1080	1080	1058
Torsional strength $N\,mm^{-2}$	855	855	833	945	945	923	1080	1080	1058
Limit of proportionality $N\,mm^{-2}$	280	261	260	302	292	305	346	345	333
0.5% Proof stress $N\,mm^{-2}$	538	533	517	628	628	615	761	734	713
Fatigue Limit (Wöhler)									
Unnotched 9.6 mm dia. $N\,mm^{-2}$	345	345	336	355	355	346	349	349	342
Notched 9.6 mm dia. root $N\,mm^{-2}$	202	202	196	170	170	166	183	183	179

easily with irons displaying a pearlitic or tempered martensite matrix structure before hardening. The time above the critical temperature is too short for C diffusion to produce a uniform matrix in a ferritic iron. Small amounts of ferrite can be tolerated if they are adjacent to graphite, as in bulls-eye structures.

Hardenability can be a disadvantage in surface hardening. The metal below the surface aids quenching. Alloying additions can increase the heating period and promote retention of austenite by delaying transformation.

Temperature control is important if surface melting

is to be avoided. Closer control is required with cast iron than with steel because the critical temperature is higher and the incipient melting temperature lower. Thus, pearlitic malleable iron is easiest to harden consistently because of its matrix structure, low critical temperature (low Si content) and low alloy content. Spheroidal irons present more problems because of their higher critical temperature (higher Si content) and alloying additions.

Several modes of operation are employed in surface hardening. The heat input is uniformly distributed over the whole area to be hardened in spot hardening. After heating for a predetermined time the surface is immediately quenched. Spin hardening is suited to the hardening of round parts on their circumference, for example, the teeth of small gears. The component is rotated as heat is applied to the surface to ensure uniform heating. The surface can be spray quenched or immersion quenched. Progressive hardening is used to treat large areas. The heat source traverses the surface and is followed immediately by a quenching spray.

Flame hardening is usually performed using an oxy-acetylene flame. Other gases, with their lower flame temperature, produce deeper hardening. Flame hardening equipment is relatively simple, but it is difficult to instrument and considerable skill is required for visual control of the process. Both unalloyed and alloyed grey irons can be flame hardened. This can be done most easily when the combined C content is between 0.5 and 0.7%.

The depth of the hardened layer and its microstructure depends on the surface temperature, the amount of C and alloying elements in solution when the surface is quenched and the quenching efficiency. The total C should be kept as low as possible consistent with the production of a sound casting for maximum hardness. Si should be less than 2.4% and Mn should be in the range 0.8–1.0% to enhance the depth of hardening. Typical surface hardness of a 3.0% C, 1.7% Si, 0.6–0.8% Mn iron is 400 to 500 BHN. Small alloying additions of Cr, Mo and Ni can increase this hardness to 550 BHN.

The surface of a flame-hardened grey iron often shows a slightly lower hardness than the material just below it. This may be due to the retention of relatively soft austenite at the surface. Stress relieving can be used to reduce residual stresses. This minimizes distortion and the risk of cracking, but it is often safely omitted. The surface stresses are usually compressive, for example, water quenching introduces compressive stresses in excess of $200\,\mathrm{N\,mm^{-2}}$ in the fillet area of a crankshaft. This increases the fatigue strength to a greater extent than can be achieved by through hardening. However, the residual stresses may be tensile if self-quenching occurs.

Induction heating is more expensive, but it is justified when a large number of components are being treated. The basic features of induction heating have been described in Chapter 2. The depth to which eddy currents, and hence heat, penetrates varies inversely with the square root of the frequency. Heat penetrates to a depth of $\sim 1\,\mathrm{mm}$ with the frequently used 9600 Hz power. Thus, the minimum depth of hardening with this frequency is $\sim 1.3\,\mathrm{mm}$. If a shallower depth is required, a higher frequency must be used. Greater depths can be obtained at the same frequency using an extended heating time at a lower power input. Induction heating is a very flexible process and complex casting shapes can be treated.

Electron beam, plasma and laser beam heating are relatively new methods of surface treatment. The general features are compared with induction heating in *Table 4.9*. Donaldson has reviewed laser beam technology[68]. A laser converts electrical energy into electromagnetic energy, a beam of light which is easily

Table 4.9 Comparison of laser heating performance with that of electron beam, induction and plasma heating (after ref. 80)

Thermal source	*Main technical features*	
	Laser compared to named source	*Named source compared to laser*
Electron beam	No vacuum required	Coating not required
	Mirrors can direct beam	Simplified beam programming
	One laser serves several stations	High efficiency of beam power generation
Induction heating	Long stand OH possible	Coating not required
	Heat pattern readily changed	Simple technology
	Less sensitive to component section or profile	Well suited to produce deep cases
	Higher intensity possible	
	No electrodynamic forces to displace melt layers	
Plasma heating	No electrode effects	Coating not required
	Heat pattern more readily changed	Very simple technology
	Higher intensity possible	

transmitted through air and directed to, and focused onto, the surface to be hardened. High power (> 500 W) CO_2 lasers are suitable for heat-treating cast irons, the surface of which must be coated with graphite or manganese phosphate to increase the surface absorption. Energy absorption rapidly increases the surface temperature. Removal of the energy source results in self-quenching by conduction of heat into the body of the component.

The particular advantages of laser treatment are:

1. treatment can be localized;
2. heat input is low, consequently there is very little distortion;
3. the depth of hardening can be controlled accurately;
4. complex shapes can be treated and inaccessible areas reached by using mirrors;
5. treatment is rapid thereby minimizing surface degradation;
6. usually no finish grinding is required;
7. the treatment is self quenching;
8. no special gaseous atmosphere or vacuum is required.

Two parameters characterize the laser and control its performance. One is the power density which is the ratio of the power of the beam to its diameter. The second is the interaction time which is the time that the beam is incident on any one point on the surface.

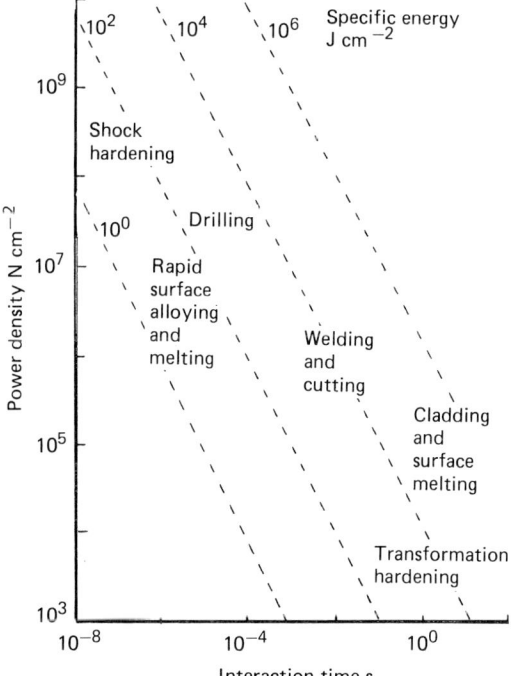

Figure 4.35 Suitable laser application areas as related to power density and interaction time

The two parameters define the specific energy. Control over each defines the type of treatment, as shown in *Figure 4.35*.

A focused beam of small diameter gives a high power density suitable for laser drilling, cutting and welding. On the other hand, surface processing requires a lower power density and wider coverage. This can be achieved with a rastering technique by using vibrating mirrors or a multifacet mirror. However it is more commonly achieved by using a defocused multimode beam. Surface melting has been used for surface alloying and cladding as well as to produce rapid solidification structures. Applications of rapid solidification structures have been made to grey iron components in the machine tool and textile industries.

One disadvantage of surface melting is that the resulting cementite layer is in tension and susceptible to transverse cracking. However, surface melting appears to be the only method of surface hardening ferritic spheroidal irons. This offers the possibility of processing a spheroidal iron casting with good impact strength to produce a wear-resistant surface.

To date lasers have been used primarily for transformation hardening[69-88]. Case depths in spheroidal and grey irons are limited to about 0.75 mm with surface hardness in excess of 800 Vickers Pyramid number (VPN). Attempts to increase the case depth often result in surface melting, which is undesirable in transformation hardening. Limited surface melting results in the formation of globules or spherical caps. A small increase in traversing speed often solves the problem. Otherwise a finish grinding operation may be required.

Surface imperfections can act as stress raisers and generate cracks. A thin layer of soft ferrite is often found at the surface when maximum case depths are produced due to decarburization. Degraphitization may occur in irons with coarse flakes or large spheroids. They leave a rough surface which requires finish grinding.

General Motors have used laser hardening on a production basis for several years. Grey iron diesel engine cylinder liners are hardened to improve scuffing resistance. Four 5 kW CO_2 lasers are used to lay a series of 19-mm wide overlapping tracks which cover the entire inner wall. A feature of laser hardening is the accurate control that can be exercised over the beam location. This has allowed spiral hardening of liners. The patent claims that this avoids cracking in service experienced due to stresses promoted by overlapping tracks. The laser surface hardening of spheroidal iron camshafts is a widely practised treatment. The camshaft lobes shown in *Figure 4.36a* are surface hardened to a case depth, defined as the depth where the hardness was 50 Rockwell hardness 'C' (HRC), of between 0.5 and 1.0 mm.

A typical hardness profile is shown in *Figure 4.36b*.

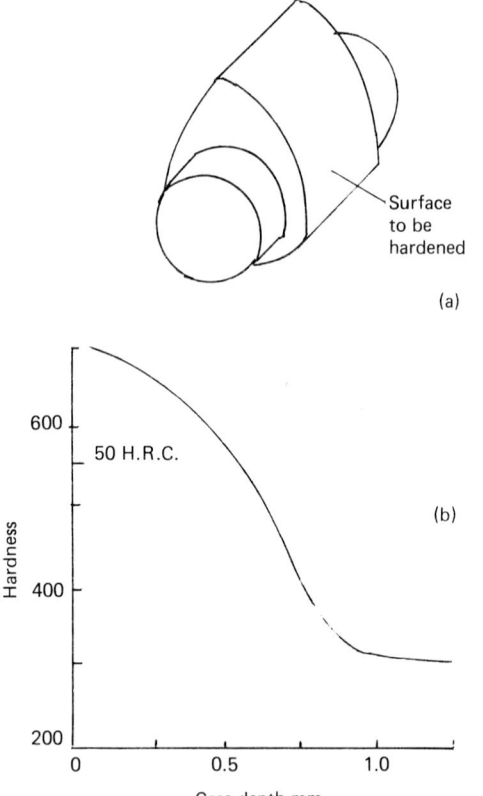

(a)

(b)

Figure 4.36 Camshaft profile (a) and hardening profile (b) after laser transformation hardening

This was achieved using a 15 kW CO_2 laser with an optical system that delivered a focused spot 10 mm in diameter to the camshaft lobe. The operational procedure was to scan this spot over a distance of 22 mm normal to the direction of processing and 25 mm in the direction of processing. The frequency of scanning was 125 Hz in the normal direction and 700 Hz in the processing direction. This formed a rectangular spot 22 mm by 25 mm on the lobe surface. The angular speed of rotation of the lobe under the laser beam must be programmed to obtain an even case because the angle of incidence of the laser beam to the workpiece changes during rotation from nearly normal incidence at the cylindrical portion of the lobe to grazing incidence at the flat portion. The profile shown in *Figure 4.36b* was obtained using a power input of 9 kW, a power density of 1600 W cm^{-2} and a linear speed of processing of 760 mm min^{-1} at the cylindrical, and 180 mm min^{-1} at the flat portion, of the lobe.

Molten salt cyaniding is a diffusion hardening process that combines the advantages of a short processing cycle (1–2 h) with low processing temperature (540 to 600 °C). It provides considerable improvement in wear and corrosion resistance and fatigue strength, including thermal fatigue. Cyanide and cyanate compounds in the molten salt bath liberate N and C which diffuse into the component. N diffuses at the subcritical temperatures to a depth of 1.2 mm, whereas C penetration is much less (∼0.015 mm). The C absorbed at the surface acts as the nucleating agent for C-bearing iron nitride, a tough ductile compound which imparts greater ductility than obtained by conventional carburizing or nitriding. The N diffuses deeper into the surface and forms a solution-hardened zone beneath the outer compound layer. This improves fatigue strength.

Joining processes for cast iron

Although one of the main attractions of the casting process is the elimination of the need for fabrication, there remain situations in which it is advantageous to join castings to other castings or to form a composite component between cast iron and a second material, usually steel. In addition, occasions arise when the repair of a casting defect can recover an otherwise scrap casting. Joining methods for cast irons fall into two main categories, brazing and welding.

Table 4.10 Characteristics of non-fusion joining processes using a filler metal

Property	Process				
	Soldering	Brazing		Braze welding	Powder welding
		Low temperature	High temperature		
Melting range of filler metal	< 450 °C	450–1000 °C	> 1000 °C	∼ 1000 °C	—
Typical filler metals	Pb-base Zn-base	Ag-base Cu-base	Ni-base Au-base Pd-base	Cu-base	Ni-base Cu-base
Joint type	capillary	capillary	capillary	large gap or external fillet	large gap or external fillet

Table 4.11 Examples of filler metal compositions

Classification	Type	Ag %	Cu %	Zn %	Cd %	Sn %	Ni %	Mn %	Liquidus	Comment
AWS A 5.8	B Ag-1	44–46	14–16	14–18	23–25	–	–	–	618 °C	Torch, induction brazing
AWS A 5.8	B Ag-5	44–46	29–31	23–27	–	–	–	–	744 °C	Furnace brazing
AWS A 5.15	RB Cu-ZnA	–	57–61	remain	–	0.25–1	–	–	899 °C	Oxy-acetylene braze welding
AWS A 5.15	RB Cu-ZnD	–	46–50	remain	–	–	9–11	–	935 °C	Good colour match
BS 1845, 1977	AG 18	45–48	15–17	21–25	–	–	4–5	6.5–8.5	625 °C	Improves wettability

Brazing

Brazing is a joining process in which capillary action draws molten filler metal into, or retains it in, the space (~ 0.1 mm) between closely adjacent surfaces of sections to be joined. The melting point of the filler is usually above 500 °C but below the melting point of the parent material. Brazing can be distinguished from other 'non-fusion' techniques such as soldering, bronze welding (braze welding) and powder welding by the melting range of the filler metal or the design of the joint. The non-fusion techniques are outlined in *Table 4.10*.

Silver-base brazing alloys are usually used for low temperature (< 850 °C) brazing. Cu-base alloys are used for high temperature brazing[89]. Some filler metal compositions are given in *Table 4.11*. Processing methods include torch, furnace, induction and dip. Process selection depends mainly on the shape and size of the assembly, the number of assemblies to be brazed and the equipment available.

Successful brazing depends on the condition of the interface between the molten filler metal and the component surfaces being joined. Factors that reduce wetability include surface oxides, high Si contents, and sand and graphite on the surfaces[90]. These factors contribute to the order of brazeability, which is malleable irons, spheroidal irons and flake irons.

The importance of surface condition means that care is required in surface preparation. Conventional chemical cleaning methods do not remove surface carbon. Abrasive blasting with steel shot or grit has

Table 4.12 Post-weld heat treatment processes

Iron	Treatment	Temperature °C	Holding time h/cm thickness	Cooling conditions
Flake	Stress relief	600–660	0.6	Furnace cool to 300 °C at 60 °C/h, air cool to room temperature
	Ferritize anneal	700 760	0.4	Furnace cool to 300 °C at 60 °C/h, air cool to room temperature
	Full anneal	730–900	0.4	Furnace cool to 300 °C at 60 °C/h, air cool to room temperature
	Graphitizing anneal	900–950	1–3 h + 0.4 h/cm	Furnace cool to 300 °C at 60 °C/h, air cool to room temperature
	Normalizing anneal	870–950	1–3 h + 0.4 h/cm	Air cool from annealing temperature to below 500 °C; may need stress relief
Spheroidal Unalloyed	Stress relief	510–560	0.6	Furnace cool to 300 °C at 60 °C/h, air cool to room temperature
Low alloy	Stress relief	560–600	0.6	Furnace cool to 300 °C at 60 °C/h, air cool to room temperature
High alloy	Stress relief	540–660	0.6	Furnace cool to 300 °C at 60 °C/h, air cool to room temperature
Austenitic	Stress relief	620–675	0.6	Furnace cool to 300 °C at 60 °C/h, air cool to room temperature
	Ferritize anneal	900–950	1 h + 0.4 h/cm	Furnace cool to 940 °C, hold for 5 h + 0.4 h/cm of thickness, furnace cool to 300 °C at 60 °C/h, air cool to room temperature
	Full anneal	870 900	0.4	Furnace cool to 350 °C at 60 °C/h, air cool to room temperature
	Normalizing and tempering anneal	900–940	2 h minimum	Fast cool in air to 600–540 °C, furnace cool to 350 °C at 60 °C/h, air cool to room temperature

been used to prepare spheroidal and malleable irons, but this has proved unsuitable for flake irons. Indeed, brazing of flake irons was considered impractical until the advent of electrolytic salt bath cleaning and the recent development of special Ag-base alloys.

Very clean surfaces can be produced using alkaline-based fused salts in the temperature range 400–500 °C. When the component is charged negatively, a reducing medium is created in the vicinity of the surface which removes oxide and sand. The reaction products, sodium silicate and sodium ferrite, are removed in a sludge collection system. If the polarity is reversed, oxidizing conditions are created which lead to the removal of surface graphite. This reaction leaves Fe_2O_3 on the component surface. This can be removed in a final reduction cycle. The outcome is a very clean component surface that produces a braze bond between the metal surfaces. The bond can exceed the strength of the parent metal.

Fluxing prior to, or during, heating chemically dissolves oxide formed below the melting point of the brazing alloy and protects the filler metal when it melts. As the brazing alloy flows through the joint, flux residues are displaced, leaving clean surfaces to bond. Recently the addition of Mn and Ni to Ag-Cu-Zn alloys has been shown to remove the need to degraphitize the iron surface[91,92]. Microstructures of brazed spheroid iron using this filler are shown in Chapter 6, *Figures 6.42 and 6.45.*

Joint design is another important consideration[93]. A narrow joint gap is essential for inducing capillary flow of the liquid brazing alloy. It ensures that, when the brazed component is loaded, the filler metal is constrained by the parent metals on both sides and prevented from deforming plastically. Thus the

apparent mechanical properties of the filler metal are better than the bulk properties of the as-cast brazing alloy. Brazed joints should be designed to be stressed in compression or to shear in service. These considerations lead to different joint designs compared to welding, as indicated in *Figure 4.37*.

Silver brazing, silver soldering and hard soldering are synonymous with low temperature brazing, which is used frequently to join cast irons. Applications include:

1. brazing of cylinder liners into engine blocks, onto steel plates and the fabrication of complex cylinder liners[94],
2. brazing of steel sprockets into cast iron hubs[95],
3. mounting of cast iron gears onto steel shafts[91],
4. fitting of steel tubes into malleable iron headers in heating installations[95] and
5. fabrication of hydraulic valves to simplify casting production by eliminating complex cores[96].

Cu base- and Ni base-filler alloys have been used in several vacuum brazing operations[96-99]. These include joining of Stellite components to the grey iron exhaust valve housings of industrial diesel engines[96] and the brazing of stainless steel rings to grey iron valve bodies and gates[98]. Ni-base fillers have been used to vacuum braze alloyed cast iron cylinder liners into engine blocks in high performance cars[99].

Braze welding is often used as a low temperature substitute for fusion welding. It resembles brazing in that non-ferrous filler metals (usually Cu-Zn base alloys) are used. Bonding is achieved without fusion of the components. Flux is usually applied to the joint surfaces, which are preheated until the filler metal wets or tins the surfaces. The joint is completed by building up weld layers. Thus, in common with welding, wider joints are made by filler metal deposition rather than capillary action.

Braze welding offers several advantages over welding. Its lower operating temperature minimizes thermal stress and distortion and susceptibility to cracking. The weld deposit is soft and ductile and displays good machinability. Dissimilar metals are readily joined and the equipment used is simple and suitable for on-site operations. The major disadvantages are a susceptibility to galvanic corrosion, the fact that the weld strength is limited to the strength of the Cu filler, and colour mismatch.

Powder welding is a non-fusion welding technique in which filler rods are replaced by powdered metal particles which are introduced into the gas stream. The process uses an oxyacetylene torch which incorporates a metering valve to regulate the flow of the powdered welding alloy into the oxygen stream flowing through the torch. The powdered alloy particles are heated progressively as they pass through the flame prior to impingement on the component surfaces. When the base iron surface temperature reaches a critical value, the deposited powder coating

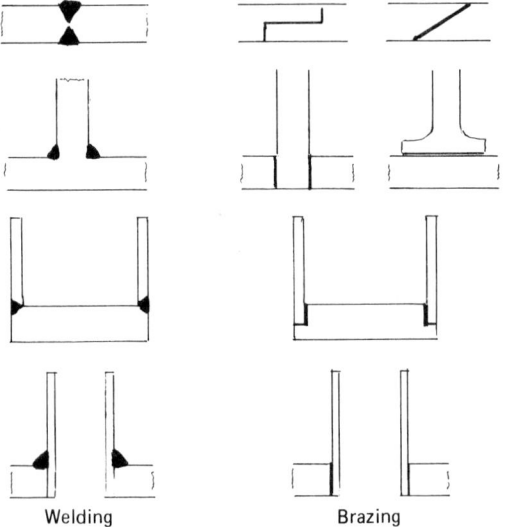

Welding   Brazing

Figure 4.37 Suitable joint designs for welding and brazing operations

melts and wets the surface of the casting. Subsequently, the weld can be built up because the preheated powder continuously melts as it is sprayed onto the wetted surface. The bond is a diffusion bond which is produced without flux. Its strength depends on thorough cleaning of the component surfaces. Either Cu- or Ni-base alloy powders are used. Although the strength of the bond is acceptable, the ductility is poor. Nevertheless, this technique is used extensively for defect repair of iron castings.

Welding

Welding is frequently used to join and repair cast irons. It differs from brazing in that the heat input is greater and the surfaces of the component are melted in addition to the filler metal.

Microstructural features of a weld

Welding is a dynamic casting process in which the base metal provides the 'mould'. The surface of the mould melts and the filler metal is the casting[100-102]. This is illustrated in *Figure 4.38* where the various structural zones in a welded cast iron are identified.

Cast irons are more difficult to weld than steels. Their high C content, coupled with the high solidification and cooling rates associated with welding, can lead to the formation of primary and eutectic carbide and martensite in the structural zones. These microconstituents make the weld hard and brittle, difficult to machine and susceptible to internal stress, distortion and cracking. Careful control must be exercised over welding conditions to minimize their formation.

The structural zone next to the base material is the heat affected zone or HAZ. The temperature in this zone during welding is high enough for austenite formation. The austenite can transform to

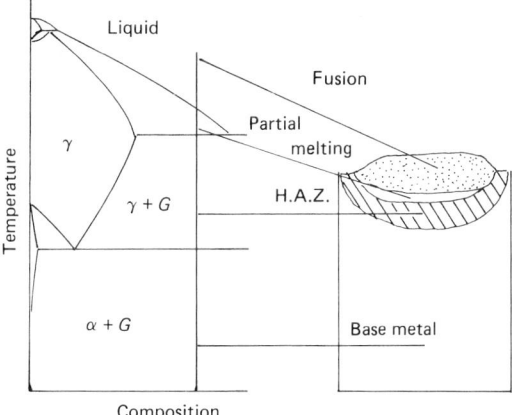

Figure 4.38 The various structural zones in a weld in relation to the temperature reached during welding

martensite on cooling depending on the iron composition and the thermal conditions. The time at temperature is relatively short. C-enrichment of the austenite by dissolution and diffusion occurs more easily in a pearlitic matrix than in a ferritic one. The higher the austenite C content, the harder is the martensite. The lower the austenite C, the higher is the M_s temperature and the smaller the volume change associated with martensite formation. Consequently, pearlitic irons present greater problems with respect to hardening in the HAZ.

Preheating may be used to reduce the thermal gradient and the rate of cooling. A preheating temperature of 400 °C will, in general, prevent martensite formation in the HAZ. Careful use of preheating can leave the weld in compression and ensure a uniform cooling rate. This will reduce cooling stresses even when light and heavy sections are being joined.

Preheating is used to greater advantage in flake irons. The preheating temperature range depends on the hardenability of the iron, the size and complexity of the weld and the type of electrode used. A typical temperature is 350 °C although higher temperatures may be used with heavier sections. The need for, and the selection of, preheating temperature for spheroidal irons is a more complicated consideration. Excessive preheating can be disadvantageous because it increases the amount of base metal melted, increases the size of the HAZ and encourages the formation of massive continuous carbides along the fusion line. The formation of massive continuous carbides is not of major concern in flake irons which already display a low ductility because of flake graphite, but it is of concern in spheroidal irons. In general, if used, preheating is below 200 °C, although higher temperatures may be necessary for production welding using high deposition rates and for welding heavy sections. On occasions a temperature as high as 600 °C may be used to counter the chill effect resulting from decreasing C content in the liquid metal when producing a spheroidal graphite fusion zone. In this case, it is advisable to reduce the heat input in order to minimize the formation of continuous carbides.

The effect of martensite in the HAZ may be reduced by tempering. This is achieved to a limited extent in multiple pass welding as previous layers are tempered. The interpass temperature should exceed the preheat temperature to benefit from the tempering effect of later deposits.

The hard structures and thermal stresses generated in welding may be eliminated by post-weld heat treatment. These treatments include stress relief, ferritizing anneal, full annealing, graphitizing annealing and normalizing annealing. Precautions necessary in performing these heat treatments have been discussed earlier in the chapter. Recommended practices are given in *Table 4.12*.

Hardness profiles shown in *Figure 4.39* illustrate the above effects[103]. The hardness in the HAZ ranges from

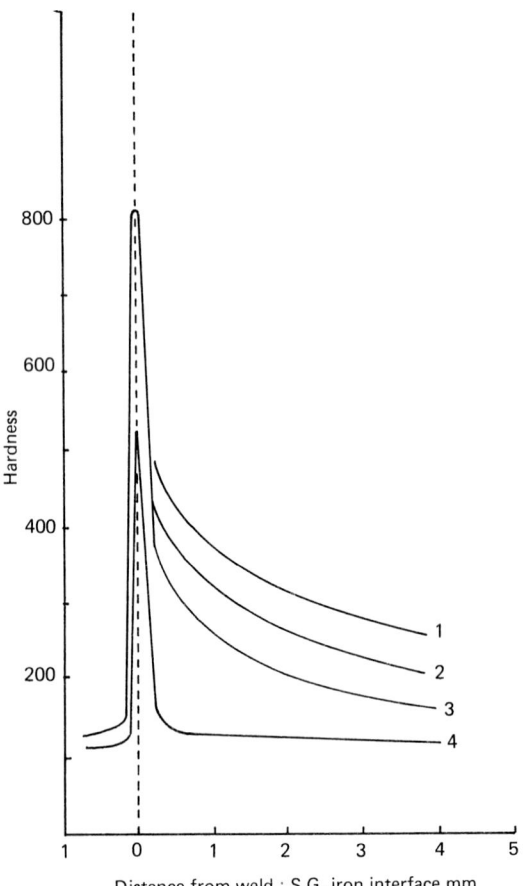

Figure 4.39 Hardness profiles in the root pass of a multipass spheroidal iron butt weld with a Ni-Fe filler; 1, pearlite content 57%; 2, pearlite content 17%; 3, ferritic; 4, sample 3; full anneal at 900 °C for 1 h (after ref. 103)

250 to 650 VPN and typical widths are in the range 0.5 to 2.5 mm.

The partially melted zone lies between the fusion zone and the HAZ (see *Figure 4.38*). It forms when the peak temperature exceeds 1150 °C and causes partial melting of the base material at the fusion line. During welding this region consists of primary phase and liquid of eutectic composition. On resolidification the high cooling rate results in the formation of ledeburite from the liquid areas and the primary phase transforms to martensite. Thus, the structure of this zone consists of a mixture of martensite, austenite, primary carbide and ledeburite surrounding spheroids or flakes of graphite. This is usually the hardest region in the weld and is susceptible to cracking.

The size of this zone can be reduced by reducing the peak temperature and the time at this temperature. This is achieved by controlling the heat input, preheat and interpass temperatures. The peak temperature also depends on the filler material used.

Thus, the selection of welding conditions must be considered carefully because they can produce a beneficial effect in one zone and a detrimental effect in another. For example, a high heat input may raise the base metal temperature of a small weld sufficiently to cause severe fusion line cracking even with no preheat. A high preheat temperature used to prevent martensite formation in the HAZ may result in fusion line cracking although a low heat input welding technique has been used.

A third zone in the 'casting' is the fusion zone. Its microstructure and properties are determined by the filler metal and the welding conditions. The fusion zone geometry depends on the heat input, welding speed and thermal properties of the base metal[102]. Its composition is primarily that of the filler metal with some dilution from the partially melted base metal and impurities introduced from the fluxes and gases. Large temperature gradients and vigorous stirring mix the weld pool thoroughly. However, as the heat source moves away, the temperature gradient drops rapidly and mixing ceases. These conditions produce directional solidification structures except in the last areas to solidify. These areas may also display segregation effects. The structural features of a multipass spheroidal iron weld are shown in Chapter 6, *Figures 6.46a* and *b*.

Welding processes

Metal arc welding is the most practised welding method for cast irons. The shielded metal arc welding (SMAW) or manual metal arc welding process joins components using the heat generated when an arc is established between a flux-covered consumable electrode and the workpiece (ref. 94, p. 75). The arc is established by momentarily touching the electrode in the weld groove and then withdrawing it to a distance of 3 or 4 mm. A schematic representation of the process is given in *Figure 4.40*.

Approximately two thirds of the heat generated is developed near to the positive pole and the remaining third close to the negative pole. It is usual to connect medium-coated electrodes and uncoated electrodes to the negative pole in order to limit their burning rate. Heavily-coated rods are usually connected to the positive pole. The electrode coating incorporates several materials. Each has a specific purpose as outlined in *Table 4.13*. Combustion and decomposition of the coating creates a gaseous shield that protects the electrode tip, weld pool, arc and the adjacent areas from atmospheric contamination. The coating also contains elements that promote ionization (this allows the use of a lower heat input), degassing, deoxidation, alloying and slagging. The slag produced provides protection for the weld pool, in addition to influencing its cooling rate.

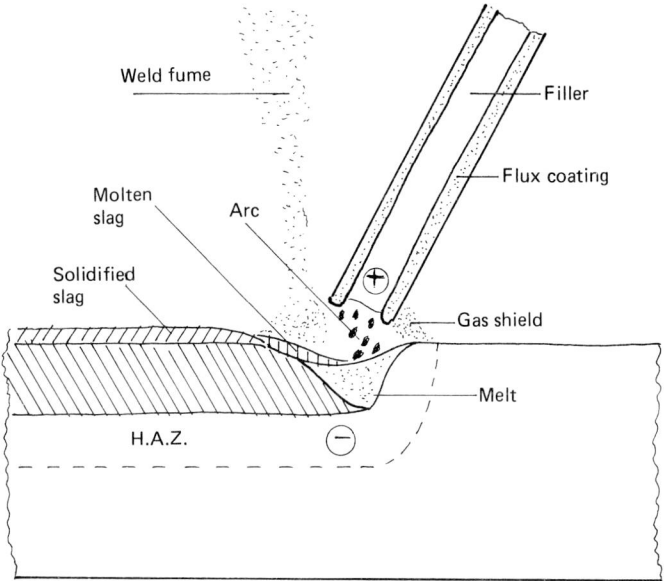

Figure 4.40 Schematic illustration of the manual metal arc welding process

The choice of filler metal depends on weld requirements and economic considerations. Electrodes designed specifically for welding cast irons are listed in the American Welding Society specification A 5.15 but many rods are produced under proprietary names[104]. Cast iron filler metals are of low cost, produce a good colour match and with careful welding can produce a weld with similar properties to the base metal. Preheat is usually applied and flake iron welds often suffer from porosity. Low hydrogen steel-type electrodes are readily available and cheap. However, the first pass deposit has a high C content due to dilution and can be brittle and crack sensitive. This restricts the use of this electrode to cosmetic repairs in non-structural areas.

Ni and Ni-Fe coated electrodes are more expensive, but are the most popular electrodes for cast irons when the weld has to be machined and is load bearing. Ni is a graphitizing element and high Ni content electrodes produce an austenitic weld deposit. Ni has a low solubility for C in the solid state. Consequently, on solidification, C is rejected from solution and precipitates as graphite. This increases the volume of the weld deposit, which offsets shrinkage stresses. It also reduces the susceptibility to fusion zone cracking. It is possible to use graphitized coatings with Ni electrodes enabling a very low welding current to be used. Ni-Fe electrodes are cheaper than Ni electrodes and offer other advantages, including:

1. the deposit is stronger and more ductile. These electrodes are used for welding higher strength grey irons and spheroidal irons;

Table 4.13 The function of electrode-coating materials

Constituent	Primary function	Secondary function
Cellulose	Shielding gas	
Calcium carbonate	Shielding gas	Fluxing agent
Fluorspar	Slag former	Fluxing agent
Rutile	Slag former	Arc stabilizer
Feldspar	Slag former	Stabilizer
Manganese oxide	Slag former	Alloying
Iron oxide	Slag former	
Iron powder	Deposition rate	
Ferrosilicon	Deoxidizer	
Aluminium	Deoxidizer	Degasifier
Strontium	Arc stabilizer	Degasifier

2. Ni-Fe electrodes are more tolerant of P;
3. Ni-Fe electrodes have a smaller coefficient of expansion. This makes them suitable for welding heavier sections without increasing the susceptibility to fusion line cracking and
4. specially formulated coatings for Ni-Fe electrodes give a soft non-penetrating arc that requires a lower heat input. A 20% Ni-Fe-Mg electrode has solidification characteristics matching those of austenitic spheroidal irons. This significantly reduces the danger of cracking in these materials.

Power settings should be in accordance with the electrode manufacturer's instructions but typical settings are given in *Table 4.14*.

Advantages of SMAW are large selection of electrodes, good availability, low cost power supplies

Table 4.14 Typical settings for shielded metal arc welding

Electrode Diameter	Ni-Fe Electrode		Ni Electrode	
	DC Current, A	AC Current, A	DC Current, A	AC Current, A
3/32 inch	40–70	40–70	40–80	40–80
1/8 inch	700–100	70–100	80–110	80–110
5/32 inch	100–140	110–140	110–140	110–150
3/16 inch	120–180	130–180	120–160	120–170

and all position welding. The main disadvantages are higher cost per unit weight of deposit, a low deposition rate and the need for a skilled operator. Welding procedure and applications are described in the literature (ref. 95, p. 307 and refs. 105–109).

Gas metal arc welding (GMAW) or MIG welding, when operated in the short circuiting or short arc mode, is suitable for welding cast irons. The arc is struck between a fine filler wire (~ 0.8 mm diameter) and the workpiece. Both are shielded by inert gas (argon or CO_2). The wire is fed at a rate that just exceeds the burn-off rate so that intermittent short circuiting occurs. When the wire touches the weld pool and short circuits the arc, there is a momentary rise in the current which melts the wire tip. A neck then forms due to the magnetic pinch effect. It melts off in the form of a droplet which is drawn into the molten pool. The arc is then re-established. It gradually reduces in length as the feed rate gains on the burn-off rate until short circuiting occurs again. This occurs several times per second depending on the arc voltage and current, type of shielding gas, diameter of the wire and power source characteristics. The mode of metal transfer is termed dip transfer.

Typical operating conditions are 20–22 V, 80–150 A, wire feed rate 12.5–15.0 cm s^{-1} and an argon flow of 12–14 litres min^{-1}. The high ratio of deposition rate to heat input means that the weld pool is smaller than in other metal arc processes. This reduces welding stresses, distortion, susceptibility to carbide formation and the size of the HAZ. This process is especially suitable for mechanization where large quantities are involved and where all positional welding is required. Bare wires of similar chemical composition to covered electrodes are used for joining cast iron to cast iron and cast iron to steel[105,108,110].

The flux-cored arc welding (FCAW) process is similar to MIG welding in that thin wire is fed continuously through a welding gun. The main difference is that the flux core produces a clean porous-free weld without an inert gas shield. FCAW combines the best features of the coated electrode and solid wire processes. Compared to GMAW operated in the spray transfer mode, the slag cover in FCAW offers better protection during cooling, an out of position welding capacity, chemical refinement of the weld deposit and a higher deposition rate. However, there is the added step of removing the slag and a lower deposition efficiency.

Filler wires are available that produce weld compositions similar to those obtained with coated electrodes. A proprietary Ni-Fe-Mn flux core wire that produces an extremely crack-resistant deposit at high deposition rates is being used increasingly. Bowen has compared the flux-cored, manual metal arc and MIG processes for welding spheroidal irons[111] and, as illustrated in *Table 4.15*, showed that the flux-cored process compares well with the other two processes. FCAW is described and illustrated in the literature[105,112–114].

Gas tungsten arc welding and submerged arc welding[115] are both expensive welding techniques and have limited use for cast irons. However, gas tungsten arc welding is the only process that can be used to repair a casting without the addition of a filler metal. Submerged arc welding is used mainly for weld overlap cladding.

Gas fusion welding uses inexpensive equipment and

Table 4.15 Comparison of spheroidal iron weld properties using metal inert gas (MIG), manual metal arc and flux cored methods (after ref. 111). U.T.S. indicates ultimate tensile strength; HAZ is the heat affected zone

Process	Hardness VPN			U.T.S. N mm^{-2}	Elongation %	Width HAZ, mm
	Parent	HAZ	Weld			
MIG	171	179	153	388	2.9	0.10
Manual arc	176	192	154	381	4.5	0.14
Flux cored	171	172	149	463	8.1	0.10

Table 4.16 A selection of filler metals used for welding cast irons

Classification	Grade	C %	Si %	Mn %	Ni %	S %	P %	Cu %	Fe %	Comment
BS 1453 (1972)	B1	3.0–3.6	2.8–3.5	0.5–1.0	–	0.15 max	1.5 max	–	remain	Gas welding, easy machining
	B2	3.0–3.6	2.0–2.5	0.5–1.0	–	0.15 max	1.5 max	–	remain	Gas welding, hard (valve seats)
	B3	3.0–3.5	2.0–2.5	0.5–1.0	1.25–1.75	0.10 max	0.5 max	–	remain	Gas welding, Ni-cast irons
AWS A 5.15	RC1/ ECI	3.25–3.5	2.75–3.0	0.6–0.75	trace	–	–	–	remain	Gas welding
	RC1A	3.25–3.5	2.0–2.5	0.5–0.7	1.2–1.6	–	–	–	remain	Gas welding
	RC1B	3.25–4	3.25–3.75	0.1–0.4	0.50	–	–	–	remain	Gas welding
AWS A 5.15	ENiCl	2.0	4.0	1.0	85 min	–	–	2.5	8.0	Arc welding; minimum preheat; easily machined
	ENiFeCl	2.0	4.0	1.0	46–60	–	–	2.5	remain	Arc welding; suitable for spheroidal iron; moderate machinability
	ENiCuA	0.35–0.55	0.75	2.25	50–60	–	–	35–45	3–6	Arc welding
	ENiCuB	0.35–0.55	0.75	2.25	60–70	–	–	25–35	3–6	Arc welding

is used for welding cast iron to cast iron, repairing damaged or worn castings and for building up on castings. The process simply involves fusion of the base metal and filler rod by the heat generated from a neutral oxyacetylene flame. In practice, the base metal joint face, after preparation, should be heated until it commences to melt and then the filler rod melted by immersion in the pool. The pool should be constantly fluxed to avoid slag entrapment.

The oxyacetylene flame does not generate as high a temperature as a welding arc. Consequently, gas fusion welding is slower than arc welding and has a greater total heat input. This produces a wider HAZ and can result in distortion. It is usual to use a high preheat temperature which means that there is less hardening in the HAZ than with arc welding. Post-weld heat treatment is often required.

Specifications for cast iron filler rods are shown in *Table 4.16*. The C and Si content are sufficiently high to allow for losses during welding. The Si content is high enough to ensure that C precipitates as graphite during solidification and to promote a soft machinable matrix as the weld cools. The composition of filler rods for spheroidal graphite formation[116] includes either Mg or Ce. Gas welding is not recommended for malleable irons[117].

Fluxes for grey irons are usually composed of borates or boric acid, soda ash and small amounts of NaCl, ammonium sulphate and iron oxide. Fluxes for spheroidal irons are rare earth-/Si-based. Some fluxes contain inoculants. Examples of the use of gas welding with cast irons are given in the literature[118].

References

1. MINKOFF, I., *Physical metallurgy of cast iron*, J. Wiley, London (1983)
2. RIDLEY, N., A review of the data on the interlamellar spacing of pearlite, *Trans. Met. A*, **15A**, 1019 (1984)
3. HONEYCOMBE, R.W.K., Steels. Microstructure and properties, E. Arnold, London
4. JANOWAK, J.F. and GUNDLACH, R.B., A modern approach to alloying gray iron, *A.F.S. Trans.*, **90**, 847 (1982)
5. FIDOS, H., Structural analysis of a graphite nodule and surrounding halo in ductile iron, *A.F.S. International Cast Metals Journal*, **7**, 54 (1982)
6. DODD, J. and PARKS, J.L., Factors affecting the production and performance of thick section high Cr-Mo alloy iron castings, *A.F.S. International Cast Metals Journal*, **5**, 15 and 47 (1982)
7. ANON., *A.S.M. Metals Handbook*, 9th edn., Vol. 4, p. 542, A.S.M. (1981)
8. ANON., *A.S.M. Metals Handbook*, 9th edn., Vol. 4, p. 529, A.S.M. (1981)
9. ANON., *The heat treatment of S.G. cast iron*, British S.G. Iron Producers' Association Ltd., London, p. 45 (1969)
10. ROY, P.L., CHAKRAHARTI, A.K. and BANERJEE, P., Effect of minor additions on the first stage graphitisation in white iron, paper presented at Third International Symposium on the Physical Metallurgy of Cast Iron, Stockholm (1984)
11. TEKAHESHI, R., Kinetics of graphitisation of white cast iron, paper presented at Third International Symposium on the Physical Metallurgy of Cast Iron, Stockholm (1984)
12. BURKE, J. and OWEN, W.S., Kinetics of first stage graphitisation in iron-carbon-silicon alloys, *Journal Iron and Steel Inst.*, **176**, 147 (1954)
13. HUSSEIN, A.A., EL-MENAWATI, L.I., KASSEM, M.A., and YOSTOS, B.I., First stage graphitisation of subcritically treated white cast iron, paper presented at Third International Symposium on the Physical Metallurgy of Cast Iron, Stockholm (1984)
14. BIRCHENALL, C.E. and MEAD, H.W., Kinetics of first stage graphitisation, *A.F.S. Trans.*, **59**, 181 (1951)
15. HILLERT, M., NILSSON, K. and TÖRNDAHL, L.E., Effect of alloying elements on the formation of austenite and dissolution of cementite, *Journal Iron and Steel Inst.*, **209**, 49 (1971)
16. APPLETON, A.S., The kinetics of first stage graphitisation in Fe-C and Fe-Co-C alloys, *Journal Iron and Steel Inst.*, **194**, 160 (1960)
17. BURKE, J., The growth of temper carbon nodules, *Acta Met.*, **7**, 168 (1959)
18. SWINDELLS, N. and AVERY, J.D., The first stage graphitisation of white cast iron, *Metallurgy of Cast Iron*, Georgi Publ. Co., St. Saphorin, Switzerland, p. 221 (1975)
19. ROY, P.L., CHAKRABARTI, P. and BANERJEE, P., The effect of NaCl addition on solidification and graphitisation in

white cast iron, *A.F.S. Trans.*, **93**, 587 (1985)

20. MAIER, R.D. and WALLACE, J.F., Literature search on controlling the shape of temper carbon nodules in malleable iron, *A.F.S. Trans.*, **84**, 687 (1976)

21. POPE, M. and GRIEVESON, P., Effect of interfacial segregation upon graphitisation in Fe-C alloys, *Metal Science*, **11**,137 (1977)

22. BOLOTOV, I.Y., Electron microscope study of the different stages of crystallisation of spherulites and floccular graphite in hard alloy Ni-C, *Phys. Met. Metallographie*, **20**, 86 (1965)

23. HULTGREN, A. and ÖSTBERG, G., Structural changes during annealing of white cast irons of high S:Mn ratio, *Journal Iron and Steel Inst.*, **176**, 351 (1954)

24. COTTRELL, D.J. and MAWSON, A.J., Modern heat treatment technique boosts production of malleable cast irons, *Metallurgia*, **52**, 70 (1985)

25. REYNOLDS, C.C., WHITTINGHAM, N.T. and TAYLOR, H.F., Hardenability of ductile cast iron, *A.F.S. Trans.*, **63**, 116 (1955)

26. DANKO, J.C. and LIBSCH, J.F., Secondary graphitisation of quenched and tempered ductile cast iron, *Trans. A.S.M.*, **47**, 853 (1955)

27. ASKELAND, D.R. and FARINEZ, F., Factors affecting the formation of secondary graphite in quenched and tempered nodular irons, *A.F.S. Trans.*, **87**, 99 (1979)

28. VOIGT, R.C. and LOPER, C.R. Jr., Secondary graphitisation in quenched and tempered ductile cast iron, *A.F.S. Trans.*, **90**, 239 (1982)

29. RUNDMAN, K.B. and ROUNS, T.N., On the effects of molybdenum on the kinetics of secondary graphitisation in quenched and tempered ductile irons, *A.F.S. Trans.*, **90**, 487 (1982)

30. MYERS, R.C., Hardening thin walled cylinders, *American Machinist*, **108**, 66 (1964)

31. DEMIDOVA, T.G. and KUNYAVSKII, M.N., Isothermal tempering of nodular graphite cast iron, *Liteinoe Proizvodstvo*, **2**, 20 (1955)

32. BRIGGS, J., *High strength irons*, Climax Molybdenum Company, Greenwich, CT, USA (1962)

33. DODD, J., High strength, high ductility, ductile irons, *Modern Casting*, **68**, 60 (1978)

34. GRINDAHI, R.B. and GENERAL MOTORS CORP., High stress nodular gears and method of making the same, US Patent 4 222 793, Sept 16th 1980

35. ROSSI, F.S. and GUPTA, B.K., Austempering of nodular cast iron automobile components, *Metal Progress*, **119**, 25 (1981)

36. HARDING, R.A. and GILBERT, G.N.J., Why the properties of austempered ductile irons should interest engineers, *Brit. Foundryman*, **79**, 489 (1986)

37. FORREST, R.D., The challenge and opportunity presented to S.G. iron industry by the development of austempered ductile iron ADI: S.G. iron – the next 40 years, paper presented at BCIRA Conference, Warwick, April 1987

38. MORGAN, H.L., Introduction to foundry production and control of austempered ductile irons, *Brit. Foundryman*, **80**, 98 (1987)

39. HARDING, R.A., Use of austempered ductile iron for gears, paper presented at Second World Congress on Gearing, Paris, March 1986

40. FISCHER G.A.G. *et al.*, Process for the bainitic hardening at least partially isothermally, of a cast iron part, International Patent Application WO 83/01959 (1982)

41. PONT-A-MOUSSON S.A. and BELLOCCI, R., Centrifugally cast tube of spheroidal graphite cast iron and its method of manufacture, UK Patent Application G.B. 2 117 000A (1983)

42. DODD, J., GUNDLACH, R.B. and LINCOLN, J.A., Advances in process technology and new applications of austempered ductile irons, paper presented at BCIRA Conference on Developments for Future Foundry Prosperity, Univ. Warwick, April 1984

43. BLACKMORE, P.A., High strength nodular irons, paper presented at BCIRA Conference on Developments for Future Foundry Prosperity, Univ. Warwick, April 1984

44. WATMOUGH, T. and MALATESTA, M.J., Strengthening of ductile iron for crankshaft applications, *A.F.S. Trans.*, **92**, 83 (1984)

45. ROUNS, T.N., RUNDMAN, K.B. and MOORE, D.M., On the structure and properties of austempered ductile cast iron, *A.F.S. Trans.*, **92**, 815 (1984)

46. DORAZIL, E., BARTA, B., MUNSTEROVA, E., STRANSKY, L. and HAVAR, A., High strength bainitic ductile cast iron, *A.F.S. International Cast Metals Journal*, **7**, 52 (1982)

47. VAN MALDEGRAM, M.D. and RUNDMAN, K.B., On the structure and properties of gray cast iron, *A.F.S. Trans.*, **94**, 249 (1986)

48. MOORE, D.J., ROUNS, T.N. and RUNDMAN, K.B., Structure and mechanical properties of austempered ductile iron, *A.F.S. Trans.*, **93**, 705 (1985)

49. BLACKMORE, P.A. and HARDING, R.A., The effects of metallurgical process variables on the properties of austempered ductile irons, *First International Conference on Austempered Ductile Irons*, Chicago, USA, April 1984

50. DUBENSKY, W.J. and RUNDMAN, K.B., An electron microscope study of carbide formation in austempered ductile iron, *A.F.S. Trans.*, **93**, 389 (1985)

51. JANOWAK, J.F. and GRUNDLACH, R.B., Development of a ductile iron for commercial austempering, *A.F.S. Trans.*, **92**, 489 (1984)

52. JANOWAK, J.F. and MORTON, P.A., A guide to mechanical properties possible by austempering 1.5% Ni–0.3% Mo ductile iron, *A.F.S. Trans.*, **91**, 377 (1983)

53. SHEA, M.M. and RYNTZ, E.F., Austempering nodular irons for optimum toughness, *A.F.S. Trans.*, **94**, 683 (1986)

54. GAGNE, M., Influence of manganese and silicon on the microstructure and tensile properties of austempered ductile iron, *A.F.S. Trans.*, **93**, 801 (1985)

55. COX, G.J., The effect of austempering time on the properties of high strength S.G. iron, *Brit. Foundryman*, **79**, 215 (1986)

56. HARDING, R.A., The effects of metallurgical process variables on austempered ductile irons, *Metals and Materials*, **2**, 65 (1986)

5.1f if draft cannot be tolerated in the hole or on the perimeter of the ring) and the designer fully appreciates the additional costs involved, the foundryman must then decide on the method best suited to the production of cores. This includes how the core box will be separated; how the cores will be located and supported; how they will be vented and how the core sand will be removed after casting. Often discussion between the designer and foundryman will produce opportunities to reduce the complexity and the costs involved.

Computer aided design (CAD)

In the past and still in many cases today, the skilled, experienced foundryman is responsible for design decisions. However, if the foundry is to remain competitive, these skills must be supplanted with a more engineering-based approach. This is possible with CAD/Computer aided machining (CAM) technology. In the 1971 Hoyt Memorial Lecture Ruddle suggested, 'we shall within two decades see the digital computer take over much of risering practice'. He continued, 'it seems likely that in a short while all one may need to do is feed into the computer a drawing of the casting plus indication of the thermal properties of the mould material and of the alloy being used, its solidification characteristics and the degree of soundness desired in the casting. The computer would then indicate just what would be the most economic risering and gating system to develop the desired degree of soundness and the location of any unsoundness'. This is one of several design functions now possible through the development of CAD/CAM technology. This technology facilitates the liaison between foundry and designer and improves the speed and accuracy with which a design concept is translated into a finished product[1,2].

Mathematical modelling or solidification simulation is central to a comprehensive CAD system. A casting solidifies as a result of heat transfer by conduction, convection and radiation. Steady state solutions of the differential equations derived from the laws governing these modes of heat transfer can be obtained analytically for uniform property problems. However, the solidification of a casting does not fall into this category because it is a transient problem requiring solution of partial differential equations relating time, temperature and position. It is complicated by the evolution of latent heat, which requires special procedures of accommodation because established differential equations for heat flow assume the description of a continuous function[3,4].

The two principal computer-borne numerical methods for solving the partial differential equations are the finite-difference method (FDM) and the finite element method (FEM)[5,6]. The FDM uses a series of regularly spaced grid points and approximates the partial differential equation (PDE) at each point. The resulting algebraic equations are then solved for the temperature at each grid point. Rapid solutions are possible because the algebraic equations are usually well structured and linear. Thus FDMs are relatively inexpensive, but they do not readily cope with complex casting geometries unless a large number of grid points are used.

FEM seeks to approximate the solution of the PDE integrated (with some weighting function) over a series of arbitrarily shaped finite elements. The temperature within each element is taken as a simple weighted average of the temperatures at a small number of selected nodes on the bounding surfaces of the elements. The result is a system of simultaneous linear equations which usually do not have a regular structure and, therefore, their solution is more time consuming and expensive. However, the number of nodes required in FEM is usually much less than the number of grid points in FDM for comparable accuracy. The FEM is advantageous when a complex casting geometry is being analysed.

Other requirements and considerations in solidification simulation include:

1. Definition of the time and temperature dependent metal-mould heat transfer coefficient. Heat transfer across the air gap that forms between metal and mould due to metal shrinkage and mould movement can be as important a consideration as the transfer across the casting and mould. Air gap formation occurs in most casting processes, but is particularly important when high thermal conductivity moulds are used and when chills are used in sand moulds. At present a mathematical description of gap formation is not available for sand castings and interfacial heat transfer coefficients must be obtained from experimental data[7–14].

2. A data base defining the thermal properties of casting alloys and mould materials. Properties required include thermal conductivity, specific heat, latent heat and density. They are required in a form amenable to computer implementation. Compiling the data is a protracted task because of the extensive range of casting alloys and mould materials in use. Thermal conductivity data for moulds presents a particular problem as it depends on sand type, sand grading, amount and type of binder, water content, compaction method, hardness and temperature[15].

3. To date, the starting point for most simulations has been a temperature field in the casting that is considered uniform and to be attained instantly on pouring. It is only recently that two-dimensional convective heat transfer models have been used to

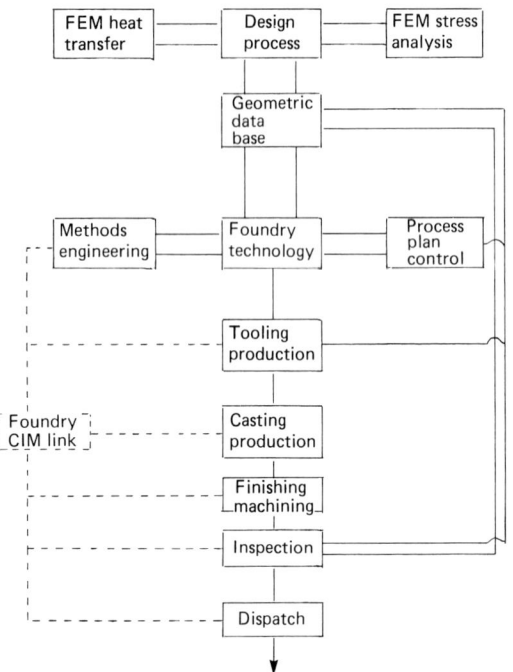

Figure 5.2 Schematic CAD system for a casting application

predict temperature losses in the gating system and casting cavity. These models have revealed the effects of a complex series of interactions of heat and fluid flow during the process[16]. These include the concept of superficial solidification and remelting of the flowing stream and an upward displacement of the thermal centre by buoyancy-driven forces of natural convection after mould filling and prior to solidification.

4. Development of software for interfacing geometric modelling and heat transfer simulation programs in two and three dimensions. Numerous studies have been reported which correlate computer simulations of heat transfer and solidification with experimental measurements of temperature and casting characteristics[17-26].

The first stage in a CAD system for castings is a geometric modeller. It serves as a user description language for defining the cast shape and provides algorithm data needed to allow enmeshment of the volume. A CAD system is of considerable benefit to those responsible for specifying the process and tooling requirements for casting production. It is an aid to improved liaison between the designer and foundryman by providing video screen output of isometric views, perspective section views etc. It represents a considerable aid for the coring of internal features. Volume, weight and surface area calculations permit section modulus and feeder calculations.

With respect to tooling, three-dimensional models overcome inherent limitations of two-dimensional drawings which under define complex shapes. Packages are available for translating complex shapes with contoured, intersected and fillet surfaces into finish machined tooling with the minimum of operator input.

CADD, computer aided draughting and design, is replacing the draughtsman's traditional tools of pencil, paper and T-square. Considerable benefits are to be gained by the foundry in using draughting packages in the preparation of pattern layouts and tooling[27]. Library features and the use of layers are particularly useful. Drawings can be constructed by combining any number of separate layers that have been constructed individually. For example, the methods engineer uses separate layers to detail features such as the pattern plate, pattern layout, core locations, running and gating system and feeding system. This technique is particularly useful for redesigns and recording of changes. A typical CAD system for casting application is shown in *Figure 5.2*.

Patterns

One of the first steps in the manufacturing process is the selection and production of suitable pattern equipment. This will include, in addition to the pattern itself, core boxes, core assembly fixtures and casting inspection equipment where needed. The emphasis is on *suitable* equipment because this decision will greatly influence both the quality and the cost of the casting that is eventually produced. Because the cost of the pattern equipment varies according to the type selected, the material used in its construction and the accuracy and quality of finish required, the final selection of the most suitable equipment will be dependent on the following criteria:

1. the design of the component and the order quantity likely to be procured;
2. the dimensional tolerance necessary in the casting;
3. the moulding process to be used and
4. an appropriate core making process, if needed.

Items 1 and 2 invariably lead to the decision made in criteria 3 and 4, and ultimately to the type of pattern equipment selected.

The greatest variety of pattern materials are used in sand moulding processes. Pine, mahogany, aluminium and grey irons are most commonly used to make working patterns. The woods are employed when a small quantity production is required. They are also used for the manufacture of master patterns from which the metal working patterns for higher volume production runs are made.

Plastics are being used increasingly, particularly when medium volumes are required. A recent de-

Table 5.2 The life expectancy of different mould materials

Pattern material	Life (numbers of moulds)
Soft wood	1 – 50
Hard wood	1 – 1 000
Epoxy resin	5 000 – 20 000
Gun metal	20 000 – 30 000
Aluminium	20 000 – 60 000
Grey cast iron	200 000 – 300 000
Low alloy steel	300 000 – 500 000

velopment in this field is the HMP process[28,29]. This process (Highly reactive machine-applied polyurethane system) produces lightweight patterns with abrasion-resistant surfaces. These characteristics make them suitable for long runs on mechanized moulding lines. Central to the process is a machine capable of mixing, metering and dispensing three different polyurethane materials involving two catalysts and a resin. The machine produces a foamed core of expanded clay pellets and a very dense, abrasion-resistant surface which is 1–3 mm thick. The surface has a cure time of approximately 45 seconds. In general, selection of pattern material is made on the basis of the amount of wear to which the equipment is to be subjected, and, in turn, its life time expectancy. *Table 5.2* is a general guide to the life expectancy of various pattern materials.

More recently, expanded polystyrene foam expendable patterns have found popularity in the colume production of iron castings by the 'lost foam' process and the 'full mould' process[30]. Patterns made from polystyrene sheet stock have been used for several years for the one-off casting situation or for design prototypes due to their relatively low cost.

The metal mould is often cast to size using a mahogany pattern in permanent moulding. The basic pattern material may be wax or plaster in ceramic moulding. The simplest and often least expensive type of pattern is referred to as a loose pattern. The moulding process utilizing this type of equipment is called simply 'loose pattern moulding'. In this case the equipment is often nothing more than a model of the finished component to be cast with added allowance for contraction and where necessary, for machining. If cores are required, projections known as core prints are added to allow for core location and clamping.

Match plates are usually selected for slightly higher volumes. In this type of equipment the pattern is segregated into two halves using a flat parting line. Each pattern half is mounted in perfect register on either side of a flat wooden board or metal plate. Metal flow channels, known as the *running and gating system*, are often attached to the pattern to eliminate the need for hand-cutting of the mould. This type of equipment can be used either in a hand squeezer or vibrating type. When the pattern reaches a size that is inconvenient for manual handling, the two halves of the pattern are mounted onto two separate plates. The half plate which produces the top half of the mould is called the *cope*. The half plate which produces the bottom half is called the *drag*. Even if the design and the two halves of the pattern are symmetrical, the cope will differ in that it will contain provision for the metal to enter the cavity gates by the inclusion of a vertical sprue.

Some foundries have developed the use of large pattern plates onto which several different casting patterns are included and frequently changed using a fixed gating system. This form of equipment is known as insert pattern equipment.

Other types of pattern methods adopted include sweep or strickle moulding. This is often used in the moulding of large simple components such as pots. The sweep or strickle pattern is a simple template which is rotated around a central spindle to shape the surfaces of the mould.

Very large castings are often produced from moulds assembled using separately hardened sand sections produced out of core boxes. This form of moulding is referred to as core assemble. Other moulding methods

Table 5.3 The colour coding used for patterns

Pattern section	Colour
As-cast surface not to be machined	red or orange
Surfaces to be machined	yellow
Core prints for unmachined openings and end prints	black
Core prints for machined openings 'A' periphery	yellow stripes on black
'B' ends	black
Split patterns – pattern joint – 'A' cored section	black
'B' metal section	clear varnish
Touch core cored shape	black legend touch
Seats of and for loose pieces and loose core prints	green
Stop offs	diagonal black stripes with clear varnish
Chilled surfaces	outlined in black, legend chill

using separately made sections include 'stack moulding' which vertically stacks separate mould cavity sections and fills them with metal via a common singular sprue. The 'H' process comprises horizontally stacked and clamp moulds with a single common horizontal runner system included in the core produced mould[31]. *Table 5.3* shows the colour coding recommended for patterns.

Running, gating and feeding

A most important aspect of the production of clean and sound castings is the method by which the iron is introduced into the mould cavity and the arrangements made to compensate for solidification shrinkage. The components of a system are shown in *Figure 5.3*. The first design step should be one of contemplation[32]. The following should be considered:

1. maximum use of the mould volume but leave adequate room for a proper gating and feeding system;
2. position the parting line to minimize the need for cores;
3. attempt to position heavy sections of the casting in the drag;
4. locate as much of the casting as possible in the cope for quiet filling;
5. aim for maximum simplicity and symmetry;
6. employ identical gating and feeding for all identical castings in multiple impression moulds and
7. consider maximum feeder utilization.

The next stage involves decisions concerning:

1. the flask size, length width and the cope height;
2. the number of casting cavities and their horizontal/vertical positioning on the pattern;
3. the number and type of feeders – side or top;
4. the shape of the feeder and feeder neck;
5. the type of gating system;
   a. runner/gate control (pressurized system) with a cope runner;
   b. sprue/runner control (non-pressurized system) with a drag runner;
   c. Inmould process – conventional system with a cope runner;
6. location of the sprue;
7. number of runners.

Dimensional calculations are the third stage and they are being performed increasingly with microcomputers. These calculations include:

1. the number of castings in the mould and the casting weight;
2. feeder and feeder contact sizes;
3. the pouring time for the weight of castings and feeders;
4. the choke cross-sectional area required to deliver the liquid iron in the desired time;
   a. for a runner/gate system;
      select the geometry and calculate gate(s) dimensions;
      calculate runner cross-sectional area;
      select the geometry and calculate the dimensions of the runner;
   b. for a sprue/runner system;

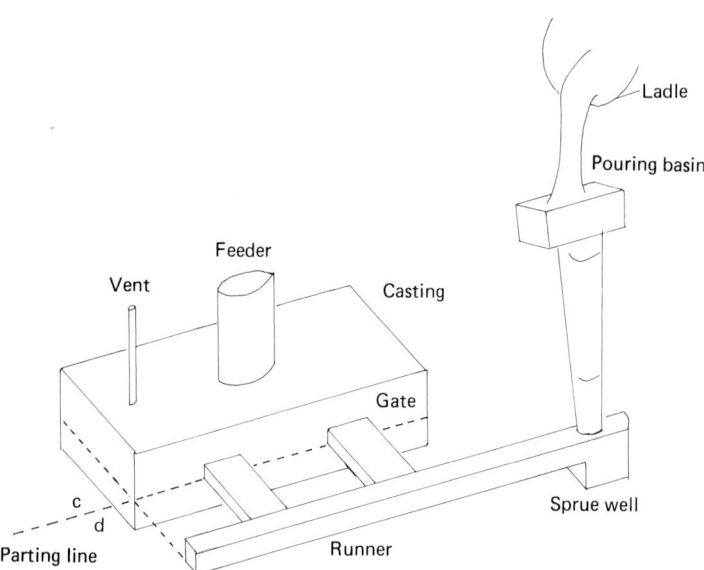

Figure 5.3 Schematic representation of the major components of a running, gating and feeding system

select the geometry and calculate dimensions of the choke;

select the geometry and calculate the dimensions of the runner(s);

calculate the total gate cross-sectional area;

select the geometry and calculate dimensions of the gate(s);

  c. for an Inmould process;

select the geometry and calculate gate(s) dimensions;

calculate the runner cross-sectional area;

select the geometry and calculate dimensions of the runner;

calculate the alloy chamber exit area;

calculate the total alloy chamber area;

calculate the weight of the treatment alloy;

calculate the alloy chamber dimensions;

5. the sprue cross sectional area;
6. the sprue diameter;
7. the minimum pouring temperature and
8. the approximate casting yield.

Although it is convenient to consider running and gating and feeding principles separately, as evident from the considerations defined above, the two cannot be treated independently in practice because they mutually interact. Situations arise in which the gating system can perform the act of feeding as described later. The location of the gates influences the temperature distribution in the liquid immediately after pouring and hence, the solidification pattern and feeding behaviour. The time elapsing between completion of pouring and gate solidification is important in the control of pressure feeding. If the design calls for the longest possible time for the feeder to remain liquid, this can be achieved by gating into the feeder and filling the entire mould cavity through the feeder.

Principles and practice of running and gating

The running and gating system is required to ensure that:

1. the flow of metal should be as free as possible from turbulence but at a rate sufficient to avoid undue delay in filling the mould;
2. slag and oxidation dross within the iron are contained in the runner system and not allowed to enter the casting cavity;
3. direct impingement of the metal stream on core or mould surfaces at a velocity sufficient to cause erosion of sand is avoided and
4. the thermal gradients required to produce sound castings are established

With respect to thermal gradients, generally, the last metal to enter the mould is the hottest. Consequently, if directional solidification is desired, iron can be run into the upper parts or heads on the

casting. This can be arranged in many cases by gating through or near the heads. This approach is top gating. However, this technique applied to other than shallow parts will result in metal falling vertically inside the mould with the risk of splashing and sand erosion. To avoid this, metal can be introduced at the lowest part of the mould cavity. This is known as bottom gating. If adverse temperature gradients result from this practice, they can be avoided by using a system called step gating. Step gating consists of a series of gates at several levels. Each becomes inoperative in turn as iron rises in the mould during filling.

Several well known physical laws govern the flow of liquids in channels and are the basis of runner and gating system design[33]. Several design features of the components in *Figure 5.3* are independent of the gating system chosen. The pouring ladle size is a compromise between minimizing differences in pouring temperature and ladle refilling. The tilting mechanism should allow easy and rapid adjustments in the pouring rate. The width of the U-shaped pouring lip should be about twice the sprue diameter. The lip should be long enough to reach the centreline of the mould to facilitate pouring from just above the mould. Pouring basins should be designed to avoid slag entry to the runners, air entrainment and vortexing.

A rectangular basin with rounded corners, as shown in *Figure 5.4*, fulfils these requirements better than a conical shape. It is easier to locate the pouring ladle lip laterally than to predict where the first liquid will touch the top of the mould. Usual basin dimensions are width twice the sprue diameter and length twice the width.

The first iron poured contains most slag and it is prevented from entering the sprue by the well of depth approximately equal to the basin width. Although this generous basin reduces casting yield, it means that there is time for the pourer to adjust pouring rate before the sprue filling commences. The constant and adequate head of metal allows the sprue to run full.

However, this will not be the case if the sprue is of uniform cross-sectional area. This is because of an increase in flow velocity, which results in a decrease in the cross-sectional area of the liquid stream. This can lead to aspiration, the introduction of gases from the mould into the iron stream. It is prevented by tapering the sprue such that

$$A_2/A_1 = (h_1/h_2)^{1/2}$$

where A_1 and A_2 are the cross-sectional areas at the sprue entrance and exit, respectively, and h_1 and h_2 are the height of metal in the system above these locations.

The sharp change in direction from the sprue to runner shown in *Figure 5.4* causes the formation of a stationary liquid pool soon after pouring has started. This diverts the liquid flow into the runner. The

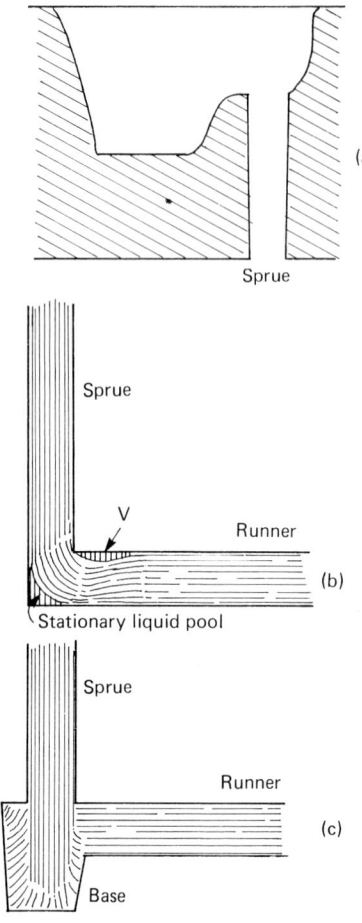

Figure 5.4 (a) Suitable shape for a pouring basin; (b) Densening of streamlines due to a change in flow direction from the sprue to runner; (c) The hydraulic effect of a sprue base. The base acts as a cushion converting downward flowing liquid into a horizontal flow with velocity $\sqrt{2gH}$

streamlines show a contraction in the liquid stream, *vena contracta*. Reduced pressure in the region marked 'V' can lead to aspiration, thus drawing air into the stream. This air can increase the turbulence further down the stream causing slag to mix with liquid iron rather than sticking to the top of the runner. Aspiration can be eliminated by using a sprue base. The effect of the base on metal flow is shown in *Figure 5.4*.

Ceramic filters are claimed to be effective in removing both dross and slag from the system[34]. Many runner systems also incorporate slag or dirt traps. A most effective form of dirt trap is the swirlgate, which is often used with horizontally parted moulds. Centrifugal force separates slag and dirt from the iron stream. However, a correctly designed runner

and gating system should prevent slag and dross from entering the mould cavity.

Foundry engineers responsible for designing runner systems refer to the gating ratio. This is the relationship between the cross-sectional area of the sprue, runner and gate. For example, a ratio of 1:2:4 describes a non-pressurized or sprue/runner-controlled system, whereas a ratio of 1:1:0.7 describes a pressurized or runner/gate-controlled system.

Several factors influence the choice of system. A satisfactory flow pattern is a prime consideration. The flow pattern is controlled to satisfy the needs of irons that form either dross and/or slag. Runner/gate control is used for slag forming irons such as malleable and flake irons. Fluidity is controlled primarily by C.E.V. and pouring temperature. Although spheroidal irons are of higher C.E.V., their higher Si content and the presence of Mg and Ce from spheroidization renders their liquid surface more susceptible to oxidation and dross formation. Spheroidal iron slag is drier, more particulate and more difficult to trap. Sprue/runner control can be beneficial in this case.

In general, although a gating system that performs well for a spheroidal iron will usually function for a flake iron, the reverse is not necessarily true. Other considerations are that a sprue/runner control produces a higher yield when a large number of small castings are in one mould. Runner/gate control is more suitable and produces a better yield with heavy castings with only a few in a mould.

A runner/gate-controlled system is shown schematically in *Figure 5.5*. The most important feature of this system is that the gating ratio decreases from the sprue to the gates which form the choke. We have seen that a tapered sprue prevents excessive turbulence and air aspiration during the initial stages of the pour. However, in horizontally parted moulds, the sprue often has a reverse taper for ease of pattern withdrawal. Such a sprue will not run full initially but the choked gate(s) will ensure that backfilling occurs quickly.

It is important to make sure that the sprue does not act as the choke. This is because a condition can be generated in which iron can leave the gating system faster than it can be poured into the pouring basin. To avoid this, at least as much liquid must be capable of passing through the sprue as through all the gates. This can be expressed mathematically using the equation for the rate of filling derived using Torricelli's equation. The velocity of liquid flow through the top of the sprue is given by the equation

$$v_s = c\sqrt{2g}\sqrt{h} \qquad (5.1)$$

where g is the acceleration due to gravity and c is a friction factor. The rate of flow, F_s, is $A_s c\sqrt{2g}\sqrt{h}$, where A_s is the cross-sectional area of the sprue. Applying the same consideration to the choke gives

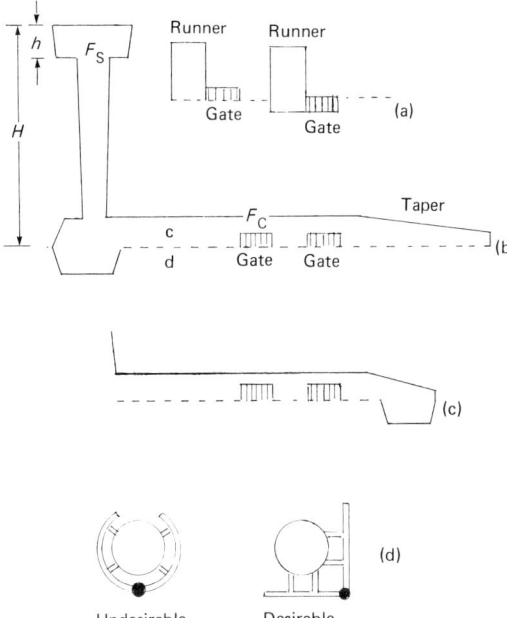

Figure 5.5 Schematic representation of a runner/gate control system: (a) The correct ways of contact between the gate and runner; (b) Running system illustrating a long distance and taper after the last gate and an approximately equal distance between the sprue and first gate; (c) Modified system using a runner well to trap slag. This is used when there is insufficient room for (b); (d) Desirable and undesirable gating arrangements

for the condition $F_s \geqslant F_c$,

$$A_s \geqslant A_c \sqrt{\frac{H}{h}} \tag{5.2}$$

This is illustrated in *Figure 5.5*. The sprue can never be too large except that the casting yield decreases; it can only be too small.

Slag and dross are trapped in the runner in a well designed system. This function can be analysed by considering two time periods:

1. from the start of pouring until the running system is full and
2. from the end of period 1 until pouring is completed.

Flow can be turbulent, but it is usually unidirectional in the runner in the first period. Unidirectional flow takes the initial slag, which contains iron, past the gate(s). If the runner is bent or curved, unidirectional flow is disturbed and turbulence is introduced. The chance of slag entering the gate is then increased. If the runner must be curved, allow a straight length before the gate. The bottom of the runner and of the gate should be in the same plane, otherwise the step

between the two can induce turbulence and increase the possibility of slag entering the gate.

The velocity of the liquid in a partially filled runner is high; so much so, that when it reaches the end it tends to fold over the on-rushing layer. It is then possible for slag to enter the last gate. This possibility is reduced if the design features in *Figure 5.5* are used, namely:

1. the runner is extended as far as possible past the last gate;
2. the runner is tapered gently, wedging the iron and preventing the folding over action and/or;
3. a well is placed at the end of the runner.

In addition, the gates should be thin and wide. A thickness to width ratio of 1:4 presents a low profile relative to the runner. The level of the iron in the runner rises rapidly so that it is well above the top of the gate before iron starts to flow through the gate. However, the width must not be excessive because this increases the possibility of slag entry. As a further aid, gates should be branched perpendicularly to the runner and not flared. Gates can be bent and as long as desired because they should be slag free. Their length exceeds their width.

Once the system is running full in the second period the only concern is for slag travelling from the sprue bottom to the first gate(s). Slag flotation and retention is aided by maintaining a good distance between the sprue and the first gate. The degree of flotation is determined by the velocity of vertical flotation defined by Stokes' Law and the velocity of liquid flow along the runner. If the latter is decreased, the degree of flotation increases. This is achieved by increasing the cross-sectional area ratio between runner and choke. This ratio is usually between 4:1 and 2:1 depending on the distance between the sprue and the first gate.

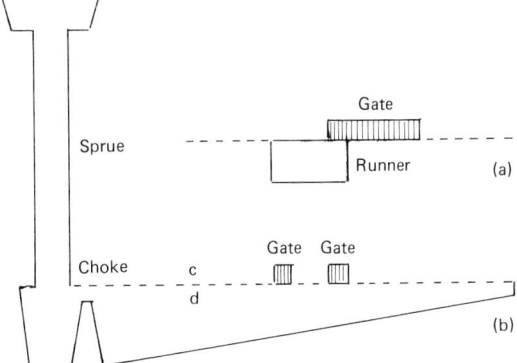

Figure 5.6 Schematic representation of a sprue/runner control system. (a) The correct way of making the contact between the gate and runner; (b) The running system illustrating gate position and a gradual uphill taper for slag entrapment

Slag flotation is aided by increasing the distance between the gate and the runner top by using a tall and narrow runner, which usually has a 2:1 height to width ratio. Choking at the gate has the advantage of producing an even distribution of flow, but can cause high gate velocities.

An example of a sprue/runner-controlled system is shown in *Figure 5.6*. The choke is at the junction between sprue base and runner, or, alternatively, at the bottom of the tapered sprue. This system shows several differences from the runner/gate system shown in *Figure 5.5*. In particular, the runner is located in the drag and the gates are located in the cope. This ensures that there is no possibility of slag entering the gates until the liquid level has reached the parting line at the end of the second period in the pouring sequence. There are three periods to consider with respect to slag entrapment:

1. this period is identical to that in the previous system and extends until the system has back filled from the choke;
2. this period is absent from the pressurized system and covers the time necessary to complete the filling of the runner;
3. this period corresponds to the second period in the runner/gate system and covers the time during which the mould is filling.

The sprue/runner system is superior to the runner/gate system in that considerably less liquid is poured in the first period. Once backfilling has been completed the amount of slag travelling down the sprue is reduced due to decreased turbulence. There is no possibility of slag entering the mould during this period. It must be confirmed that the top of the sprue is not acting as a choke using the procedure defined earlier.

Slag separation in the second period during which the runner is being filled is aided by several design features. The straight but tapered runner traps the initial slag-bearing iron in the wedge at its end. The uphill taper minimizes turbulence to give quiet mould filling and helps to hold the slag. There is a possibility of slag entering the casting cavity towards the end of this period when the runner has been filled. This depends on the ratio of the total runner/gate overlap area and the horizontal surface area of the runner.

The horizontal surface area is increased by using a wide and thin runner; a ratio of 2:1 at the mid-length is common. As the runner is tapered, the width:thickness ratio increases from the sprue to the end of the runner. The gate cross-section is approximately square and the overlap should be slightly in excess of the gate thickness.

The slag retention features during the mould filling period are similar in both systems. A relatively large runner/choke cross-sectional area ratio (e.g. 4:1) produces a low velocity flow in the runner which helps

to promote slag separation. This is aided by maintaining a long distance between the sprue and the first gate. The total gate cross-sectional area must be greater than the choke area. The larger this ratio, the quieter the mould filling.

Combinations of the two systems offer the benefits of both and a better guarantee against slag inclusion. Combinations of sprue height, choke area and gate area can be chosen for the system in *Figure 5.6* such that it operates initially as a sprue/runner system. However, the mould filling rate is controlled by a runner/gate choke. The casting yield decreases with hybrid systems, but this is justified if the rejection rate is reduced. They are used mainly on large pattern plates with multiple castings.

Several factors influence the choice of mould filling time. They include production schedule, freedom from misruns and sand expansion defects, fluidity and feeding considerations. Average pouring times for spheroidal irons are less than for flake and white irons. Calculation of the filling time is based on Torricelli's equation. If the casting is entirely in the drag (see *Figure 5.7a*), the velocity at the gate is

$$v = c\sqrt{2g}\sqrt{h}$$

and the flow rate through the gate, F, equals velocity × gate area,

$$F = Ac\sqrt{2g}\sqrt{h}$$

However, the flow rate equals casting volume, V_D, divided by filling time, t. Hence:

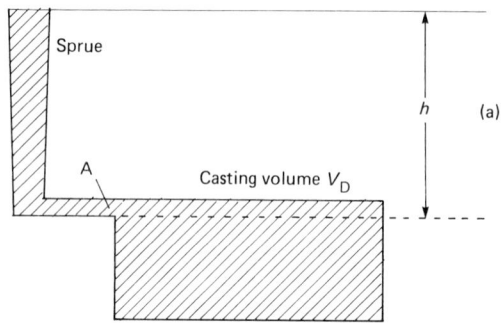

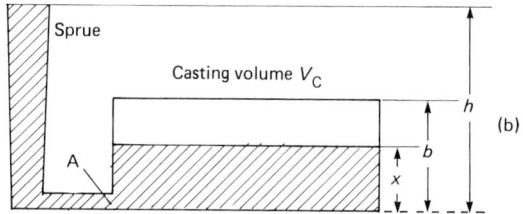

Figure 5.7 Cavity filling for (a) Casting entirely in the cope; (b) Casting entirely in the drag

$$t = \frac{V_D}{Ac\sqrt{2g}\sqrt{h}} \tag{5.3}$$

If the gate remains on the parting line, but the casting of volume V_C is entirely in the cope, the head of liquid, and hence the gate velocity, decreases during filling. With reference to *Figure 5.7b*, the gate velocity at any instant of filling with liquid height x in the casting is

$$v = c\sqrt{2g}\sqrt{h-x}$$

The average velocity is:

$$v = \frac{c}{b}\sqrt{2g}\int_0^b (h-x)^{1/2}\,dx$$

$$= \frac{2\sqrt{2gc}}{3b}[h^{3/2} - (h-b)^{3/2}]$$

Hence the filling time is:

$$t = \frac{3bV_C}{2Ac\sqrt{2g}[h^{3/2} - (h-b)^{3/2}]} \tag{5.4}$$

If the casting is split between the cope and drag,

$$t = \frac{1}{Ac\sqrt{2g}}\left[\frac{V_D}{\sqrt{h}} + \frac{3bV_C}{2(h^{3/2} - \{h-b\}^{3/2})}\right] \tag{5.5}$$

Observations described by Webster and Young[35] concerning the running and gating system for the refrigerator housing casting shown in *Figure 5.8* illustrate the use of these equations. *Table 5.4* details the flow rates through possible choke areas. The volume of the casting is 1701 cc in the drag and 3213 cc in the cope. Using Equations (5.3) and (5.4) gives a filling time for the drag of 5 seconds and for the cope of 11.65 seconds. This compares with a measured time of 31 seconds. The calculations for the filling times did not include a friction factor, c. However, a value of 0.54 would have to be assumed for calculated and measured times to coincide. This is considered to be low. Closer examination of the pouring process revealed that the basin was only a quarter full during pouring and was only topped up at the end of the pour.

Table 5.4 shows that if the head of liquid is reduced to 1.25 cm in the basin, the top of the sprue becomes the choke. Under these conditions the filling time increases to 27.1 seconds and the c value is 0.9. It is likely that both the runner and the top of the sprue operated as the choke during filling with flow surging between them depending on the height of liquid in the basin. The gate area is greater than is required with a filling time of 31 seconds. The average flow rate is $4914/31 = 158.5\ \text{cm}^3\text{s}^{-1}$. The time to fill the drag is $1701/158.5 = 10.7$ seconds leaving a filling time for the cope of 20 seconds. Assuming a friction factor of

Table 5.4 Flow rates through possible choke areas in the refrigerator casing casting in *Figure 5.8* (after ref. 35)

Possible choke	Dimensions		Calculated flow rate $v = A\sqrt{2gh}$ cm^3/s
	Area cm^2	Liquid head cm	
Top of the	3.66	5.0	363
sprue		1.25	181
Narrow section of runner	1.88	16.5	338
Gates	4.18	16.5	752

0.7 gives a gate area of $1.6\ \text{cm}^2$ for a 20 seconds filling time. This example indicates the benefits of calculations in redesigning gating systems. Webster and Young demonstrate the use that can be made of computers for calculating gating system dimensions and illustrate an animated sequence that shows the pattern of flow during filling.

Examples of gating and running systems for vertically parted moulds are shown in *Figure 5.9*. System A is sprue/runner-controlled and system B is runner/gate-controlled. The second system offers a higher casting yield, but there is a greater possibility of slag entry to the casting because of the mould taper required on the runner. Dimensional calculations for these systems are similar to those described in the previous section.

The calculations for system C are more complicated because the impressions are on different levels. Conditions at each level must be analysed and the influence of pouring at one level on the conditions at other levels considered. Karsay has detailed[36] the calculation for a system C design which contains

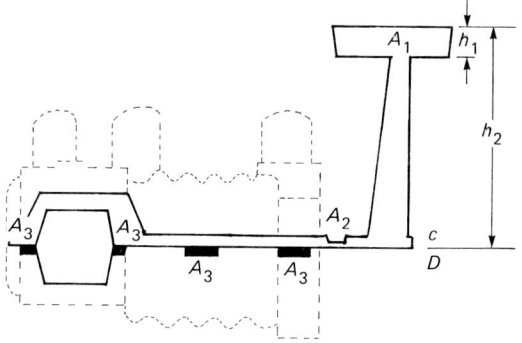

Figure 5.8 The refrigerator casting described by Webster and Young; h_1 = depth of metal in the pouring basin; h_2 = head of metal above the parting line; A_1 = cross-sectional area of the sprue top; A_2 = cross-sectional area of the runner choke; A_3 = cross-sectional area of the gates

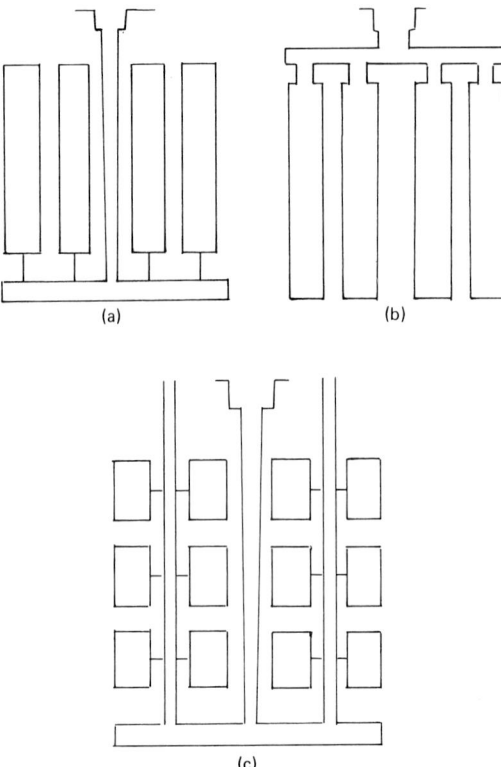

Figure 5.9 Examples of vertically parted moulding systems; (a) A sprue/runner controlled system; (b) A runner/gate controlled system; (c) A multilevel system

impressions at six levels with four impressions at each level. The system is reduced to that shown in *Figure 5.9* by considering three groups of impressions. The bottom group contains three levels, the middle group two levels and there is a single level in the top group.

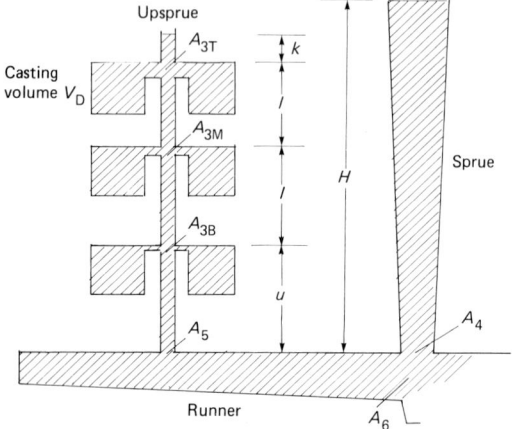

Figure 5.10 The bottom group in one half of a vertically parted gating system (after ref. 36)

The groups fill sequentially from the bottom, but the levels within a group fill simultaneously. The filling time, t, is the same for each casting. Initially the system is non-pressurized with the bottom of the sprue acting as the choke. However, as shown in *Figure 5.10* for a single group (the top level in the bottom group) the liquid rises in the upsprue to above the gate before filling commences. This pressurizes the system at this instant. If each casting is considered to be in the drag, Equation (5.3) can be used to calculate the gate area, A_{3T}, for the top level in the bottom group. With reference to *Figure 5.10*:

$$A_{3T} = \frac{V_D}{tc\sqrt{2g}\sqrt{k}}$$

The head of liquid increases by l, the interlevel distance for each lower level in the group. Thus the gate area A_{3M} for the middle level in the bottom group is:

$$A_{3M} = \frac{V_D}{tc\sqrt{2g}\sqrt{k+l}}$$

The gate areas calculated for the three levels in the bottom group in Karsay's analysis are given in *Table 5.5*. The area of the sprue base, A_4, can be calculated because the rate of liquid transfer through the base must equal the rate of mould filling of all the cavities in the bottom group. The rate of mould filling in a group is $m \times n \times V/t$ where m is the number of castings in one level, n is the number of levels in the group, V is the volume of the mould and t is the filling time. The velocity of flow through the bottom of the sprue, v_4, is governed by the head of liquid which is the difference in liquid heights in the sprue and upsprue. Consequently,

Table 5.5 Calculation of the various gate, sprue and runner areas in *Figure 5.10* (after ref. 36)

The example considers a mould containing four cavities at each level and six levels in three groups. The bottom group contains three levels, the middle group contains two levels and the top group contains one level. The system contains a central sprue and two upsprues. The casting volume = 38.6 cm^3. The head of liquid = 100 cm, $t = 4 \text{ s}$, $c = 0.5$, $k = 3.8 \text{ cm}$, $l = 12.7 \text{ cm}$ and $u = 5 \text{ cm}$

Calculations

Gate location		Gate area cm²
Top group	Top level	1.1
Middle group	Top level	0.67
	Bottom level	0.32
Bottom group	Top level, A_{3T}	0.56
	Middle level, A_{3M}	0.27
	Bottom level, A_{3B}	0.20
Area at bottom of the sprue, A_4		1.61 cm^2
Area of the upsprue, A_5		0.80 cm^2
Runner cross sectional area, A_6		3.20 cm^2

$$v_4 = c\sqrt{2g} \times \sqrt{H - [u + k + (x - 1)l]}$$

where x is the total number of levels from the bottom to and including the uppermost level which is being filled. H, u and k are defined in *Figure 5.10*. The flow rate, F_4, is given by

$$F_4 = v_4 A_4$$

Hence:

$$A_4 c\sqrt{2g} \times \sqrt{H - [u + k + (x - 1)l]} = \frac{mnV}{t}$$

Therefore:

$$A_4 = \frac{mnV}{ct\sqrt{2g} \times \sqrt{H - [u + k + (x - 1)l]}}$$

The value calculated for this area is given in *Table 5.5*. The next step is the calculation of the remaining gate cross-sectional areas group by group and inside the groups, level by level. The rate of liquid transfer through the sprue bottom is reduced as each group is filled because the liquid level in the upsprue is increased by nl after each group is filled. Thus:

$$A_4 c\sqrt{2g} \times \sqrt{H - [u + k + (x - 1)l]} = A_3 mnc\sqrt{2g}\sqrt{k}$$

Hence the area of the gate at the top level in the middle group is:

$$A_3 = \frac{A_4\sqrt{H - [u + k + (x - 1)l]}}{mn\sqrt{k}}$$

The same equation is used to calculate the gate areas at lower levels in the group with the last parameter in the denominator increased by l for each lower level. Gate areas calculated for the middle and top groups using this equation are given in *Table 5.5*. The sprue and runner are tapered as shown in *Figure 5.10*. Although these calculations are tedious, they are

necessary because simultaneous filling at all levels without correct gate area selection to counter the much reduced gate velocity at higher levels leads to misruns in the upper half of the mould.

A pressurized system is usually advocated for the Inmould technique as shown in Chapter 2, *Figure 2.37*. However, the non-pressurized system in *Figure 5.11* is an alternative. There is a primary choke leading into the reaction chamber and a secondary choke. The area of the secondary choke is less than that of the primary choke. It is calculated to satisfy the requirements of the Inmould technique. This system mixes the first, possibly undertreated, iron with treated iron in the runner before entry into the casting.

The principles and practice of feeding cast irons

Feeding is the practice of providing sufficient liquid metal to compensate for volume changes that occur during the cooling and solidification of liquid iron in the mould cavity. The design of a feeder system involves calculation of feeder size and shape and choice of number of feeders and their location in order to ensure that directional freezing proceeds from casting extremities along identified feeding paths

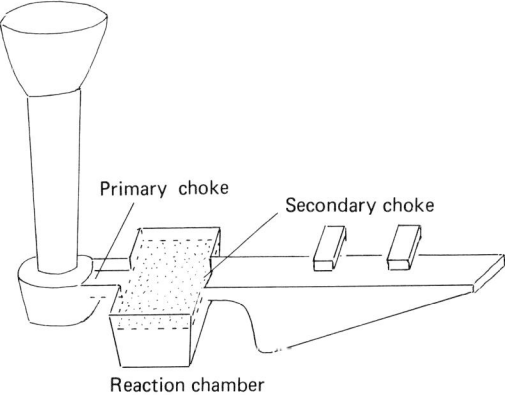

Figure 5.11 A sprue/runner control system suitable for Inmould Casting

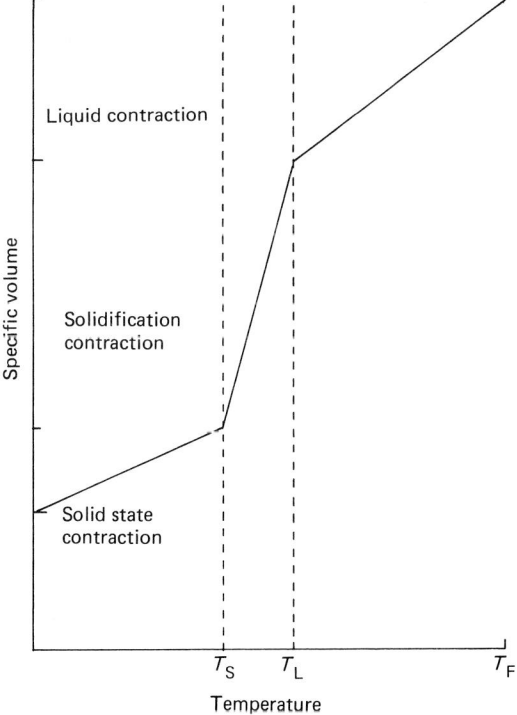

Figure 5.12 The volume–temperature characteristic of a white iron; T_F = temperature of the liquid after mould filling; T_L = liquidus temperature; T_S = solidus temperature

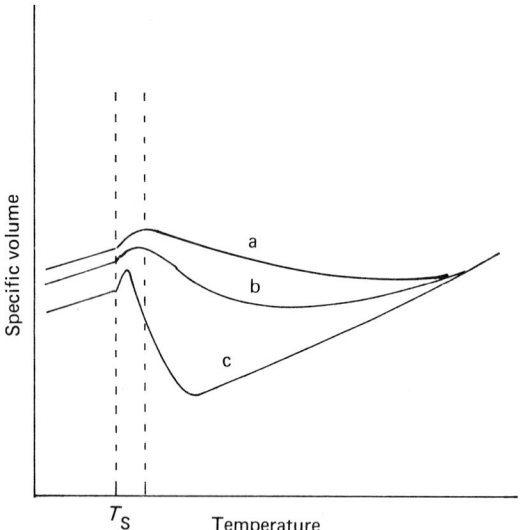

Figure 5.13 The different patterns of volume change in grey irons

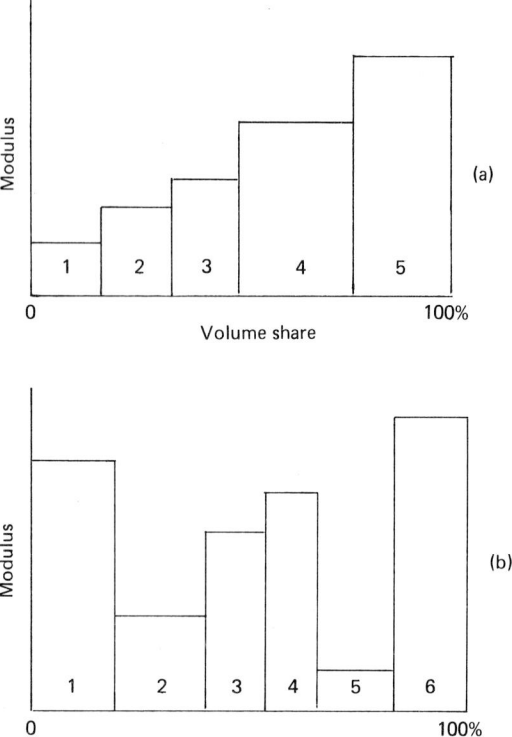

Figure 5.14 Two examples of modulus – volume share diagrams

through the feeder connection and into the feeder itself. *Figure 5.12* shows the pattern of volume change displayed by many casting alloys including white irons. The liquid shrinkage is approximately 1 vol % per 55 °C and the solidification shrinkage lies between 4 and 5.5 vol %, depending on the white iron composition. This pattern of volume change reflects conventional behaviour and the practice is termed conventional feeding.

Figure 5.13 shows the pattern of volume change displayed by grey irons. This differs from the pattern in *Figure 5.12* in that it consists of liquid contraction, expansion and secondary contraction followed by solid state contraction. The solidification shrinkage in grey irons depends on the type and quality of the iron and the solidification conditions. This coupled with the fact that the liquid expansion can cause mould dilation, which creates an extra volume to be fed, makes the application of conventional feeding practice to flake and spheroidal irons a more difficult and uncertain process. However, under certain conditions, the expansion can be used to compensate for the secondary contraction (applied feeding techniques) and so reduce the amount of feed metal required.

Conventional feeding

The first step in feeder design is to determine the freezing order in the casting and to establish the freezing path(s). This is often decided on the basis of experience but is the first step in computer programs for feeding[19]. One of the exercises in geometrical modelling is the division of the casting into a series of subshapes, either in two or three dimensions, and the calculation of the section modulus for each subshape. Section modulus is defined as the casting volume divided by its surface area, which transfers heat out of the casting. The freezing order calculation is based on Chvorinov's rule, which states that the solidification time of a casting or portion of a casting is a function of its section modulus.

The section modulus–cumulative volume diagram in *Figure 5.14a* is typical of a step bar in which freezing is unidirectional along the path 12345. A feeder would be attached to subshape 5 and its modulus must exceed that of subshape 5. The diagram in *Figure 5.14b* describes a casting in which freezing is not unidirectional. The freezing order is 523416. Several freezing paths are identified, 21, 56, 234 and 54. Consequently, feeders must be placed on subshapes 1, 4 and 6. *Figure 5.15* shows a casting divided into subshapes and the section modulus–cumulative volume diagram including that of the feeder.

The feeder must be large enough to solidify last and to deliver sufficient liquid iron to compensate for the total solidification shrinkage. There isn't a universal criterion for determining feeder size even for steel which is the most examined casting alloy. Many of the criteria used for cast iron are based on experience with steel. They include:

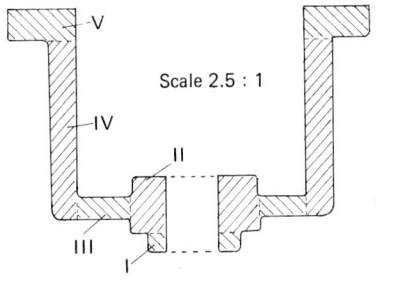

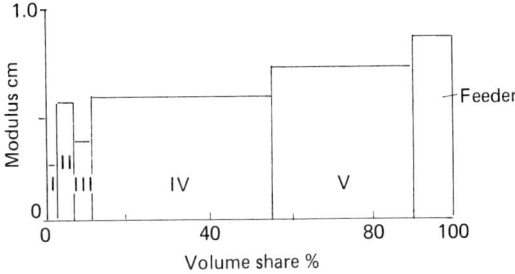

Figure 5.15 A hub casing casting and its modulus – volume share diagram

1. *Empirical relationships.* These are the result of experience gained with specific types of castings and mould materials. For example, it has been suggested[37] for top and side feeders for spheroidal iron in green sand moulds:

feeder diameter = casting section + 5 cm

feeder height = 1.5 – 2 × feeder diameter.

2. *Shape factor.* The easily calculated casting shape factor $(L + W)/T$ (L = length, W = width and T = thickness) is used to represent casting geometry instead of modulus. Empirical graphs relating the shape factor and V_c/V_f (V_c = casting volume, V_f = feeder volume) are used to determine the required V_f knowing V_c. This method was developed for steel[38] and extended to cast iron[39]. It is the basis of a recent computer program for malleable and spheroidal irons[40].

3. *Modulus concept.* The basis of this popular method is that the feeder modulus should exceed that of the casting. A single and constant factor of 1.2 has been found successful for steels. However, the unconventional shrinkage behaviour of grey irons means that it is not possible to define a single factor that describes all combinations of mould rigidity, casting section/modulus and metallurgical condition. A wide band of M_f/M_c ratios has been found for spheroidal irons[41]. One disadvantage of the modulus concept is that a feeder designed on the basis of cooling rate considerations for a casting with a high surface area/volume ratio may not have sufficient volume to satisfy the

casting's liquid and solidification shrinkage. This problem has been addressed recently by Ruddle[42].

4. *Volume controlled feeding.* Gough and Clifford[43] have shown that when composition, mould material and pouring temperature are constant and only casting weight and design vary, there is a well defined relationship between the ratio V_f/V_c and soundness for spheroidal iron castings of a size typical of the small jobbing castings market.

5. *Geometric risering technique.* This method was developed by Heine and coworkers[44-51] and is used extensively in the USA. The method is computerized[19,52,53] (RISERING Program) and when used in tandem with the SWIFT (Solidification Wavefronts in Foundry Technology) Program goes a long way to fulfilling Ruddle's 1971 prediction. The geometric RISERING Program commences by using geometric modelling to determine casting volume, surface area and section modulus to define the freezing order, freezing path(s) and, with a knowledge of freezing distances, the number of feeders required. The program proceeds to calculate the feeder dimensions.

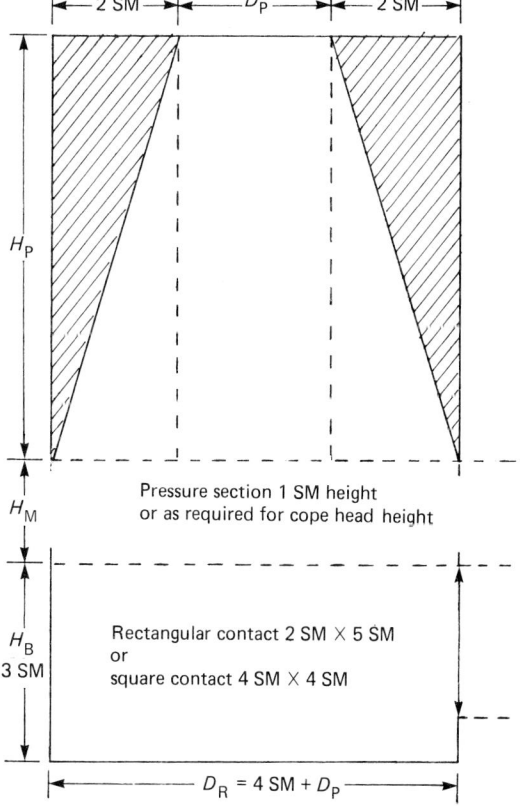

Figure 5.16 Dimensions of a feeder and connection for a full size cylindrical or tapered feeder geometrically modelled; $2 \times SM$ is the effective section thickness of the feeding path (after ref. 53)'

The ideal feeder shape in cooling modulus terms is a sphere, but moulding and pouring difficulties mean that cylindrical or tapered cylindrical feeders are used. The design of a conventional feeding system provides a reservoir of feed liquid in the centre of the feeder and a continuous feeding path. Both remain liquid and open to atmospheric pressure sufficiently long for feeding to occur progressively from the extremities of the feeding paths through the feeder connection. This leaves a well formed pipe in the feeder on the completion of solidification. To achieve this, the top or side feeder is designed with three sections as shown in *Figure 5.16*.

The feeding section of height H_p provides the feed metal from the central pipe cavity of diameter D_p and of volume equal to the feed metal required to satisfy the shrinkage volume. This information is input into the program. The pressure section of height H_m is chosen so as to ensure that as the liquid level in the pipe drops during solidification, there remains a sufficient head of liquid to move liquid to the highest level to be fed in the cope. The base section is designed to ensure that it does not freeze until all the feed metal has passed through the connection. Dimensions of these sections found satisfactory for malleable irons are indicated in *Figure 5.16*. A H_p/D_p ratio of 2.5 has been found to give optimum yield for white irons.

The computer program produces several alternative feeder D_p and H_p values. A specific combination is chosen taking into account mould requirements. The tapered cylindrical feeder increases the yield compared to a straight walled cylinder. Yield can be increased significantly with use of feeding aids. Chills and padding can be used to improve directional freezing. Hot topping compounds and feeder sleeves

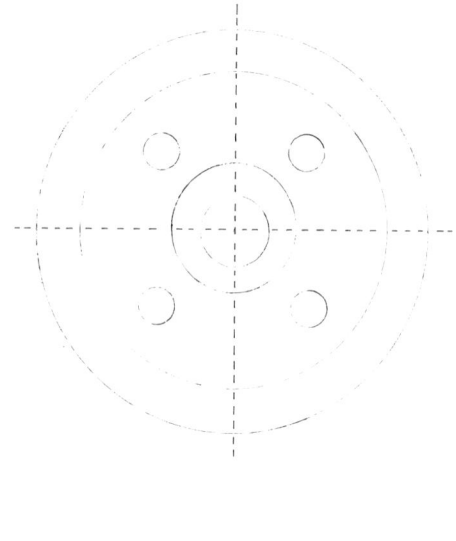

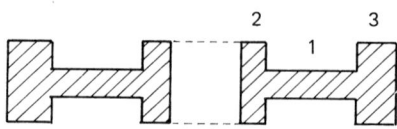

Figure 5.18 Gear blank casting; scale 8.3:1. Section modulus values define the freezing sequence. 1, modulus 1.85 cm; 2, modulus 2.08 cm; 3, modulus 2.25 cm; Casting modulus 2.15 cm (after ref. 19)

can be employed to increase the efficiency of feeders. These benefits are illustrated in *Figure 5.17*.

The SWIFT program calculates section moduli and feeder dimensions. It also provides a graphic display of solidification wavefronts, which highlights areas of potential shrinkage. Temperature gradients across and between sections are predicted for directional solidification monitoring. Mould dimensions can be superimposed round the casting and allowance made for superheat and mould material variations. A SWIFT program analysis of the gear casting shown in *Figure 5.18* illustrates the use of this technique. Although this example is detailed for steel, it is representative of a conventional feeding system. The section modulus and freezing order are shown in *Figure 5.18*.

The wavefront analysis for the gear blank is shown in *Figure 5.19*. The cross identifies the last area to freeze and any area of potential shrinkage. The effect of attaching a feeder in three different positions on one of the flanges is shown in *Figure 5.20*. The top feeder is preferred from a yield consideration, though it can present moulding problems. The second position in *Figure 5.20* delays solidification longer than when there is no feeder attached and shrinkage will occur in the flange. The side feeder attached so

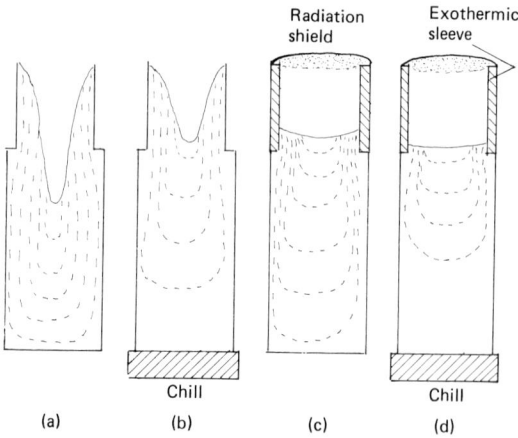

Figure 5.17 Schematic illustration of the effect of insulating sleeves and chills on the efficiency of a top feeder; (a) Open feeder; (b) Open feeder with chill; (c) Feeder surrounded by insulating or exothermic material; (d) Same as (c) but with a chill

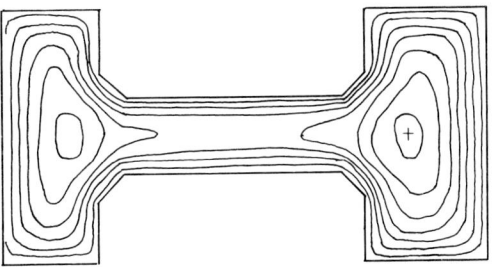

Figure 5.19 A wavefront analysis for one half of the gear blank shown in *Figure 5.18*. The cross indicates the last area to solidity (after ref. 19)

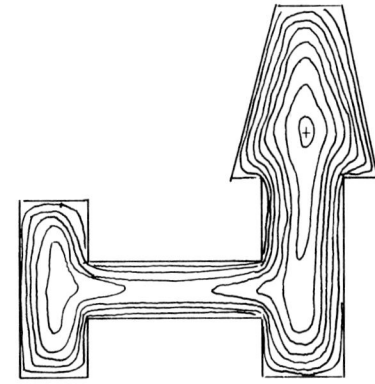

that the casting is in the drag, as shown in the lowest diagram in *Figure 5.20*, produces favourable directional solidification. The influence of using a tapered feeder with the two acceptable systems in *Figure 5.20* is shown in *Figure 5.21*. In both cases the final freezing position is moved closer to the feeder contact location.

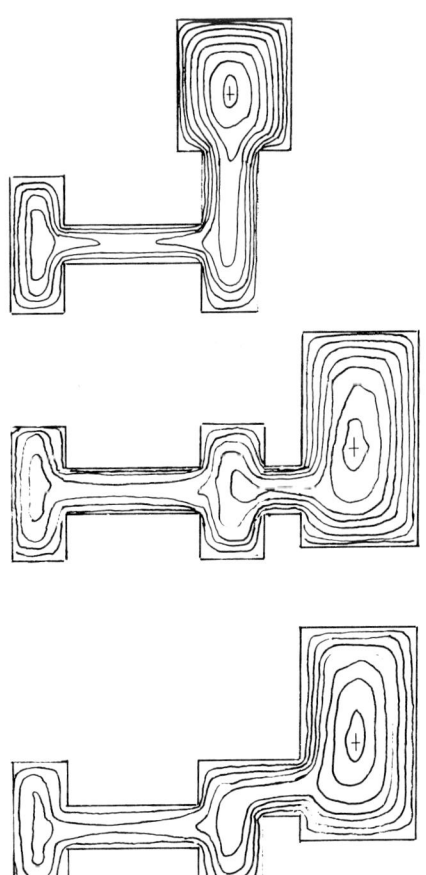

Figure 5.20 The effect on the solidification wavefronts shown in *Figure 5.19* of attaching a top feeder and a side feeder in different positions (after ref. 19)

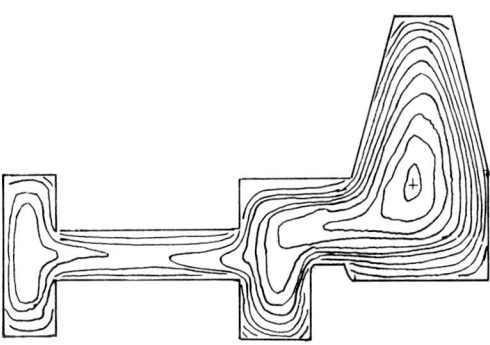

Figure 5.21 The change in solidification wavefronts when a tapered feeder is used in the two acceptable feeding positions in *Figure 5.20* (after ref. 19)

Unconventional feeding

Several factors combine to make designing of feeding systems for flake, compacted and spheroidal irons more difficult than for steels and malleable irons. Irrespective of the volume and temperature characteristics, the mode of solidification influences the ease with which feed metal can pass from the feeder to the casting. This effect has been measured in terms of the centreline feeding resistance (the ratio of the time during which solid is forming at the centre of the casting to the total solidification time).

Progressive solidification from the mould wall (skin formation) favours directional solidification and promotes easier feeding. Cast irons do not behave in this way. Increasing in the sequence white, flake, compacted, spheroidal, solidification is less directional and more pasty. The pasty mode of solidification is characterized by a higher centreline feeding resistance and greater difficulty with feeding.

The mode of solidification is influenced by cooling rate; by the type of mould for an iron of fixed composition and size; by the casting size for an iron of fixed composition poured into a certain type of mould. Both flake and spheroidal irons with small modulus, < 0.6 cm, cool fast enough to approach a skin-forming mode of solidification. As the modulus increases, solidification becomes more pasty and feeding difficulties increase. Inoculation of a flake iron changes its solidification mode closer to that of a spheroidal iron and, in this respect, inoculation makes feeding more difficult.

The second factor influencing feeder design is the pattern of volume changes. As shown in *Figure 5.13* unconventional behaviour consists of liquid contraction which has been measured as being between 0.3 and 2.33 vol % per 100 °C followed by an expansion. This expansion is attributed to the density difference between C-containing liquid and graphite or, alternatively, according to the gas bubble theory, precipitation of numerous minute CO bubbles. A secondary contraction occurs after the expansion period and lasts until the entire casting is solid. The secondary shrinkage amounts to 0.1 to 0.5 vol % in good quality spheroidal irons, but can exceed 1.0 vol % in low C.E.V., high strength flake irons. Strengthening of grey flake irons is preferably achieved through alloying with respect to feeder design.

The rate of liquid shrinkage, the temperature at which liquid shrinkage changes into liquid expansion, the total specific expansion and total secondary shrinkage are all variables influenced by the rate of cooling, chemical composition and metallurgical quality. Increasing the rate of cooling has the undesirable effect of promoting the type c curve. However, increased cooling rate also favours a more skin-forming mode of solidification which compensates for this effect. Chilling also increases the cell or spheroid count. This improves the metallurgical quality which favours the type a curve. These factors account for the overall beneficial effect that chilling can have on feeding. However, careful use of chills is necessary to ensure that surface carbides do not form.

Improving the metallurgical quality by inoculation favours the type a curve. However, inoculation increases the tendency for the pasty mode of solidification. The balance of these effects depends on the iron type. Promotion of the pasty mode of freezing can predominate in flake irons and make feeding more difficult. However, spheroidal irons (depending on modulus) already solidify in the pasty mode. Increased inoculation can decrease the tendency to form secondary shrinkage defects. Increasing the C.E.V., particularly the Si content, favours the type a curve. Elements that decrease the graphitization potential increase secondary shrinkage. Intercellular carbides are particularly detrimental in this respect. They can negate the effect of parameters that promote a type a curve.

The third factor of significance is mould dilation due to heating and due to the expansion period in the volume shrinkage pattern. Volume changes in the mould due to heating are significant immediately after pouring, but are readily compensated for by the feeding system. The amount of dilation depends on mould material but is usually less than 0.9 vol % in a green sand mould. It can be increased by moisture and sand fines and decreased by coal additions and sawdust.

Expansion effects associated with graphite formation are of concern because they occur later in the solidification sequence. This expansion transmits pressure to the mould wall. The amount of mould dilation depends on the mould rigidity (this increases in the sequence green sand, CO_2 silicate, cold-set resin, cement) and the thickness of the skin of the casting. Spheroidal irons solidify in the pasty mode with a thin skin that offers little resistance to the expansion pressure. Flake irons develop a thicker skin, which not only protects the mould from the expansion pressure, but provides a means for the expansion pressure to be transmitted to the liquid phase. Spheroidal irons with a modulus exceeding 1 cm produce sufficient pressure to exceed the compression strength of green sand moulds ($\sim 0.6 \, N \, mm^{-2}$), whereas a modulus exceeding 1.5 cm is required before mould deformation with flake irons.

These factors complicate considerations when applying conventional feeding practice to grey cast irons. Design changes are necessary to satisfy the different pattern of volume change and the increased difficulty in promoting directional solidification. The casting yield is often low with conventional feeding practice. The expansion period offers the designer the possibility of using alternative 'applied' feeding methods, in which the expansion period is used to compensate the secondary shrinkage. Although these methods have a much higher yield, stringent conditions must be satisfied for their successful operation.

Heine has discussed the application of the geometric risering technique to grey irons[50,51] and detailed the changes necessary for feeder conversion from malleable to spheroidal iron[54]. Overall design principles are similar to those for conventional feeding. The feed metal requirement is not as easily defined. It varies between 2 and 8 vol % depending on the casting section thickness and the mould type and characteristics. A large percentage of the feed metal required is to satisfy mould dilation when green sand or shell moulds are used.

Difficulty with piping is often experienced with spheroidal irons. This is partly due to the pasty mode of freezing and partly due to the sprue providing the early feed metal in a normal gating system. The feeder will not pipe until the sprue itself, or the connections from the runner to the feeder, solidify. When the feeder fails to pipe, it often shrinks into the casting at the feeder connection.

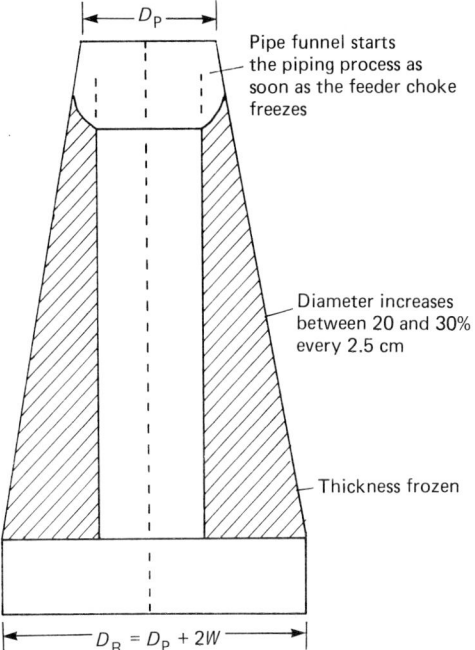

Figure 5.22 The design features of tall tapered feeders (after ref. 51)

Design modifications that promote early freezing of ingates and the use of tall feeders minimize sprue feeding and promote early piping. Thin and wide ingates are advocated. An increased number of such gates may be necessary to meet the pouring time required. Cylindrical feeders with H_p/D_p in the range 4–8 are used. Tapered feeders are more efficient. The taper is designed to retain the pipe cavity as the liquid feed level drops and the thickness frozen increases. The thickness frozen at the top of the pipe is almost zero if piping begins early. The top diameter must not solidify before the sprue and, therefore, should be D_p or 10% larger than the sprue diameter. The thickness frozen at the bottom should equal the effective section of the casting fed, W.

The feeder taper obtained by joining the top and bottom diameters depends on the H_p/D_p ratio. The smaller this ratio, the greater the yield. However, a feeder with too small a ratio will not feed casting sections in the cope and may not generate the pressure head and solidification gradients required for piping. On the other hand, if the ratio is too large, there is the possibility of the feeder freezing across the pipe. Recommended dimensions are shown in *Figure 5.22*.

Once the feeder has been designed for correct piping it is necessary to design the base and feeder connection. The design principles are the same as for conventional feeders, but the solidification mode of grey irons means that the size of the feeder connection can be reduced. Suitable dimensions are 0.5 to 1 SM (section modulus) by 4 SM for rectangular

connections and 0.5 to 1 SM by 0.5 to 1 SM for square connections. These reductions may be achieved by notching the larger connections designed for conventional feeders. Notching facilitates fettling. Selection of the correct connection size is a significant factor in the incidence of shrinkage on the casting side of the connection. Too small a contact permits shrinkage in the casting just inside the contact area. Shrinkage can occur in a similar position if there is excessive metal flow through the contact. If this occurs and the feeder is piping well, the hot spot can be removed by introducing metal directly into the casting using knife gates from the runner.

Applied feeding methods are listed in *Figure 5.23*. They utilize the expansion associated with graphite formation to compensate for secondary shrinkage. The design for the first of these methods, directly applied feeding, must provide feed metal to satisfy the liquid contraction. Once this has been achieved, the feeder connnection must freeze before the liquid expansion period. On further cooling, the expansion deforms the mould. Provided this deformation is elastic, the mould can spring back as the pressure against the wall decreases. The spring back is sufficient to maintain the remaining liquid above atmospheric pressure and prevent porosity.

Factors that influence the amount of expansion and mould dilation have been discussed. For green sand moulds this technique is applicable to relatively thin castings with an arbitrary modulus limit of 0.4 cm for spheroidal irons and 0.6 cm for flake irons. This limit increases as the mould rigidity increases.

The feeder contact design begins with the identification of the significant modulus, M_s, from the modulus-cumulative volume diagram. The significant segment is defined as that with the smallest modulus, but whose freezing expansion is capable of keeping all the remaining liquid above atmospheric pressure in the heavier segments until they commence freezing and expanding. The feeder contact modulus, M_n, is calculated in terms of the significant modulus, M_s, and pouring temperature by equating the heat extracted per unit surface area of the significant section to the start of liquid expansion with the heat extracted to solidify the feeder contact. The resulting calibration curve is shown in *Figure 5.24*.

Care must be exercised to ensure that the freezing of the connection is not delayed by heat from the casting. To avoid this, the connection should be long. It should be at least four times the shorter dimension of the connection. The connection should be horizontal to avoid convection heating. This means that side feeders are usually used.

Applied pressure feeding is popular with thin castings. If the significant modulus is only 0.4 cm, the contact modulus is 0.36 cm. This can be satisfied with a contact 3×0.9 cm and more than 3.6 cm long. This gate is capable of acting both as a runner/gate choke and a feeder connection. Thin castings may be

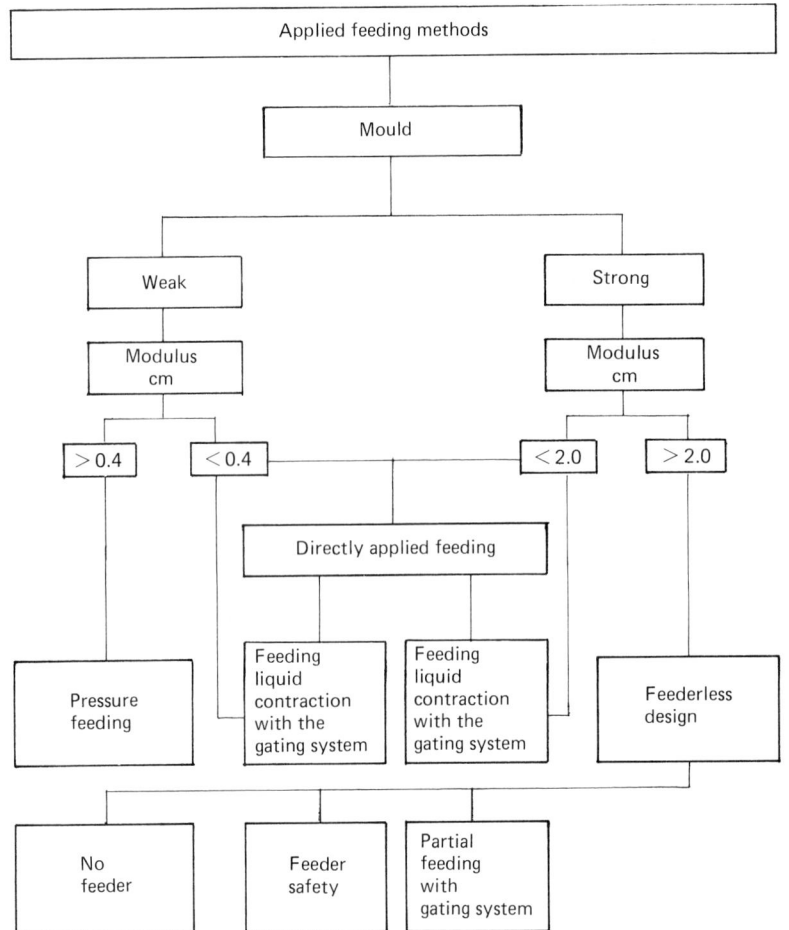

Figure 5.23 Applied pressure feeding systems; modulus values quoted apply to spheroidal irons

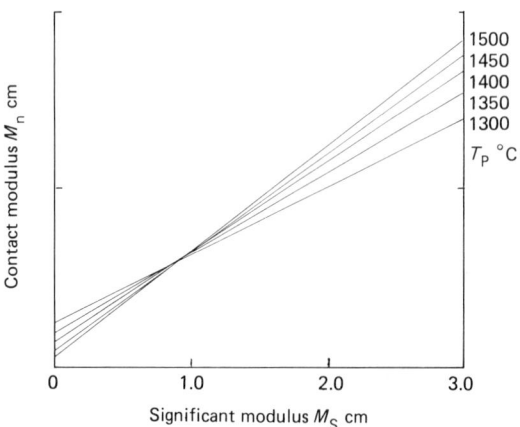

Figure 5.24 The relationship between contact modulus, M_n, and significant modulus, M_s, as a function of pouring temperature, T_p, for directly applied feeding.
$M_n = M_s \times (T_p - 1150/T_p - 900)$ where T_p is in °C
(after ref. 36)

pressure fed through the gating system in this way. Thin and light castings lend themselves to multiple castings in one mould for which sprue/runner control gives the best yield. This system separates the function of choke and feeder connection and allows the method to be used with a significant modulus up to 1 cm.

Difficulties encountered in using the gating system for pressure feeding have been discussed by Heine[50]. A common cause of shrinkage is too cold an iron after pouring. The calibration in *Figure 5.24* cites pouring temperature, but is based on temperature after pouring. Temperature after pouring may be lower than pouring temperature with thin castings. If this occurs, the contact modulus calculated using the pouring temperature and *Figure 5.24* will be smaller than required. The connection will freeze early and the expansion pressure will have to satisfy liquid contraction. It may not be great enough to compensate secondary shrinkage. Defects often take the form of draws or depressions on the cope surface. Directly applied feeding is successful provided the modulus is small, the mould strong and adequate control can be maintained over pouring temperature. The method offers good casting yield, freedom in system design and minimum fettling costs.

Feederless design offers maximum yield, but stringent conditions must be satisfied for its success.

1. The pouring temperature must be less than 1350 °C to ensure that liquid expansion commences immediately after pouring.
2. The modulus of the major part of the casting must be greater than 2.5 cm for the same reason.
3. The metallurgical quality must be good enough for the iron to display the type a curve characteristics shown in *Figure 5.13*. This is promoted by a high C.E.V. (~4.2), low Mn and Cr, good inoculating practice.
4. The gate design must promote freezing immediately after, but not during, pouring. Suitable gates are 13–16 mm thick and rectangular, with width approximately four times the thickness and length exceeding the width. Multiply gates may be required to satisfy pouring time. They will prevent the development of hot spots.
5. Fully rigid moulds are required to prevent plastic deformation of the mould by the high liquid expansion pressure developed. The cope and drag must be held firmly together by bolting or clamping.
6. The runner system should be designed to equalize temperature gradients in the mould during solidification. A fast pour is desirable to prevent temperature loss during pouring and to avoid heating of the mould top by radiation. A fast pour is facilitated by vents that fully penetrate the cope.

Although there are many examples of the successful application of feederless design[36], it is restricted to a limited class of casting design and is extremely demanding on metallurgical and mould quality control. Some dilation is experienced with all but cement-bonded moulds. The selection of pouring temperature must consider other factors in addition to the reduction of liquid contraction. Failure to satisfy conditions 3 and 4 are the most common cause of internal porosity. Surface depressions often result when the other conditions are not fulfilled. They may be avoided by locating a small blind feeder (volume ~0.02 V_c) on top of the casting.

Blind feeders are used in the third applied feeding technique, pressure relief feeding. This approach is used when conditions for the other applied feeding techniques cannot be satisfied. This is the case with a combination of weak mould and large casting modulus. (see *Figure 5.23*). Pressure relief feeding is used to produce porosity-free, medium to heavy grey iron castings in green and shell sand moulds. The principle of the technique is to use a blind feeder(s) to relieve liquid expansion pressure that would otherwise deform the mould plastically but to leave sufficient liquid pressure to ensure the liquid remains above

atmospheric pressure during the final stages of solidification. This is achieved by filling the mould cavity and blind feeder complex using a gating system designed for gate closure immediately after pouring. Transport of liquid from the feeder into the casting satisfies the liquid shrinkage that occurs as the casting and feeder cool.

Feeder placement must ensure that the required feed metal (2–3 vol % depending on modulus) is in a portion of the feeder that is above the topmost level of the casting. Top feeders are preferable for this reason. The void created in the blind feeder is used to relieve liquid pressure during the expansion period by liquid movement from the casting into the feeder. Pressure relief ceases once the feeder is refilled. Consequently the size of the void must be controlled to produce the desired level of liquid pressure for the final stages of solidification.

The varied and often unpredictable nature of the combined influences of the many factors that influence the volume temperature characteristics of grey irons made the design of a pressure relief system a matter of trial and error in the past. However, Karsay has recently detailed[36] a design procedure. The analysis designs for maximum casting yield and minimum feeder removal costs. It is based on the criterion that no further liquid transfer is required once the desired liquid pressure has been achieved and the feeder and feeder contact should freeze at this instant in time.

Feeder dimensions are calculated by equating expressions based on Chvorinov's equation for the time necessary to reach the range of safe expansion

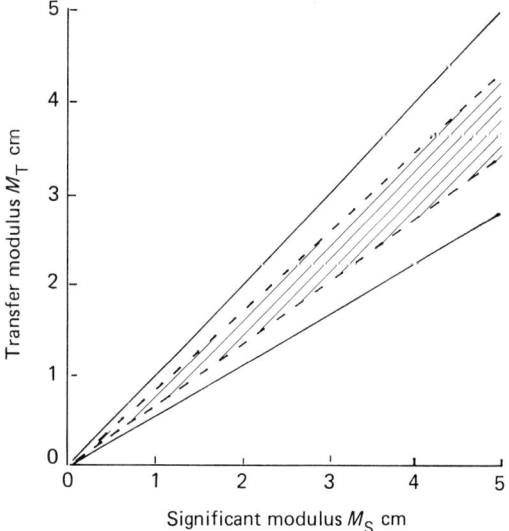

Figure 5.25 Relationship between modulus necessary for liquid transfer, M_T and the significant modulus, M_s. Low values in the shaded area can be used with irons of high metallurgical quality (after ref. 36)

pressure and that of liquid permeability in the feeder system. This results in an expression of the form $M_T = f M_S$ shown in *Figure 5.25*. M_T is the modulus of the feeding system corresponding to the limit of liquid transfer considered to occur when the system is 75% solid. M_S is the modulus of the significant (heavy) section. The factor f is given by:

$$f = \frac{1 + X}{1.75}$$

where X is the fraction of solid present when the desired liquid expansion pressure is attained in the significant section. The value of X will vary from 0 (the iron is still completely liquid when all transfer has taken place) to 0.75 (liquid can no longer permeate in the casting/feeder complex). Thus f varies from 0.57 to 1.0 depending on the solidification characteristics of the iron, which depend upon the iron type and metallurgical quality. The feeder modulus, M_f, should equal the transfer modulus M_T.

The shaded area in *Figure 5.25* identifies the working range. Low M_T values should be selected for irons of good metallurgical quality and vice versa. Selection of a higher modulus increases design safety but increases production costs. The feeder contact should freeze at the same time as the feeder. The acting contact modulus, M_n, should equal the transfer modulus. However, the absence of cooling surfaces at the casting and feeder junctions means that the acting modulus is 1.5 to 2 times the modulus of a body of the same size and shape cooled on all sides. Consequently, the contact modulus can be reduced to 0.6 M_T. M_n depends on iron quality through the dependence of M_T and practice has shown that an M_n value in the range 0.35 to 0.55 M_S is suitable.

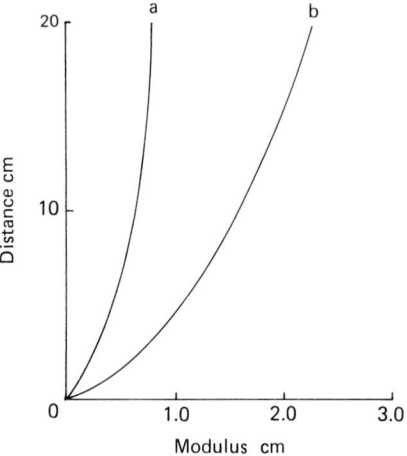

Figure 5.26 Liquid transfer distances as a function of modulus and iron quality in spheroidal iron plates: a, good metallurgical quality; b, poor metallurgical quality (after ref. 36)

In contrast to directly applied feeding, the feeder contact should be as short as moulding permits. It is possible to make the feeder contact to a segment that is thinner than the significant one, provided the modulus of the segment separating the significant one from the feeder contact equals or exceeds M_T. Liquid must be able to transfer from casting to feeder.

Factors that influence the limit of liquid transfer include iron type, modulus, iron quality and geometrical shape. Heine[50] has discussed these factors, in particular, limitations due to changes in geometry to bar and plate shapes. *Figure 5.26* shows transfer distances in plate-shaped castings. In common with other feeding methods, a constant pouring temperature is desirable. However, unlike the previous method, a high pouring temperature should be used (1370 to 1430 °C). A sound feeder usually indicates a sound casting in a pressure relief feeding system. However, this is not the case if delayed gate freezing allows the sprue to feed the liquid contraction. If this occurs, the void formed in the feeder will be too small to allow the pressure relief required. The mould will dilate and the liquid pressure will be reduced below that required to prevent secondary shrinkage. Under these conditions the feeder will be sound but the casting porous.

The design procedure described by Karsay can be used with open feeders. However, if the value of M_T selected for an open feeder is too large and results in the feeder and feeder contact being permeable to the end of the expansion period, all the expansion pressure will be used in transporting liquid from the casting to the feeder and a secondary shrinkage defect will occur in the casting. Blind feeders avoid this possibility by limiting the amount of liquid transport possible. If the selected M_T is too small, the feeder freezes prematurely and a deep shrinkage pipe occurs in the feeder. The liquid expansion pressure dilates the mould, resulting in an enlarged, porous casting.

Figure 5.27 shows a hub wheel casting used to illustrate[56] an instance where a change from conventional feeding, which resulted in unacceptable porosity, to pressure relief feeding eliminated shrinkage defects. The component was made in good quality spheroidal iron to specification ASTM A536-80 grade 65-45-12 or BS 2789 (1985) grade 420/12.

The following steps were used in following the design procedure for a pressurized running system and a pressure relief feeding system. The casting was divided into three segments with modulus values $M_1 = 1.25$ cm, $M_2 = 2.2$ cm and $M_3 = 1.95$ cm. The modulus exceeds 1 cm. Because green sand moulds were used, pressure relief feeding was the applied feeding technique selected. The significant modulus was 2.2 cm. A transfer modulus of 1.25 cm was selected from *Figure 5.24*. A blind feeder of standard shape and with the dimensions shown in *Figure 5.27* was used to satisfy a modulus of 1.25 cm. The feeder was connected to segment 1. The contact modulus

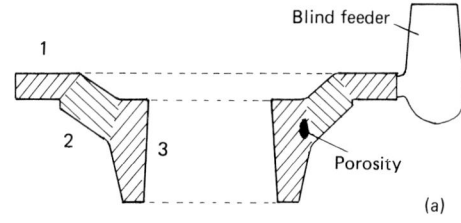

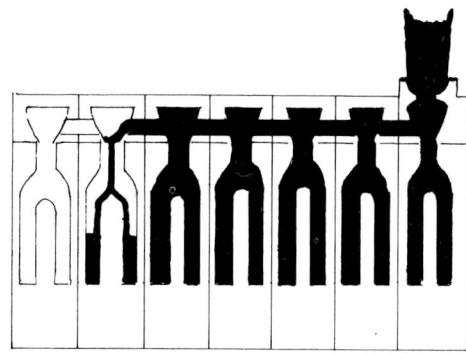

Figure 5.28 The pouring and feeding sequence in the 'H' Process

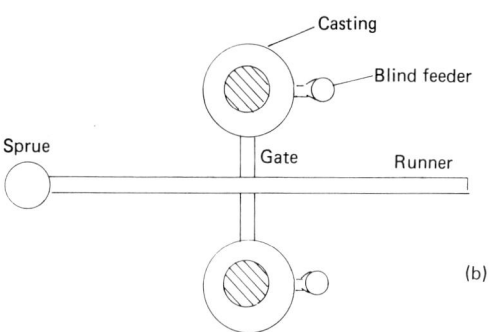

Figure 5.27 (a) Cross-sectional view of a hub casting and attached blind feeder, scale 5:1; (b) Schematic representation of the running and feeding system (after ref. 56)

used was $0.67\ M_T$ and this was satisfied with a rectangular contact 5 by 2.5 cm. Sufficient feed metal (3 vol %) was present above the top of the casting to satisfy liquid contraction. The choke cross-sectional area required in the running system was $4.2\ cm^2$. This was satisfied with one gate per casting and of dimensions 4 by 1 cm. The gate modulus was 0.4 cm, which was approximately one third of the contact modulus. It was small enough to ensure gate closure soon after pouring. In keeping with the design principles of a pressurized running system, the runner bar cross-sectional area varied from 2 to 4 times the choke area and the height was twice the width, 7 by 3.5 cm at the mid point. A sprue diameter of 4.35 cm filling two castings satisfies Equation (5.2). The volume of two castings was 19 680 cc, feeder and contact 1050 cc, sprue and basin 1020 cc, runner 820 cc, gate 16 cc, sprue well 164 cc. This gives a pattern yield of 86.5 %.

The controlled pour and flow feed of the 'H' process represented a breakthrough in the gating and feeding of small castings of the highest integrity. Cores are produced which carry a half impression of the casting on each side and on one or both sides, the ingate or runner. The cores are assembled vertically. They are located each on the next one by complete location pieces and clamped together by bolts. As shown in *Figure 5.28* the holes made by the runner/feeder section provide a means of pouring a complete line of castings.

The design of the system balances the rate at which it fills the casting against the rate at which it fills the blind feeder above each casting such that the feeder and casting itself must be nearly full before metal passes over the weir into the next casting. Thus the castings fill sequentially and the feeder above each casting always contains hot metal. This hot feeder promotes directional solidification. By heating the adjacent sand it allows the use of a smaller ingate which facilitates fettling. Loper has described a similar casting technique[57]

Moulding and coremaking practice

The design freedom offered by the family of cast irons does not demand an in-depth knowledge of moulding materials or methods of the designer. Indeed, except in a 'tied' foundry situation, the designer generates a casting geometry without a specific moulding process in mind. This is chosen by the casting engineer on the basis of metallurgical, economical and environmental considerations. The process selected can influence pattern equipment and may require modification to the casting design. Examples of moulding processes are given in *Table 5.6*.

Sand moulding

Green sand moulding

'Green' or 'moist' sand is one of the oldest moulding materials[58, 59]. Its cheapness and relative freedom from environmental effects guarantee it remains a basic method of producing castings up to about 100 kg in weight, especially by machine moulding techniques (see *Figure 5.29*). The demand for more dimensional accuracy, good surface finish and freedom from unsoundness has led to the use of higher ramming pressures. Coupled with mechanization this has made green sand a favoured material for the production of high quality, mass-produced moulds for automobile castings[60].

Table 5.6 Some moulding processes used with cast iron

Moulding process	Pattern material	Mould material	Binder	Scope	Casting size kg	Cost Labour	Cost Equipment	Cost Mould
Sand	wood, plastic metal	silica, zircon, chromite, olivine sand	clay, cement, CO_2-silicate, resin self-set, silicate self-set	limited only by pattern	Green < 500, dry sand > 500, others to & above 500	L	L	L
Shell cronig	metal	silica or zircon sand	PF resin, NOVOLAC, hexamethylene tetramine + pattern heat at 260 °C	better surface finish and accuracy than sand	up to 150	L	MH	M
Full mould	low density EPS	silica sand	Na_2SiO_3, resin or unbonded sand	one off castings complex shape	up to 5000	L	L	L
Replicast fm.	medium density EPS	refractory paint, sand + vacuum	–	complex shape more reliable than Full mould	up to 5000	L	M	M
Ceramic mould	wood, plastic metal	molochite, mullite sillimanite	ethyl silicate + gelling agent 50/50 NH_4OH/H_2O	complex shape fine detail	up to 180	H	MH	MH
Replicast cs.	high density EPS	ceramic shell + sand + vacuum	–	complex shape very fine detail	up to 300	M	M	M
Permanent mould	–	cast iron, Cu alloys	–	Casting design must allow removal from the die		L	H	H
Centrifupal casting	wood	metal die, conventional sand mould	–	used to produce pipes	pipes 1.2 m dia. × 11 m long	ML	H	H
Continuous casting	–	water cooled metal die	–	semi-finished bar	up to 40 cm dia.	L	H	H

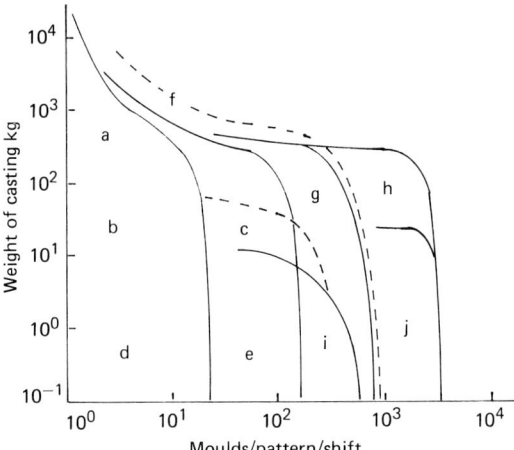

Figure 5.29 Summary of sand moulding processes.
Legend: a, cements; b, resin silicates; c, shell; d,
continuous mixers; e, high speed mixers; f, mechanized
cold set; g, mechanized green sand; h, high pressure; i,
manual and squeeze jolt green sand; j, flaskless high
pressure

Flowability and green strength are required in a
sand for successful moulding. Other characteristics,
such as refractoriness and permeability, promote
casting soundness. Flowability measures the ability of
the sand to be compacted to a uniform density. Green
strength is important in preventing mould collapse
when the pattern is withdrawn. However, an increase
in green strength is accompanied by a reduction in
flowability.

A balance between these properties is struck
depending on the mould compaction method. These
range from a pneumatic ramming hand-held gun, in
the case of hand moulding, which is also referred to as
floor moulding, to jolt squeeze machines or plain
squeeze machines to high pressure moulding[61–63],
including jolt squeeze machines with higher squeeze
pressures, plain squeeze machines for boxless
operation and shoot-squeeze operations. Air impulse
moulding is a recent development. It allows moulding
sands with a wide range of properties to be compacted
successfully[64].

A high flowability is desirable when ramming is
non-selective and the energy for compaction is transmitted through the sand mass. The concept of flowability has been discussed by Moore[65]. Several
methods are used to measure flowability or compac-
tibility. They are based on the compaction of a
standard cylindrical test specimen, on the extrusion of
sand through an aperature in a tube under a standard
load and on measurement of hardness differentials on
the sloping base of a cylindrical core rammed in a
modified tube mould.

There are two main types of sand, naturally bonded
and synthetic sands. Naturally bonded sands contain
up to 12% clay. When blended with 4–6% moisture
they produce a moulding sand of reduced permeability and refractoriness. Natural sands are used for light
castings when good surface finish is essential, but they
are not suitable for the production of high density
moulds or for recirculating sand systems. Synthetic
sands contain little or no binder in the natural state.
Mould properties are developed by separate
additions. Moulding properties can be achieved with
less clay (5–7%) and less moisture (3–3.5%) giving a
higher permeability and more refractoriness.

Important features of the base sand are composition, mechanical grading and grain shape. Most
foundry sands are based on quartz silica which
displays refractoriness as well as a high coefficient of
thermal expansion and a large transformation
expansion at about 600 °C. Parts of a mould surface
may be heated above this temperature by local,
uneven or prolonged heating during mould filling (for
example, metal impingement on the drag surface
adjacent to an ingate). The transformation expansion
can cause the hot surface layers of sand to buckle,
crack or even spall. This leads to casting expansion
defects such as scabs, buckles and rat-tails. Susceptibility to these defects can be reduced in several ways,
including modifying the gating and running system,
reducing the moisture content and adding cushioning
agents to improve the thermal shock resistance.

Special sands based on zircon, olivine and chromite
are used instead of quartz when cost is justified and
where dimensional accuracy is a prime consideration
and for selective chilling. The grain size and distribution of the sand influences permeability, surface
fineness and strength of the bonded sand. Coarse and
uniform grading are associated with high permeability, high flowability and refractoriness. Fineness,
which prevents metal penetration and produces a
smooth casting surface and low permeability is
associated with fine-grained or continuously graded
sands. It is desirable to combine fineness and permeability. This can be achieved by using a highly
permeable sand and mould coatings to provide
surface fineness[66].

The green compression strength of bonded sand is
inversely related to the grain size of the base sand[67].
Grain shape influences flowability and strength.
Angular sand particles reduce flowability. Rounded
grains produce a higher strength at lower ramming
densities, but angular grains produce a higher
strength at higher ramming densities. Strength is
generated in green sand by a silica–water–clay–water–
silica(clay) linkage[68].

The principal mineral constituents of clays are
kaolinite, montmorillonite and illite. Natural mont-
morillonites occur in two forms, sodium and calcium
bentonite. These clays have a high capacity for water
absorption, which is retained up to 550–700 °C. They
also develop favourable bonding characteristics,
which means that strength can be developed with low

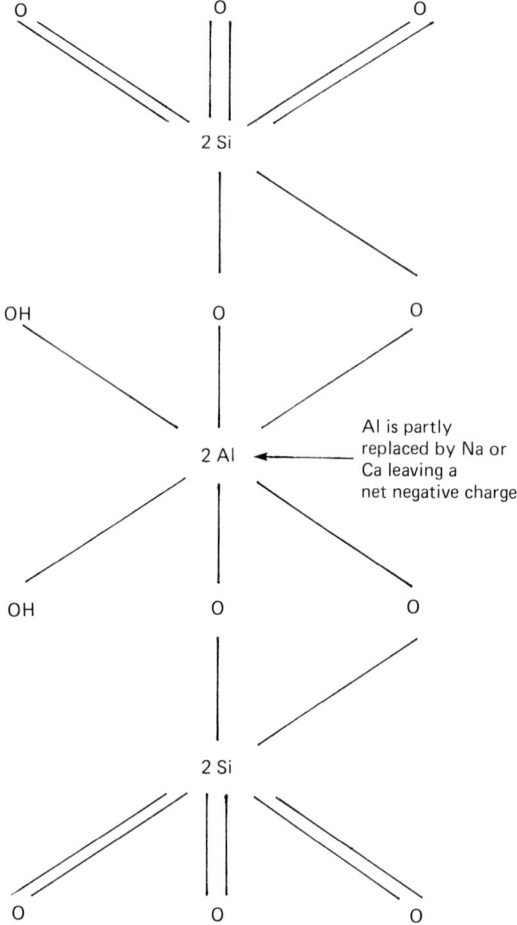

Figure 5.30 The layer structure of montmorillonite

clay additions. A high green strength consistent with a minimum clay addition and a relatively lower dry strength to improve knockout after casting are desirable sand properties. These may be achieved by mixing sodium and calcium bentonite or by affecting base exchange of the calcium ion in calcium montmorillonite with additions of soda ash or sodium carbonate (treated Fuller's earths).

The SiO_4^{4-} tetrahedron is the basic structural unit of the various forms of silica and silicate clays. The silicon oxygen bond is partly covalent and partly ionic. Each oxygen shares two electrons with a given silicon. The silicon bonds are satisfied by four silicon electrons and four electrons from the surrounding oxygen ions. Each oxygen ion must acquire an additional electron, either by sharing or by donation, from an ionic bond with a metal in order to form a stable outer eight electron configuration. Thus, there is a mixture of covalent and ionic bonds in silicates and a metal donates an electron to satisfy the require-

ment of the additional oxygen bond. The endpoint in the silicate series is silica itself, in which each oxygen atom is attached to two silicon atoms. Thus SiO_2 is covalently bonded. The three structural forms, quartz, tridymite and cristobalite differ in the orientation and density of packing of the tetrahedra.

The tetrahedra are linked to each other in two dimensions only in clays. They form layers with foreign atom bonds in the third direction. There is a sheet of oxygen ions with a single free bond per oxygen ion on one side of the layer. The other three oxygen ions in the tetrahedra have their bonds satisfied by oxygen ions from adjacent tetrahedra. The excess oxygen bonds are satisfied by a layer of aluminium ions. The additional aluminium bonds are satisfied by OH^- ions. These and oxygen ions are of a similar radius and can be accommodated in the same layer.

Tetrahedra are present on both sides of the aluminium ions in montmorillonite clays (*Figure 5.30*). Some of the Al^{+++} ions are replaced with Mg^{++} ions on an ion to ion basis. This creates a capacity for absorption of exchangeable cations, such as Na^+ and Ca^{++}, to which the properties of the clay are very sensitive.

Bonding develops when water molecules are attracted between the layers in the clay structure. The water molecules are polar. As the first layer adjusts to the electric field of the clay, a second layer is attracted, but less strongly. Water molecules are attracted to quartz because of the unequal balance of charges at fractured quartz surfaces. As the development of bond strength depends on hydration of the clay, the green strength increases with increasing moisture content up to an optimum value, which depends on the clay content. Above the optimum, additional free water causes the green strength to fall. Dry strength continues to increase to much higher original moisture contents. This is possibly due to improved binder distribution and the higher bulk densities obtained. Thus, if the optimum moisture content is determined, required strength properties can be obtained with minimum clay additions.

The relationship between green strength, clay and moisture content is shown in *Figure 5.31* for sodium bentonite, calcium bentonite and a 50:50 mixture[69, 70]. In order to obtain satisfactory moulding properties and a constant sand condition, it is usual to operate a sand with a moisture content just above the maximum green strength, that is, 2.5 to 3.5% with a sand containing 5–7% clay. If operated at the peak of the green strength or below, moisture control must be very precise. Otherwise large fluctuations will occur in green strength and other properties.

This principle should apply equally to high pressure moulding sands. This is not always the case because of the belief that a high strength and a low moisture content are required for high pressure moulding. These characteristics can be obtained only by using a high clay content and operating below the optimum moisture range.

This belief developed because of the necessity to achieve a compromise between a high degree of compaction necessary to give dimensionally accurate and sound castings and the compaction necessary to allow successful separation of mould from pattern. Machine design, mould spring back and pattern damage limit the moulding pressures used to about $120 \, N \, mm^{-2}$. This pressure will produce maximum compaction in a sand with a low clay content. This produces a rigid mould that will not separate successfully from the pattern. The required deformation for pattern removal can be obtained either by increasing the moisture content or by increasing the clay content. Increasing the moisture content is not desirable because the green strength will fall.

Several tests are used for measuring the mechanical properties of moulding sand[71]. Hardness measurements are convenient to use, but are often used erroneously in the belief that mould hardness measures mould rigidity. Mould hardness measures the degree

to which a sand has been compacted. It is directly relatable to green strength of all sands when they are compacted to the same extent. However, if the composition of a sand changes due to a change of sand type, moisture or clay content, mould hardness values are not comparable, either in terms of bulk density or mould rigidity (see *Figure 5.32*).

Hardness measurements are usually made with a spring-loaded spherical indentor. The depth of penetration from the flat reference surface of the instrument corresponds to an empirical scale from 0 to 100 units. The basic disadvantage of this method is that geometrically similar indentations are not achieved as the penetration increases. As a result the relationship between hardness and depth of penetration is not linear. This renders the measurements unsuitable for high density sands.

This has been analysed by Kupcruns and Szreniawski[74] who show that geometrically identical indentations are produced with a conical indentor with an α angle of $120°$. In this case the scale is based on the actual pressure exerted by the indentor on the

Table 5.7 Examples of typical moulding sands and mixtures used for green sand moulding

	SiO_2	Al_2O_3	Fe_2O_3	CaO	Na_2O + K_2O	Loss on ignition	AFS GFN	AFS clay	ZrO_2	TiO_2	MgO	Cr_2O_3
					Typical moulding sands							
Natural sand												
Bramcote	87.5	4.41	2.51	1.0	1.49	2.55	75	8.4				
Swynnerton	84.1	7.69	1.91	0.11	3.27	1.80	115	11.4				
Mansfield Red	82.5	4.96	1.27	2.39	2.30	2.15	137	13.6				
Washed silica sand												
Redhill 65 (F)	99.5	0.14	0.06	0	0.08	0.14	66	–				
Chelford Fine (95)	94.7	2.73	0.20	0.1	1.76	0.4	98					
Non-siliceous sand												
Ama Zircon	33.19		0.14			0.16	70		63.4	0.08	48.75	
Olivine No. 2	41.35	0.6	6.25	1.51								
FW chromite	3.1	15.6	26.5		10.4	under N_2 0.23						42

Examples of mixture for Green sands

	Sand	Clay	Additions	Added moisture %
Green sand				
Thin castings	60A 37B	–	3M	6
Medium castings	76A 20C	–	4M	6
Thick castings	66A 28C	–	6M	7
Synthetic sand				
Thin castings	91A	3D	6M	3.5
Thick castings	54A 36E	4F	6M	3.5

A – old sand renovated
B – natural sand of very fine grain size e.g. Mansfield Red
C – natural sand of fine grain size e.g. Bramcote
M – special additions, coal dust, blacklead, blacking
D – added clay i.e. bentonite
E – silica sand of fine grain size e.g. Redhill 65
F – Fuller's earth.

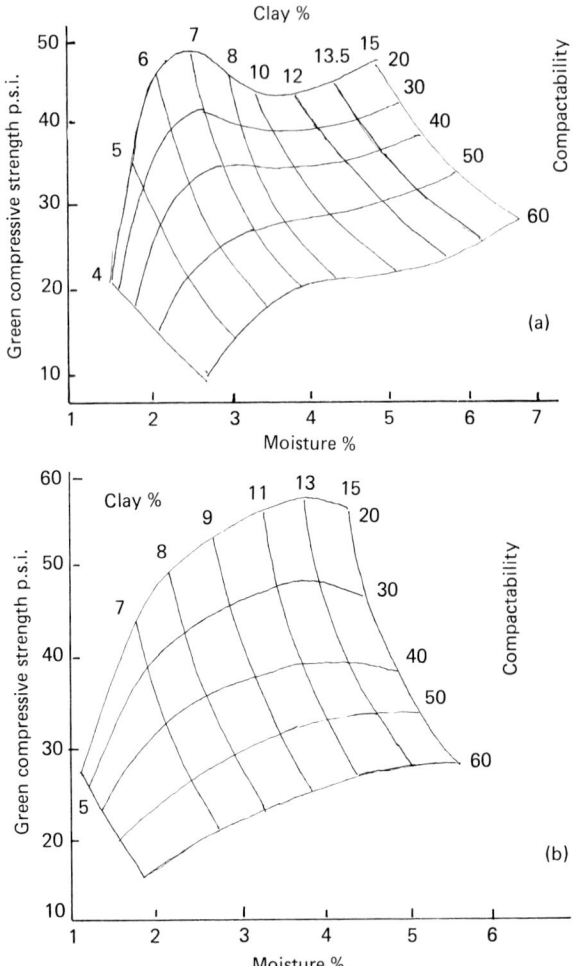

Figure 5.31 The relationship between green compressive strength, compactability and moisture for the stated clay additions for (a) Na bentonite; (b) Ca bentonite; (c) 50:50 mixture; (d) Na bentonite with seacoal (after refs. 69, 70)

specimen. This scale is compared with existing scales. Relationships between the new scale and properties such as compressive strength, permeability and flowability are defined for several sands. The extent to which hardness can be used as an indirect measurement of sand properties is demonstrated.

Severe burn-on and poor surface finish often result when iron is poured into a green sand mould that is neither coated nor contains a suitable carbonaceous additive. These defects are attributed to a rapid reaction which takes place between silica and the oxidized surface of the molten iron to produce a fusible iron silicate. This promotes mould penetration and results in sand adhesion and poor surface finish. Seacoal or suitable substitutes such as Gilsonites and cellulose flours are added to moulding sands to alleviate the problem.

Different explanations have been given for their action. The gas cushion theory suggests that metal mould interactions are prevented either by a layer of discrete bubbles or evolved gas which gathers between the sand and molten iron. The reducing gas theory considers that thermal reduction produces a mould atmosphere of reducing gases which prevents oxidation of the molten iron. The lustrous carbon theory considers that protection results from the formation of volatiles which then pyrolyze and deposit as a lustrous carbon layer at the metal mould interface. This provides a physical barrier to iron silicate formation.

The amount of coaldust added varies between 2 and 3 weight % for small castings to 7 to 8 weight % for large castings. Suitable coaldust contains a minimum of 30% volatile matter, less than 10% ash, very low S and should be of a sieve grading below that of the sand. If the coaldust is too fine, permeability is

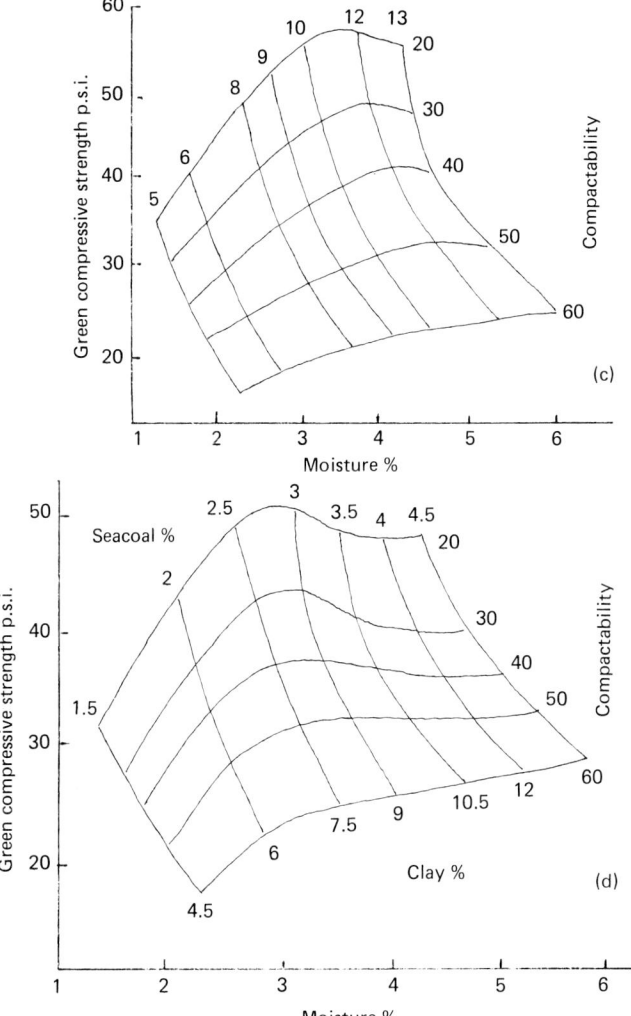

Figure 5.31 Continued

reduced. If it is too coarse, it can produce localized areas of high gas content, particularly in thin castings. This leads to pitted surfaces. The influence of seacoal on the properties of a clay bonded sand are illustrated in *Figure 5.31d*. Sands used for moulding and typical sand mixtures are given in *Table 5.7*.

The quality and properties of the base material must be maintained in any green sand system in which the sand is reused continuously. Sand must be replenished to prevent the buildup of dead clay and dead carbonaceous fines and treated to take into account loss of moisture and temperature increase[73]. A monitoring and control system is essential to regulate these procedures. Sand tests identified by Bethke[74] as being the best indicators of sand condition are compactability, permeability, green compressive strength, moisture, mulling efficiency, methylene blue clay determinations[75] and dry compressive strength. The analysis, which has been computerized[76], converts averaged sand test results run in consecutive groups into statistically operational trends as shown in *Table 5.8*. This informs the muller operator of the most logical direction in which the system is heading and enables him to make corrections in accordance with the problem source and correction guide shown in *Table 5.9*.

Hard sand moulding

High pressure green sand moulding is very suitable for mechanized quantity production of small iron castings. However, hard sand moulds are used for larger castings to obtain mould rigidity. They are a common replacement for green sand for jobbing or

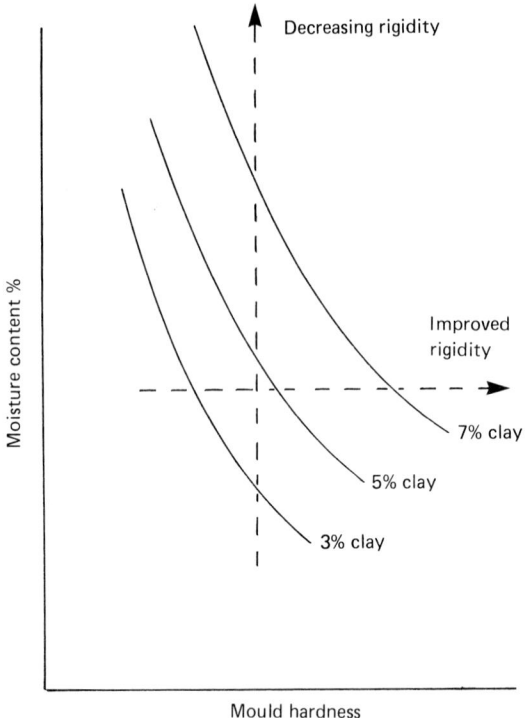

Figure 5.32 Change in mould rigidity with sand properties

medium volume production, where flexibility is as important as productivity. The hardening of moulds has changed from the traditional methods, which employed sands bonded with clay and pitch and hardened by drying in an oven or stove, to the use of chemical binders which can be hardened by gassing techniques or by liquid chemicals. These methods are referred to as the no-bake systems.

As indicated in Chapter 1 there are many advantages in using chemically bonded systems. These include:

1. lower sand metal ratio, ranging from 1:1 to 4:1 compared to 5:1 to 8:1 for green sand;
2. better dimensional accuracy and stability;
3. lower energy requirements;
4. more versatility, various metals can be cast with a common sand mix;
5. higher compressive strength;
6. applicable to stack moulded castings i.e. piston rings and
7. a wide range of binder systems, each with its particular advantages.

The sodium silicate process is an important inorganic system that has recently regained popularity after being one of the first systems to offer an alternative to green sand, dry sand and oil-bonded sands[77-90]. Its particular advantages are ecological, good mould rigidity, freedom from nitrogen, compatibility with green sand and cost. However, prior to

Table 5.8 Trend logic table (after ref. 76)

		Compactability	Permeability	Green compressive strength	Moisture	Mulling efficiency	M.B. clay	Dry compressive strength
Poorer mulling	10							
Lower sand/metal ratio	9							X
Higher new sand	8	X				X		
Higher dead fines and ash	7	X						X
Lower bonding	6	X						
Higher bonding	5					X	X	X
Lower fines	4		X					
Higher fines	3			X	X			
Lower moisture/bond ratio	2			X				
Higher moisture/bond ratio	1	X						

(Increasing ↑)

		Compactability	Permeability	Green compressive strength	Moisture	Mulling efficiency	M.B. clay	Dry compressive strength
Higher moisture/bond ratio	1			X				
Lower moisture/bond ratio	2	X						
Higher fines	3		X					
Lower fines	4			X	X			
Higher bonding	5	X						
Lower bonding	6					X	X	X
Higher dead fines and ash	7					X		
Higher new sand	8					X		
Lower sand/metal ratio	9	X	X					
Poorer mulling	10			X		X		

(Decreasing ↓)

The reaction proceeds through various intermediate stages to produce a glycol in an acidic medium[96]. The formation of the glycol is the rate-controlling step. This reaction induces a change in the SiO_2/Na_2O ratio which causes gelation. Commonly used esters include glycerol diacetate (fast cure), ethylene glycol diacetate (medium cure) and glycerol triacetate (slow cure). These esters, and others, can be blended to give a wide range of setting times (20 to 120 minutes). The silicate additions are limited to 2.5 to 3.5 weight % and silicates of medium ratio (2.5 to 2.85) are used. Hardener additions are between 10 and 12.5% of the silicate addition.

The optimum curing temperature is 20–25 °C. Lower temperatures retard the curing action and create stripping problems. Higher temperatures cause dehydration which results in premature hardening and friable moulds. As with CO_2-gassed silicates, additions such as carbohydrate polymers improve the properties of self-set systems. Do's and dont's of the self-set silicate process and some problems encountered are given in *Tables 5.11* and *5.12*. Some typical liquid hardening characteristics are shown in *Figure 5.37*.

Organic no-bake binders offer a wide range of complementary hard sand mould and core making processes. These resin-based systems may be self-setting, vapour-cured (cold box) or thermosetting (warm and hot box). They are more versatile than inorganic sodium silicate with respect to curing times. They develop good strength, produce rigid moulds when through cured, exhibit good breakdown behaviour and are easily reclaimed. However, they are expensive and can present more environmental problems than inorganic binders. Characteristics of the more common no-bake systems are given in *Table 5.13*.

Self-set or cold-set systems can be grouped as furan/acid, phenolic/acid, alkyd/isocyanate and phenolic/isocyanate. Furan binders are derived from urea formaldehyde/furfuryl alcohol (UF/FA), phenol formaldehyde/furfuryl alcohol (PF/FA) and furfuryl alcohol/formaldehyde (FA/F) condensates. They harden at room temperature under the influence of acid catalysts. The bond reaction is dependent on the catalyst addition and the sand temperature.

The type of acid catalyst depends on whether the resin is urea- or phenolic-based. Phosphoric acid is suitable for highly reactive binders (urea-based). Otherwise, curing can be slow, for example, with zero nitrogen binders based on furfuryl alcohol blends with reactive compounds, acetone formaldehyde and

Table 5.11 Do's and don'ts with sodium silicate self-set moulding (after ref. 96).

1.	Catalyst should be selected in relation to mould size or core application, type of mixing equipment, transport distance.
2.	Sand temperature should be within limits 20–30 °C, otherwise change catalyst.
3.	Always allow the system to stand for the recommended cure time, otherwise it may be surface cured only.
4.	When support rods are used in a mould or core ensure that they are at room temperature, otherwise curing rate will vary.
5.	Regularly calibrate pumps that deliver material into the sand mix. Otherwise the system will get out of control.
6.	Any change in the supply of aggregate addition may result in the binder level being too low or too high.
7.	The order in which additions are made is important. The carbohydrate polymer should be added to the sand mixture first, followed by the catalyst and finally the sodium silicate. This is to ensure that the sand grains are adequately coated with the carbohydrate polymer first to ensure a complete reaction occurs between the catalyst and the sodium silicate. If the order of addition is changed, the mould often shows partially cured areas and areas that have not cured at all.
8.	Ingredients should not be mixed prior to delivery of sand.
9.	It is possible to increase the cure rate by stripping the core or mould from the box and passing it through a low temperature oven. Also CO_2 gas may be passed through a core that has had the catalyst system applied.
10.	Water-base washes can be applied to cores. If the wash is applied efficiently and then passed through a core oven, a good core surface is obtained.
11.	If external heat is applied to the mould or core it should be for a limited time, otherwise the core will degradate due to premature breakdown of the shakeout additive.

Table 5.12 Mould and core self-set silicate — problems and solutions (after ref. 81)

Defect	Cause	Solution
Friable surface	Catalyst has partially cured sand prior to ramming	Select a slower catalyst
	Sodium silicate has partially cured sand prior to ramming	Select a different silicate
	Rough pattern	Improve equipment
Sand sticking to pattern	Sand mixture has not cured completely	Allow mixture to stand longer in box or select a faster catalyst
	Poor ramming	Improve ramming method
	Rough pattern	Improve equipment
Sagging and through cure	Undercured sand mixture	Allow sand mixture to stand longer or select faster catalyst
	Insufficiently mixed sand	Improve mixing method
	Insufficient binder level	Increase or decrease binder level. Check metering equipment
	Insufficient catalyst level	Increase or decrease catalyst level check measuring equipment

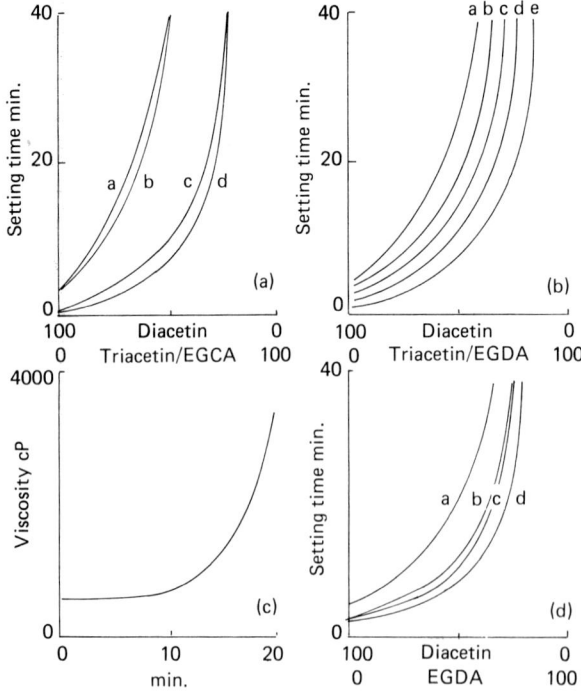

Figure 5.37 (a) Effect of setting time of ester blends a, silicate 100S with diacetin/EDGA; b, silicate 100S with diacetin/triacetin; c, silicate 95 with diacetin/EGDA; d, silicate 95 with diacetin/triacetin. (b) Effect of temperature on setting time: a, 5 °C, b, 10 °C, c, 15 °C, d, 20 °C, e, 25 °C. (c) Typical setting reaction. (d) Effect of some breakdown agents on setting time: a, corn syrup; b, powdered dextrose; c, cane sugar; d, no addition (after ref. 97)

Table 5.13 Characteristics of some no-bake binders

	Relative tensile strength	Rate of gas evolution	Degree of thermal plasticity	Collapsibility speed	Ease of shakeout	Moisture resistance	Curing speed	Strip time, min	Resistance to overcure	Optimum temperature, °C	Clay & fines resistance	Flowability	Air drying rate	Pouring smoke
Thermosetting inorganic														
silicate (warm box)	H	L	L	M	P	F	H		F	120	P	G	M	M
Self-setting inorganic														
Sodium silicate – Ester	M	L	H	L	P	P		5 – 60		24	F	G	M	N
Sodium silicate – FeSi	H	L	L	L	P	P		30		24	F	G	H	N
Sodium silicate – 2 CaOSiO$_2$	L	L	L	L	P	P		30		24	F	G	H	N
Cement – Hydraulic	M	L	L	L	P	F		45		24	F	F	L	N
Cement (fluid) – Hydraulic	L	L	L	L	P	F		30 – 60		24	P	G	L	L
Phosphate – Oxide	M	L	L	M	G	P		30 – 60		32	F	F	H	N
Vapour cured inorganic														
Sodium silicate – CO$_2$	L	L	L	L	P	P	H		P	24	F	F	H	N
Thermosetting organic														
Shell dry blend	H	M	L	M	F	G	H		G	260	F	E	N	M
Shell warm coat	H	M	L	M	F	G	H		G	260	F	E	N	M
Shell hot coat	H	M	L	M	F	G	H		G	260	F	E	N	M
Furan Hot box	H	H	L	H	G	F	H		F	230	P	G	M	M
Phenol hot box	H	H	L	M	G	G	H		F	230	P	G	M	M
Core oil	M	M	M	H	G	G	L		P	205	F	F	H	M
Self-setting organic														
High, medium, low N$_2$														
Furan – acid	H	M	L	M	G	G		1 – 45		27	P	G	H	M
Phenol – acid	M	M	L	M	G	G		2 – 45		27	P	G	H	M
Alkyd	H	M	L	M	F	G		2 – 20		32	F	F	H	H
Phenolic – Pyridine	M	H	L	H	G	F		2 – 20		27	P	G	H	M
Vapour cured organic														
Phenolic – Amine	M	H	L	H	G	F	H		G	24	P	G	H	M
Phenolic – SO$_2$	M	H	L	M	G	G	H		G	24	P	G	M	M
Solvent evaporation – air	M	H	–	–	–	G	H		G	–	–	G	H	–
FRC – SO$_2$	M	M	–	M	G	G	H		G	24	–	G	M	L
Phenol – Ester	M	L	M	M	G	G	H		G	24	–	G	M	L

phenol modified binders. The commonly used aryl sulphonic acid catalysts are toluene- (TSA), benzene- (BSA) and xylene- (XSA) sulphonic acids.

The catalyst initiates further condensation of the resin and affects the cure by cross-linking reactions. The condensation reactions produce water, which dilutes the acid catalyst and tends to slow the rate of cure. Consequently, it is necessary to use concentrated acid catalysts to ensure a reasonable rate of cure and good deep-set properties.

Curing characteristics are enhanced by the addition of silane to the resin. Its role is to act as a coupling agent between the inorganic sand grains and the organic binder. It also acts as waterproofing to prevent moisture degradation of sand cores and moulds.

The selection of the resin system depends on the iron's tolerance to nitrogen. Furan binders are available with zero, low, medium and high nitrogen[98]. The nitrogen content of a furan depends on the amount of urea or urea formaldehyde used. Cost is reduced as more urea is used. In general, urea-based resins are avoided when spheroidal irons are cast because of pin-hole type defects caused by nitrogen emissions.

Continuous mixers of the trough and high speed type are suitable for cold set processes. Constant temperatures and humidities are essential for consistent setting times. It is usual to change catalyst additions to compensate for temperature change.

Temperature and humidity effects have been shown to be interrelated[99]. The rate of cure is directly proportional to temperature and inversely proportional to the relative humidity. Sand strength is highest at low temperature and low relative humidity. The humidity effect is most apparent during the initial cure with acid-cured furans and phenolics. With alkyd and phenolic urethane systems the effect is more pronounced after 24 hours.

The phenolic/acid systems use resole phenolic resins, relatively cheap polymers of phenol and formaldehyde (PF) dissolved in a water solvent and catalysed with TSA. Sulphonic acid is used with phenolics because of their sluggish reactivity. The acid catalysed curing reaction proceeds by further condensation of methylol and methylene ether groups with reactive phenolic ring positions in the resin. Alkyd resins are reaction products of a vegetable oil, a dibasic acid and a polyol blended with petroleum hydrocarbon polymers and dissolved in a suitable solvent.

The alkyd/isocyanate system is a three part system consisting of the alkyl resin, a polymeric isocyanate and an amine metallic catalyst. The system is easy to use and offers working times in excess of one hour. The curing action occurs in two distinct stages. The isocyanate reacts with part of the resin under the influence of the amine in the catalyst. The alkyd resin completes the cure by reaction with O_2 present in the air. The second reaction occurs slowly at room temperature. The hot strength of this system is poor and 1–3% of iron oxide is often added to overcome this problem[100].

The fourth type of cold set resin system is the phenolic/isocyanate or phenolic urethane system. This is a three part system consisting of a phenolic resin in a solvent, a liquid pyridine catalyst and a polymeric isocyanate in a solvent. The PF resin provides hydroxyl groups, which chemically combine with the isocyanate in the presence of the catalyst by a cross-linking reaction. This sytem offers the characteristics of organic cold-set systems with strip times shorter than a minute when utilized with a high intensity mixer and core blower. This characteristic derives from a delay in the initiation of the curing reaction. There is little change in the flowability of the sand after mixing until curing begins. Towards the end of the work time there is a sudden and rapid increase in the tensile strength. The pattern should be removed from the cured sand before the tensile strength buildup. Otherwise severe sticking ·may occur. There is no oxidation so the curing reaction is completed shortly after removal from the pattern.

Although the ambient temperature, vapour or gas cured binder process dates backs to the silicate-CO_2 process, the major impact on core and moulding technology was made in the 1970s with the introduction of the cold box process, in particular, the phenolic urethane system[103]. This is similar to the cold set system described above except that the catalyst is in gaseous form.

The three part system consists of phenolic resin in

solvent, polymeric isocyanate in solvent and an amine catalyst in a CO_2, N_2 or air carrier. The phenolic resin provides hydroxyl groups, which combine with isocyanate groups to form a rigid bond. The reaction is accelerated in the presence of the amine. Two amines are used commonly, TEA (triethylamine) and DMEA (dimethylethylamine). The amount of binder is determined by the sand type, grain size and distribution and the minimum core strength required. The two resins are usually mixed into the sand in equal quantities. Total addition varies between 1.5 and 2.0% by weight of sand.

Cold box systems engineering has been described on several occasions[102, 103]. It is important to prevent high temperatures developing during sand mixing and contamination of the mixed sand with moisture. Excessive use of catalyst should be avoided. CO_2 or N_2 is often preferred as the carrier because of increased permeation of the sand and reduction in the risk of expolsion. Total gassing and purging time for medium-sized, simple-shaped cores may be less than five seconds. It is essential to take the precautions necessary to cope with the handling and environmental problems associated with phenolic resins, isocyanate and amine materials[104]. This is a rapid curing cold resin system which is particularly suitable for high volume production techniques but which can be adapted for small volume runs.

The Sapic or SO_2 cold box process is a comparable technique that was introduced in the 1970s[105]. Silica sand of average grain size 250–300 μm is coated with 1–1.5% of phenolic or furan resin and 0.4–0.8% of organic peroxide using either a batch or continuous mixer. Mixing time should be short ($\sim$ two minutes for batch mixers) and care taken to avoid heat generation. The resins usually used are low-water, nitrogen-free or low-nitrogen furan polymers with silane additions. Methyl ethyl ketone peroxide (MEKP) or occasionally hydrogen peroxide, if shorter bench life can be tolerated, are used.

The coated sand has a bench life of up to six hours and is easily blown into core boxes using conventional equipment. SO_2 gas is injected for about a second and the core is purged with air for about ten seconds. Alternatively, a separate gassing chamber may be used. Precautions must be taken to prevent the corrosion and environmental effects of SO_2 and the combustion of MEKP. The SO_2 is scrubbed from exhaust purge air with a wet scrubber containing 10% NaOH in water. The reaction between SO_2 and peroxide is immediate. The SO_3 generated dissolves in the water present in the binder to form sulphuric acid, which induces the rapid exothermic polymerization of the resin. The optimum curing temperature is 25 °C. The rapid curing action produces a strong handlable core within a few seconds. Consequently, production rate is determined by handling time.

This system is comparable to other organic no-bake

systems and has several outstanding characteristics. Stripping ability is good, final compression strength is about $700 \, N \, mm^{-2}$. The tensile strength on stripping is about $125 \, N \, mm^{-2}$ for a resin content of 1.3%. It increases as the resin content increases. Intricate cores can be produced and breakdown characteristics are good. The hot strength is greater than that of the amine gassing process but gas evolution and casting blowhole defect tendency is not as good[106].

Problems associated with the process include limiting the amount of SO_2 in the working area, buildup of resin on core box surfaces and S pickup from the cored face. Poor economic and critical process control requirements have combined to limit the use of this process compared to the phenolic urethane cold box process.

Although the introduction of epoxy chemistry into the SO_2 cold box process has partially redressed this situation[107, 108], other recently developed gassing systems offer equally attractive characteristics. The FRC cold box system falls into this category[109-111]. This system consists of a binder (low viscosity blend of vinyl unsaturated resin), an organic hydroperoxide initiator (2–5% of the binder weight), an adhesion promoter (3–5% of the binder weight) and SO_2 (1–5%) diluted in a N_2 carrier gas. Although the system and gassing procedure are similar to the SO_2 cold box process, the curing mechanism is different. On gassing the SO_2 breaks down the peroxide into free radicals, which initiate the cross-linking of the binder components to produce a cured solid polymer.

The first generation of FRC binders have shown that the process offers the following characteristics in comparison to other cold box systems[110].

1. the fastest curing binder system available;
2. adaptability to presently rigged cold box tooling;
3. can be used with most types of foundry sand;
4. non-acidic or basic curing mechanisim;
5. extremely long mixed sand bench life;
6. lower consumption of curing gas;
7. reduction or elimination of veining defects;
8. improved hot distortion properties;
9. improved dimensional accuracy and
10. improved shakeout characteristics.

The chemically neutral curing action means that the binder is not sensitive to sand acid demand value (ADV) or pH. This contributes to the long bench life properties and also means that reclaimed FRC sand should not have any deleterious effects on any acid- or base-cured resin system that may be operated elsewhere in the same foundry. The amount of SO_2 required is considerably less than in the acid SO_2 process. All these features contribute to the significant cost savings that can be achieved with this system. *Figure 5.38* shows results of a BCIRA hot distortion study on several no-bake binder systems.

This test gives an indication of how a system will react to the initial heat of a casting and its stability under the hot metal conditions. All the systems go through an initial expansion period, which is related to the thermal characteristics of the sand. However, whereas the majority of other systems then distort, which can lead to mould cracking and casting veining, FRC moulded sand shows thermal stability after the initial expansion. This reduces susceptibility to veining and also promotes good dimensional accuracy.

A second development of the 1980s is the ester cured phenolic no-bake system[112-115]. This system consists of an alkaline phenolic resin (1–2% of the sand weight) and an ester (0.25–0.50% of the sand weight). The curing reaction is a two stage process. The polymerized resin remains thermoplastic until heated. Once thermoset, the bond is very rigid. This characteristic introduces a flexibility which permits the moulded sand's expansion on casting to be absorbed before the mould becomes rigid. Thus, a resistance to thermal shock is created and this reduces the susceptibility to casting defects. The rate of cure can be adjusted by blending different esters. The characteristics of this system are indicated in *Table 5.14*. They show that it cures like a phenolic urethane, casts like a silicate and shakes out like an alkyd.

The BCIRA CO_2-Polidox process[116, 117] can be compared to the two preceding processes. It uses CO_2 gas to harden sand bonded with a polyacrylate resin in the presence of CaOH. Foundry trials[118] have confirmed the rapid cure rate of this binder. The process, as developed to date, produces good quality

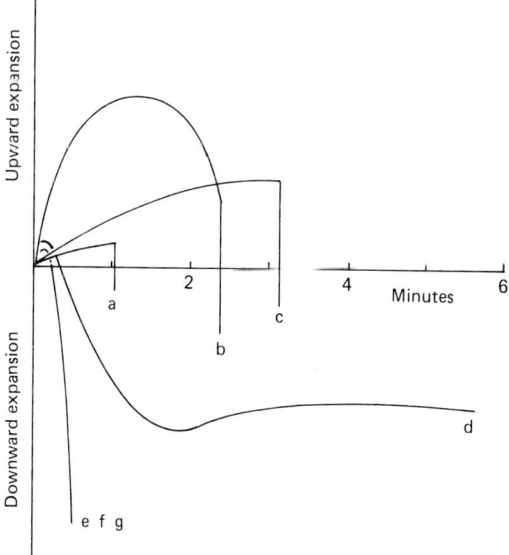

Figure 5.38 Typical hot distortion test results for various binder systems: a, FRC binder; b, Phenolic; c, Furan; d, Ester phenolic; e, Phenolic urethane; f, Alkyd; g, Silicate

Table 5.14 Comparison of the characteristics of phenolic ester binder with other no-bake binders

Characteristic	Alkyd	Phenolic urethane	Furan acid	Phenolic acid	Silicate ester	Phenolic ester
Odour on mixing	Moderate	High	High	High	None	None
Work to strip time ratio	25	50 +	20	20	20	50 +
Pattern release	Good	Poor	Poor	Poor	Poor	Good
Core storage	Average	Poor	Good	Good	Poor	Good
Sand temperature effect	High	Moderate	High	High	High	Low
Humidity resistance	Moderate	Low	High	High	Low	High
Nitrogen present	Yes	Yes	Yes/No	No	No	No
Sulphur present	No	No	Yes	Yes	No	No
Pour off smoke	High	High	Moderate	Moderate	Low	Low
Carbon pick-up	High	High	High	High	Low	Low
Gas defect potential	High	High	Moderate	Moderate	None	None
Hot strength	Low	Low	High	High	High	High
Hot tear potential	Low	Moderate	High	High	Low	Low
Shake out	Good	Good	Moderate	Moderate	Poor	Good

cores with good storage properties and with excellent breakdown properties after casting. The system has the advantage that the CO_2 gas used is environmentally acceptable. The principal requirements for success with this system are:

1. correct measuring and weighing of sand and additions;
2. control over mixing time to prevent overmixing and loss of properties;
3. use of vented core boxes to obtain good gas distribution through the cores;
4. control of CO_2 gas flow rates and consumption;
5. use of air drying mixtures, preferably if cores are to be coated;
6. application by painting or spraying of isopropyl alcohol-based coatings if required and
7. well ventilated conditions if cores are to be stored for prolonged periods.

Other well established endothermic core and mould making processes are shell, hot box and core oil. These processes are still practised extensively today despite predictions that they would become obsolete when cold box processes were established. Traditionally cores were produced from sand bonded with linseed oil ($\sim 2\%$), corn flour (2–3%), 3–4% H_2O for green strength and 1–2% Fe_2O_3 to improve hot strength and surface finish. The processes used ranged from hand production in simple wooden core boxes, sometimes referred to as 'bench' core making to more mechanized systems for volume production. The sand core is removed from its box and placed on a core plate to be baked hard in a special oven. This is known as the oil sand process. Alkyd and resole phenolic oils are also used to give a range of drying times.

The shell core process has become extremely popular in recent years. When high volume requirements apply it utilizes a free flowing coated sand principle. The sand is coated with a thermal setting resin, which is dumped or blown into a heated metal core box. The sand against the heated core box surface cures sufficiently for the core to be removed, usually in less than a minute. The uncured sand in the centre can be removed and reused. After inspection and dressing the core is ready for use in the mould. Shell cores have high strength with superior finish and dimensions.

The process known as the hot box process also uses a heated metal pattern. Sand which is blended with a resin binder and hardener is blown by compressed air into the core box. The hot box causes the hardener to vaporize and in turn, harden the resin. Early hot box systems were based on furan binders with acid catalysts. More recently, phenolic based hot box systems have become more popular. Cores produced by this process are used after inspection and dressing.

Hughes[119] has emphasized the considerable influence that coatings can have on core properties. *Table 5.15* shows that they may increase or decrease the strength of hot box cores.

The hot box process continues to be used because of the substantial capital investment in manufacturing and ancillary equipment and a reluctance to give up the expertise gained over the years. However, as shown in *Table 5.16* it has several disadvantages. Developments recently have been towards the use of a warm box process which reduces these disadvantages as shown in *Table 5.16*.

Casting defects attributable to cores include:

1. *Misalignment of internal cores* caused by low strength cores fracturing during the pouring operation, incorrect core sand mixture and badly fitting cores and core joints;
2. *Metal penetration.* This defect can result from low hot strength, poor coating of the core with refractory paint or incorrectly rammed or compacted sand. It may be caused by high sand expansion characteristics;

Table 5.15 Change in strength of hot box cores when cold (room temperature) and hot (1000 °C) as a result of applying various coatings (after ref. 119)

Coating	Liquid carrier	Refractory	Percentage change in strength	
			Hot	*Cold*
A	water	zircon	+ 48	+ 25
B	water	talc	+ 52	− 7
C	water	zircon	+ 88	+ 9.8
D	2-propanol	zircon	+ 32	+ 2.8
E	2-propanol	talc	+ 4	+ 5.7
F	2-propanol	graphite	− 16	− 5.7
G	2-propanol	coke/graphite	+ 56	− 4.1
H	2-propanol	zircon	+ 44	+ 16
I	1:1:1 trichlorethane	graphite/zircon	+ 4	+ 5.7

Table 5.16 Characteristics of cold box, hot box oil sand and shell core making processes; benefits of warm box over hot box and the advantages of the cold box process

Characteristics	Cold box	Hot box	Oil sand	Shell
Cure conditions				
Core box temperature °C	Room temp	175–250	Room temp	200–250
Cure time	V. rapid	Mod. to slow	Slow, sep. cycle	Slow
Extent of cure at strip	Through cure	Skin cure	Green	Skin cure
Strength at strip (% ult.)	60–75%	30–40%	Green	35%
Ultimate strength, $N\,mm^{-2}$	170–250	350–400	200	310
Core scrap	Low	Medium	High	Low

Hot box	vs	Warm box
Formaldehyde odour		much reduced
Phenol odour		none
Poor sand flowability		Improved due to lower viscosity resin and lower addition rates.
Poor through cure		Improved due to reactivity of binder
Poor storage humidity resistance		Improved by new catalyst
Gas defects due to resin decomposition		Low viscosity, high quality binder means lower additions and less gas evolved.
High box temperature		Lower box temperature 150 °C.
Box distortion		Reduced distortion
High energy consumption		Reduced by 25–33%.

Advantages of the cold box process

Elimination of heat, rapid cure, high productivity
Dimensional accuracy, fast start-ups
High out-of-box tensile strength reduces breakage
Greater versatility in core design.
No over-curing
Can pour within five minutes
Reduction in core wash, good casting finish
Highly flowable sand mix, excellent core shakeout
Lower tooling costs and longer life.

3. *Finning*. This is often the result of poorly graded sand or premature collapse of the binder;
4. *Blowholes*. This is a defect found just below the casting skin or on the casting surface. It can result from high binder content, poor drying control, poor ventilating practice, blocked vents, undried core paste or low quality core binder;
5. *Core lift*. This is not in the true sense a defect caused by the core, but the result of poor foundry practice and pattern equipment;

6. *Hot tears.* Cracks in the casting can result from too slow a core collapse. If this occurs the hot strength of the core or the positioning of any core reinforcement should be investigated and

7. *Pin holes.* This defect is often not discovered until the casting is in the machine shop. It can be caused by insufficient drying of the core, wet core gas or a binder that yields a high nitrogen content during casting.

The shell process is used extensively for mould making. This casting method was originally developed in Germany during World War II. It is still known by its original name, the Cronig Process, in Europe. Its original claim to importance was the relatively simple equipment required to produce a castable mould. In its basic form this is still true, but in recent years developments have led to increasingly sophisticated and expensive equipment used for production and casting of the shell moulds.

As in the original process, the moulds are made from a refractory medium, usually silica sand, which is coated with a resin system that is thermoplastic initially. Tricon, Chromite or Olivine sands are also used as a filler material. The choice of the material, apart from the obvious one of economics, is often based on the need for specific properties such as refractoriness and cooling requirement. The coated medium is normally dumped on to a heated metal pattern and this causes the resin to convert to a thermosetting condition such that on removal from the pattern the shape is retained without distortion. This shape subsequently forms the basis for the shell mould.

To produce a satisfactory shell mould with the optimum economic resin usage the sand used should exhibit certain defined properties. Because the strength of the shell mould depends on a relatively strong resin bond between the sand grains, any material which interferes with the bond will require extra resin to counter its effect if the shell strength is not to be impaired. Clay or peat, commonly found in association with sand deposits needs to be removed to render sand suitable for economic bond strength development. The sand should be free of chemically-bonded impurities which impair refractoriness. Sands with a rounded grain shape are preferred to ensure the free flowing property desirable for this process.

The distribution of grain size must be controlled to achieve good shell mould strength and density. The size of sand grains selected is usually based on the quality of surface finish required. This decision also controls the quantity of resin used. Most modern shell facilities incorporate sand reclamation systems to offset the high cost of the mixed sand.

Having selected the sand system, the next consideration is the most suitable bonding agent. Resins in common use are of the phenolformaldehyde type. These base resins usually contain insufficient formal-

dehyde to render them thermosetting. An extra quantity is supplied by breakdown at temperature of an additive made to the resin. Hexamine is used for this purpose. After the sand has been invested on to the pattern, the resin retains its viscous thermoplastic condition. It then hardens gradually as the hexamine breaks down and releases formaldehyde. This reaction is time–temperature dependent. A third component is added to the majority of resin systems. This is normally natural or artificial wax. Its main purpose is to act as a release agent, but it also functions as a lubricant to the sand grains improving flowability and hence, casting definition and quality.

Originally the sand mix was prepared simply by stirring dry resin into the dry silica sand. More recently the technique of precoating the individual sand grains has replaced the original process. The two most popular processes for precoating sands are the warm coating and hot coating methods. The warm coating technique, as the name implies uses the lower temperature. Hexamine and the release agent are milled together with the sand at a temperature in the range 40–70 °C in a modified core mixer which has a means of blowing warm air through the sand as it is mixed. The resin is solvent-based (novolak) and is introduced as a viscous liquid. The solvent is driven off by the warm air passing through the sand aggregate. As milling proceeds, the lumps of sand breakdown and the coated sand are finally discharged through screens into bags for storage until required for use.

Temperatures of 125–130 °C are used in the hot coating system, which is usually used for high volume production. The heating is carried out in fluid bed systems heated by oil or gas. The hot sand passes into the mixer where the resin is introduced as flakes of solid resin. At this temperature it melts and coats the sand grains. The hexamine addition is then made as an aqueous solution. The water evaporates off and also serves to cool and harden the resin again. As with the warm coating process, screening takes place before storage.

Because the shell moulding process is a thermal process, some form of pattern heating must be provided. The trend at present appears to be in favour of gas heating rather than electric heating. This is probably due to the physical size of the heating units required and the complexity of safety precautions which render gas more reliable, safer and faster as a heat source. Gas heating is normally applied to the pattern plate in the form of a multi-burner system mounted below the plate. It may or may not move with the pattern plate during the production cycle. Burners are located so that critical areas of the pattern may be heated more than others in order to ensure an even sand shell thickness when dumping takes place.

A difficulty commonly encountered in rapid production cycles is temperature maintenance in the pattern and plate to produce satisfactory shells consistently. Typical working temperatures are between 230 and 240 °C. Once a series of mould halves are produced they have to be fitted together, normally in pairs, to form complete moulds. If the shell moulds are to be cast unsupported, glueing is the normal method of holding shells together whilst they are cast. When some backing medium is used for support, shells may be glued or clamped together lightly using metal clips around the edge of the shell. In some cases, it is possible to rely on the backing medium to hold them together without any preliminary fastening. Glueing is by far the most common method of fastening an assembly.

Advantages and disadvantages of shell moulding are

1. extremely good detail definition retained;
2. better surface finish is possible than with any other casting process;
3. shell moulds have a slower cooling rate than solid moulds, so backing materials with a high thermal conductivity may be necessary to avoid metallurgical problems. Analysis correction or the use of zircon sands may have to be considered.
4. economics vary considerably with the type of casting to be made and
5. knockout properties of shell moulds are excellent and the casting requires only a light cleaning operation.

Shell moulding offers a means whereby accurate castings possessing fine detail may be produced rapidly in large quantities. However, high tooling costs are involved and the choice of this process should be made with due consultation with the casting experts.

Many of the advantages of shell moulding and the cold box processes have been combined recently in the development of the Anatoly Michelson process[122].

This new process consists of the following steps; closing the core box, charging it with sand binder mixture, sealing the investment orifice, flushing air out, introducing pressurized catalyst gas through a permeable pattern to the outer sand mixture layer, purging residual gas out of the formed core and the core box, discharging and collecting unhardened sand from the inside of the core, opening the core box and ejecting the shell core on a receiving surface.

An area of major concern with chemically-bonded sands is environmental pollution[123]. Increased legislative pressure has resulted in studies to identify, evaluate and control pollutant hazards within recently imposed limits[124]. *Table 5.17* details some of the foundry atmosphere contaminants generated during the core and moulding operations described above. The contaminant of most concern with the hot box process is formaldehyde. Shell moulding generates phenol from the resin and ammonia as a result of thermal decomposition of the catalyst, hexamethylene tetramine. Many older hot box and shell installations do not meet current environmental control requirements. Such installations have been modified successfully by installing local exhaust ventilation, such as ventilated core tables, and an air input above the operator's work station.

Emissions of environmental concern with cold setting systems are formaldehyde, furfuryl alcohol and phenol derived from the resin and some aromatic hydrocarbons liberated from sulphonic acid[125]. Well ventilated working areas and control over sand temperature help to prevent atmospheric pollution. Although the SO_2 and amine gassing processes are potentially troublesome, these processes have been introduced to the foundry relatively recently and are supplied with enclosures, fans and scrubbers which are capable of meeting current environmental control requirements.

Table 5.17 Some foundry atmosphere contaminants evolved during various mould and coremaking processes (after ref. 123)

Process	Resins	Resin raw material	Catalyst	Contaminant
Shell	PF resin	Phenol Formaldehyde	Hexamine	Ammonia, phenol, hexamethylene tetramine, stearates, fatty acids, formaldehyde,
Hot box	PF resin PF/FA PF/UF UF/FA	Phenol Formaldehyde Urea Furfuryl alcohol	NH_4Cl + occasionally ammonium salt of pTSA	Phenol urea furfuryl alcohol
Cold-set (air-set) (self-set) (no-bake)	PF-resin PF/FA PF/UF/FA UF/FA	Phenol Formaldehyde Urea Furfuryl alcohol	pTSA, XSA BSA, H_3PO_4 H_2SO_4 and acid mixtures	Formaldehyde, furfuryl alcohol Phenol, benzene, Toluene Xylene
Cold box SO_2-gassed	PF PF/FA	Phenol Formaldehyde Furfuryl alcohol	Sulphur dioxide MEKP	Formaldehyde, furfuryl alcohol, sulphur dioxide, MEKP
Cold box amine-gassed	Part 1 PF Part 2 MDI	Phenol Formaldehyde MDI pre-polymer	TEA DMEA	Carbon dioxide, triethyl amine Dimethyl ethyl amine, MDI, phenol, resin solvents naphthalene and homologues

Table 5.18 Foundry atmosphere contaminants resulting from mould casting and cooling for different moulding processes (after ref. 123)

Process	Atmosphere contaminant
Shell	Ammonia, aromatic hydrocarbons (benzene, toluene etc.), phenol and homologues (phenol, cresol etc.), hexamethylene tetramine, other amines (trimethylamine), hydrogen cyanide
Hot box	Aromatic hydrocarbons, phenol and homologues, ammonia, chlorinated hydrocarbons, hydrogen cyanide.
Cold set	Sulphur dioxide, hydrogen sulphide, Mercaptans (methyl, ethyl mercaptan), aromatic hydrocarbons, phenol and homologues, furan and homologues (furan, methyl furan), carbonyl sulphide, carbon disulphide, aromatic sulphur compounds, methyl ethyl ketone ⎱ from SO_2-gassed acetone ⎰ systems only
Cold box	Hydrogen cyanide, phenol and homologues, aromatic hydrocarbons, aniline and homologues (aniline, toluidine), aliphatic amines, resin solvents, naphthalene and homologues, Isocyanates (methyl, phenyl isocyanate), Benzoquinolines

Different pollutants may be encountered during mould casting and cooling. They derive from the high temperature pyrolysis of the chemical binder which produces a complex of free radicals which recombine to form a wide range of chemical compounds. Foundry experience, together with pyrolysis in conjunction with gas chromatography/mass spectrometry, has identified the pollutants associated with different processes. These are shown in *Table 5.18*. Quantitative information can be obtained from a particular mould by examining the gases evolved using an infrared gas analyser and a transportable quadrupole mass spectrometer.

The concentration of compounds present in air from the sampling hood is measured as a function of time. The area under the curve generated with this data estimates the total amount of a component liberated during the time interval from casting until the mould is cool. Additional information about when emissions maximize and how long they take to decay can be deduced from the curve. Pollutants of concern may be identified by normalizing the measurements with respect to the CO exposure limit of 50 p.p.m..

Table 5.19 shows estimated atmospheric concentrations normalized for a cold setting PF/FA resin with a *p*-toluene sulphonic acid catalyst. It identifies SO_2 as a pollutant of concern and suggests that in a foundry using this binder with a CO concentration at the exposure limit, the SO_2 concentration would be just below its limit. However, if catalyst additions were increased to account for a low sand temperature, they would almost certainly exceed the exposure limit.

SO_2 levels are of concern in foundries with insufficient local exhaust ventilation, particularly as levels may build up with reclaimed sands. Methods suggested for reducing SO_2 emissions include using an

Table 5.19 Estimated atmosphere concentrations normalized to carbon monoxide at its exposure limit for a PF/FA resin with a p-toluene sulphonic acid catalyst (after ref. 123)

Compound	Exposure limit 8 h T.W.A. p.p.m.	Estimated concentration p.p.m.
Carbon monoxide	50	50
Sulphur dioxide	2	1.84
Carbon disulphide	10	0.05
Benzene	10	0.27
Toluene	100	0.84
Furfuryl alcohol	5	0.77
Phenol	5	0.12
Cresol	5	0.18

alkali mould wash, using a renewable absorbent such as an activated charcoal mould-covering cloth, using a thin mould covering of sand containing an alkali, use of minimum quantities of the acid catalyst or the use of a phosphoric acid catalyst.

The discussion above has been concerned mainly with gases and vapours. However, of recent concern, but as yet not fully examined, is organic matter of higher molecular weight which condenses on, and may be carried by, dust particles into the body.

Vacuum has been found to be a valuable aid in moulding. The Q-ATSU process[126] for gas curing systems is illustrated in *Figure 5.39*. It is a two stage process of sand filling and gas filling. The flask is set on a vented pattern board in the moulding process. A coverless hopper with a slotted bottom plate and which contains mixed gas setting sand is located over and sealed against the flask. Vacuum is applied to complete the seal, draw sand into the flask and compact it against the pattern. A film or plate can be placed over the upper sand surface and the vacuum reapplied if a high density mould is required.

The hopper is replaced with a gassing case and the curing gas is drawn into the mould by vacuum in the second stage. After curing, and before pattern removal, air is drawn through the system. The core making process is similar. The use of suction rather than blowing, as in conventional cold box processes, eliminates gas leakage and the need for heavy clamping. As a result, the machine is more compact and of lighter construction. Sand is more easily and more uniformly compacted. Thus, uniform and through curing is achieved more rapidly. These features mean that complex shape and size do not limit the process as much as they do with conventional cold box machines.

The 'V' process is a second, similar, more established Japanese process that uses vacuum to aid moulding[127-130]. A thin, heated ethylene vinyl acetate film is vacuum-formed to the pattern. The flask is positioned on the pattern and is filled with loose, dry, unbonded sand which is compacted using moderate vibration. The sand is then rigidized by covering the flask with a second film and drawing a vacuum in the sand. The mould and flask can be stripped from the pattern by releasing the vacuum applied originally to the pattern and plastic sheet. Sand rigidity is retained by maintaining the vacuum within the packaged mould halves. Vacuum is maintained during pouring. When solidification is complete, the vacuum is

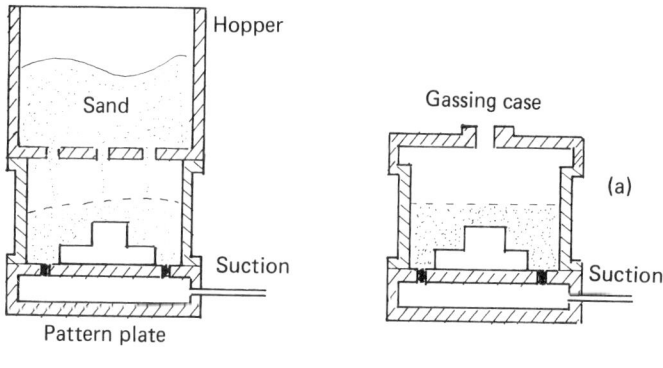

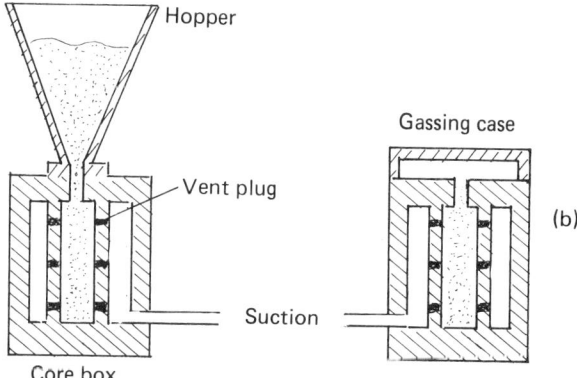

Figure 5.39 (a) Mould and (b) Core making processes using the Q-ATSU Process (after ref. 126)

removed and the sand allowed to free fall away from the casting.

This process eliminates the need for high energy moulding equipment, sand binders and additives reducing cost, pollution and combustion hazards. Metallurgical advantages include:

1. the possibility of producing thinner sections;
2. improved casting yield;
3. improved surface finish;
4. a slower cooling rate due to a different heat extraction process, (*Figure 5.40*). This influences iron structure, i.e. minimizes carbide formation.

Casting processes using expendable patterns do not require parting lines and consequently offer much greater freedom of design. Cast-in inserts, clearance holes to size, less machining stock, zero draft, back draft, undercuts, pilot holes and keyways are some of

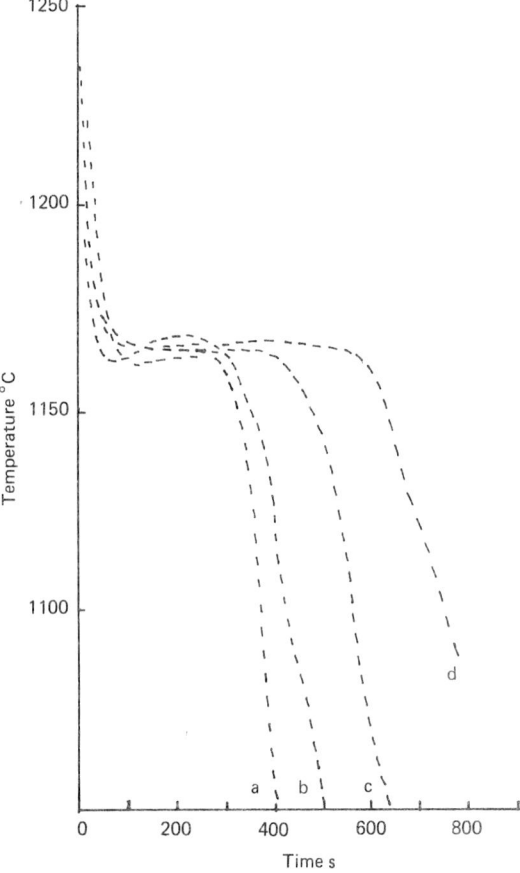

Figure 5.40 Cooling curves for two-inch grey cast iron sections poured in a, Full mould, unbonded sand, screen size AFS 25; b, Full mould, unbonded sand, screen size AFS 45; c, CO_2 mould, sand AFS 45; d, V Process, unbonded zircon sand. The cooling curve for a green sand mould was similar to that for the CO_2 mould and a V Process mould made with a silica sand was similar to (d) (after ref. 129)

the design features possible with expendable patterns. There are two casting processes that fall into this category. The lost wax, precision or investment casting process is well established. The lost foam, full mould or evaporative pattern casting process is being used increasingly.

A separate pattern is made for each casting produced in these processes. Patterns are produced from wax in the lost wax process by injecting the material into a metal die. Wax gates and runners are then attached to the pattern and assembled into a compilation called a tree. This tree is then dipped into a silica-based ceramic slurry. Repeated dipping produces a thick and strong shell of hard refractory. The wax pattern is removed by superheated steam in an autoclave and the shell is then fired in an oven.

The second process was patented in 1958 by Shroyer who used a polystyrene pattern and a self-hardening resin-bonded or silicate-bonded sand. Variations have included the use of unbonded, compacted sand as the mould. In 1971[131] Kubo used a free flowing magnetizable moulding material. This was magnetic shot, 2–4 mm diameter. It was rigidized immediately before pouring by a powerful magnetic field.

The evaporative pattern casting process has been developed in slightly different forms over the past decade[132-139]. It is becoming increasingly popular, particularly for the casting of automobile components. Typical of these developments are the Replicast full mould (FM) and Replicast ceramic shell (CS) methods developed by the Steel Castings Research and Trade Association (SCRATA)[140-142]. *Table 5.20* describes and compares these processes with the lost wax process.

The expanded polystyrene (EPS) pattern production method is similar for all evaporative pattern casting (EPC) processes. It is usually made with an Al die. Depending on the complexity of the component, a one-piece pattern, with the feeding/gating system as an integral part, or with individual parts which can be rapidly assembled using a spray contact adhesive, is made. As received, EPS beads are pre-expanded to a prescribed density, either by using a vacuum pre-expander or by exposing them to steam when the blowing agent, usually pentane, expands. After drying the pre-expanded beads are blown into the Al die. Application of heat, usually steam, causes the beads to soften and expand further, thus completely filling the cavity. Prior to pattern ejection the mould is cooled with water sprays to ensure that the pattern does not continue to expand after ejection.

Polystyrene patterns shrink with age[143] to the extent of about 0.8 cm/m. Approximately 75% of the shrinkage occurs in the first seven days. The shrinkage is completed after about 30 days. High density patterns are desirable because they have good surface finish and are strong. This allows large patterns to be handled with minimum risk of damage and distortion.

Table 5.20 A comparison of the Replicast FM, Replicast CS and lost wax processes of mould making (after ref. 142)

Replicast FM	Replicast CS	Lost wax ceramic shell
EPS pattern Blow partially expanded EPS beads into Al die at low pressure and complete expansion by injecting steam. *Result*: low-density pattern dimensionally accurate and with good surface finish, light and readily handled	*EPS pattern* Blow partially expanded EPS beads into Al die at low pressure and complete expansion by injecting steam. *Result*: high density patterns dimensionally accurate and with very good surface finish, light and readily handled	*Wax pattern* Inject softened wax at high pressure into metal die. *Result*: wax prone to shrinkage and deformation as size and section thickness increases wax heavy and expensive but reclaimable to a degree.
Mould production Apply refractory coating by painting, spraying flow coating or dipping. *Result*: only one or two coats required, handling manually acceptable even on large components possible to automate	*Mould production* Investment with ceramic by applying successive coats of refractory slurry and stucco. *Result*: only 3–4 coats required, relatively light, manual handling even on large components possible to automate Removal of EPS and shell firing by heating in a furnace at 100 °C for 5 minutes to remove EPS and fire binder *Result*: light moulds readily handled in cold condition ready for pouring	*Mould preparation* Investment with ceramic by applying successive coats of refractory slurry and stucco. *Result*: 5–10 coats required, often heavy and difficult to handle manually, possible to automate Removal of wax by subjecting to high pressure super-heated steam in autoclave. *Result*: unfired heavy shell that is vulnerable Shell fired in furnace at 1000 °C for 20 minutes to remove residual wave and fire binder *Result*: heavy mould can be difficult to handle especially in the hot condition in which it is poured
Pouring of mould Coated EPS surrounded by loose sand, vibrated to maximum bulk density and vacuum applied during pouring. *Result*: products from the combustion of EPS removed and support provided for the thin refractory coating to prevent break out and maintain dimensions	*Pouring of mould* Ceramic shell surrounded by loose sand, vibrated to maximum bulk density and vacuum optionally applied during pouring *Result*: possible to provide good support to ensure dimensions held and to avoid break out from the thin ceramic shell	*Pouring of mould* Metal usually poured into hot mould. *Result*: often unsupported and break out possible

The conventional full mould process requires a pattern of low density for it to be consumed satisfactorily by the molten metal. The pattern is removed prior to pouring in the CS process. The use of vacuum during pouring in the FM process means that high density patterns can be used in the CS process and higher density patterns than are used in conventional full mould processes can be used in the FM process. The pattern in the Replicast FM process is coated with a 1-mm layer of refractory paint. After drying the pattern is transfered to the flask, unbonded silica sand is compacted round the pattern.

Although this technique of mould production has been used successfully it always carries the risk of mould collapse. This is caused by sand movement during pouring. This can be due to inadequate compaction in difficult locations or to gas generated on pattern breakdown escaping through the paint layer and fluidizing the sand. The volume of gas produced is greater with the higher pouring temperatures used with cast irons.

The FM process overcomes this difficulty by applying a vacuum just before pouring and for a short time after. The vacuum increases mould rigidity as in the 'V' process. The risk of mould collapse is substantially reduced as a result of a pressure drop created across the refractory coating in the zone between the progressively evaporating EPS and the advancing metal interface. As a result, the coating is held in position against the rigid sand. This preserves the integrity of both the sand and the mould as the pattern is consumed by the liquid metal. The vacuum also removes the products of combustion of the pattern. This renders the process environmentally acceptable. The thickness of the coating must allow the gases from the pattern to be drawn off at the desired rate. However, the coating should not be too permeable in order to prevent the pressure drop across the coating. It should also be sufficiently strong to prevent metal penetration under vacuum.

A pattern is coated with a slightly thicker refractory shell in the CS process, using a similar technique to that in the lost wax process. The pattern is first coated with a ceramic slurry by dipping or flow coating. It is then stuccoed with a granular refractory three or four times until the required thickness is achieved. In contrast to the lost wax process, the pattern is removed during the shell firing at 1000 °C. This is possible because the EPS collapses away from the shell, whereas wax expands when heated and would stress the coating. This shell is much stronger than that in the FM process. Although there is not the same need for vacuum it is usually applied during the casting operation.

Permanent moulding

Permanent moulding processes such as die casting, squeeze casting and continuous casting are energy saving processes that offer the opportunity of producing cast products with high integrity provided their geometry is simple. They can be cheap in the case of volume production. Currently this is proving a successful production route for lower melting point alloys but, hitherto, developments have been much slower with higher melting point alloys, such as cast irons[144]

Gravity die casting

Interest in this process for cast irons has been stimulated recently because of the considerable energy savings that can be made compared to sand casting. The savings accrue from the elimination of sand processes such as mixing, conveying, compacting, cooling and extraction. Higher mould yields also mean reduced melting and material costs. The elimination of feeders reduces finishing costs. The elimination of sand minimizes environmental problems. The die casting of cast irons has been limited to a few small automobile components. It is typically no more than 5% of any country's iron casting output, despite

the advantages of the process listed in *Table 5.21*. This has been attributed to difficulties experienced in filling moulds, their cost and limited life. Details of plant have been given in the literature[145, 146]

The mould, normally in two halves, is usually parted in the vertical plane. Opening, closing and clamping of the mould halves is engineered using pneumatic or hydraulic systems. The mould may be cooled by water, oil, forced air or even naturally. Some form of mechanical ejection system is incorporated into the mould construction. Vents must be included in the mould design and construction. Moulds have been made from ferritic flake iron, Cu and Cu alloys, alloy steels and graphite.

It is common practice to preheat iron moulds and to use a mould coating to improve surface quality, reduce mould chilling and to extend mould life. Coatings vary from silica, alumina and mixes of refractory oxides. They are applied by spray coating as a mixture with silicate and water to a hot mould face. They are dried by evaporation or sometimes by stove drying. A final coating of acetylene sooting, which acts as a lubricant or release agent and assists ejection, is usually applied.

Thermal fatigue is a significant factor in determining mould life. Consequently, for a selected mould, factors determining its thermal behaviour must be carefully controlled and co-ordinated for optimum economic operation. These include the chemistry and pouring temperature of the liquid iron, the 'freeze' time, which is the time between mould filling and casting ejection, ejection efficiency and the mode of cooling that controls the upper temperature reached by the mould.

Henych and Gysel[147] have discussed the properties required for a die casting mould and analysed the thermal performance of various moulds. They emphasize the merits of a high performance mould made from a Cu-base alloy with the properties shown in *Table 5.22*. The active layer is the depth of mould influenced by heat flow at any time. The heat removed from the casting is proportional to the growth enthalpy in the mould.

With respect to the growth enthalpy the table shows that the chilling tendency of the HP mould is similar to that of a sand mould. That of the grey iron is several times higher. The HP mould also self-heats rapidly. This means that a good surface quality is obtained without the need for mould preheating and coating. The high mould thermal conductivity sustains a high temperature gradient between casting and mould wall and a low gradient in the mould itself. This reduces thermal stressing and increases mould life, as shown in *Table 5.22*. The high temperature gradient promotes rapid skin formation, which allows a reduction to be made in the dwell time of the casting. A pour to pour time of approximately 300 seconds can be achieved for small cast iron components using

Table 5.21 Some of the advantages of die casting cast iron compared to sand casting

Characteristic	Sand cast	Die cast	
Geometry of the casting	+		
Universality of process	+ +		
Degree of difficulty of casting	+		
Surface quality		+	DC no blasting
Liquid metal yield		+ +	SC 40–65%
			DC 70–98%
Product uniformity		+ +	
Mould costs per casting		+ +	DC to 80% lower
Total energy consumption		+ +	DC can be up to
			70% lower
Specific investment costs		+ +	
Floor space requirement		+ +	DC to 10^4 kg/m^2
Operation productivity		+ +	DC to 6×10^5
			kg/man-year
Environment burden		+ +	DC no sand
Manufacturing costs		+ +	DC 10–40% lower

+, Advantage   + +, Important advantage.

Table 5.22 The properties of a Cu alloy high performance permanent mould; the active layer, average mould temperature increase and the growth enthalpy of different moulds 1 and 5 seconds after the liquid iron contacts the mould, and a comparison of grey iron and high performance mould lives (after ref. 147)

	Cu alloy high performance permanent mould					
	Specific weight 8.9 g cm^{-3}					
	Tensile strength 550 N mm^{-2}					
	Modulus of elasticity 1.5×10^5 N mm^{-2}					
	Thermal conductivity 352 W m^{-1} K^{-1}					
	Specific heat 0.38 J K^{-1}					

Time after liquid iron touches mould wall	*1 s*			*5 s*		
Material	*Active layer mm*	*Average temperature increase °C*	*Growth of enthalpy kJ*	*Active layer mm*	*Average temperature increase °C*	*Growth enthalpy kJ*
copper	31.0	11.3	11.7	75.0	25.0	62.8
H.P. mould	29.0	14.0	14.3	70.0	26.6	62.5
Grey iron	15.0	81.2	46.9	28.0	89.0	96.0
Sand mould	1.5	505.0	14.3	4.0	312.0	23.5

Cast material	*Weight of casting kg*	*Grey iron mould*		*H.P. mould*	
		Mould consumption kg/t	*Numbers of cast cycles*	*Mould consumption kg/t*	*Numbers of cast cycles*
Grey ion	1.6	13.1	3000*	0.40	> 150 000
	1.6	6.6	7000$^+$		
	100	33.3	150*		
High-Cr iron	30–500			0.35	15 000

* air cooled, $^+$ water cooled.

a cast iron die. A typical rotary type gravity die casting plant would be equipped with between six and twelve stations. The same casting, with a pour time of 15 seconds, would require a dwell time of about 10 seconds in a HP die. The casting rejection and mould preparation time would be about 40 seconds. Consequently, a significant reduction in pour to pour time is possible. This means that the same casting performance can be achieved with four stations and a third of the number of moulds.

Although the cost of each HP mould is significantly higher than a grey iron mould, its greater life more than compensates for this, especially in high volume production situations. Structural studies[148-154] have shown that considerations for structural control are similar to those described for sand castings. However, particular attention must be given to the increased chilling tendency, particularly with cast iron moulds (see *Figure 1.9* in Chapter 1). Mould temperature and pouring temperature have a minor influence on chilling. The critical factors are control over inoculation and the iron composition.

Experiences with the commercial production of ferritic spheroidal irons in grey iron moulds has been described by Godsell[155]. Important factors were chemical composition, spheroidizing practice, inoculant choice and practice, mould face temperature, metal pouring temperature, casting temperature at ejection and mould coating. A two-stage inoculation procedure was adopted in the commercial process developed. One third of the total Si was added as FeSi on tapping the furnace. A final inoculation was performed as a late stream addition. A granular 7% rare earth, containing magnesium ferrosilicon alloy, was used to spheroidize. Pouring temperature was maintained between 1320 and 1380 °C and the mould face temperature was between 150 and 300 °C.

Figure 5.41 shows the influence of iron composition on various iron properties. These effects resulted in varous guidelines. Residual Mg content was kept below 0.025% to avoid chill. Mn was restricted to 0.3% for all section sizes but below 0.1% for sections of 25 mm or less. The final Si content was about 3% for sections exceeding 50 mm and up to 4% for sections less than 12 mm. The minimum C level was 3.6%. The C.E.V. was increased from 4.5, for sections over 50 mm, to over 5 for sections of 12 mm and less. P was maintained at levels less than 0.04%. Cu was used to promote pearlite in as-cast irons. The casting was ejected in the austenite range to enable matrix structure control by controlled cooling after ejection.

A variety of spherical iron components, including brake carriers, crankshafts, conrods and thin walled manifolds were cast. Their structures displayed high spheroid count, good spheroid shape, freedom from chill and intercellular carbide and various matrix structures to satisfy different specifications. *Figure 5.42* shows the structure of a brake carrier. Increasing the Si content and the amount of pearlite has the disadvantage of reducing the fracture toughness of the casting[156,157]. Benefits claimed for permanent mould castings include good surface finish and definition, close dimensional tolerances, good structural integrity and a reduced need for machining. Where required, there is better machineability. Experience has shown that cutting speeds for turning operations on permanent mould castings can be increased compared to sand cast components. Drill life is also increased considerably. The two reasons for

these differences are derived from the absence of sand inclusions in the casting surface and the different graphite morphologies resulting from the different solidification rates.

Centrifugal casting processes

The centrifugal casting of irons, particularly of the spheroidal graphite types is still a relatively recent development in the history of the iron casting industry. The first recorded patent was issued to Anthony Ekhardt, an Englishman, in 1809. During the next 100 years development was slow and erratic. The first significant growth in this process occurred during the first World War with the horizontal casting of pipe. Known as the horizontal axis centrifugal process, pipe sections are produced by pouring iron, usually at one end only, into a mould which is spinning at a rate that will produce a G force of 70 or more on the inside diameter of the finished tube. During pouring metal distributes itself uniformly along the length of the mould. Control of pouring

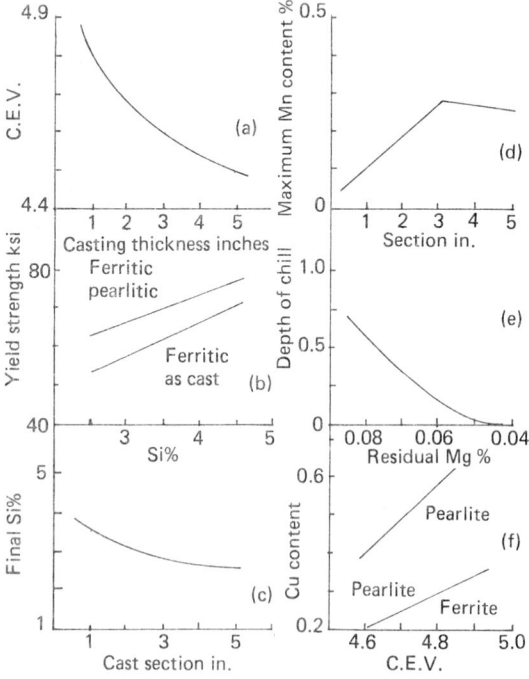

Figure 5.41 (a) Recommended carbon equivalent for different casting thicknesses; (b) Effect of Si on the yield strength of spheroidal iron permanent mould castings; (c) Recommended final Si as a function of cast section; (d) Limits of Mn content to avoid carbide formation; (e) Influence of residual Mg on chill depth of one inch section of an iron of C.E.V. = 4.8; (f) Recommended Cu content for pearlite formation in as-cast irons (after ref. 155)

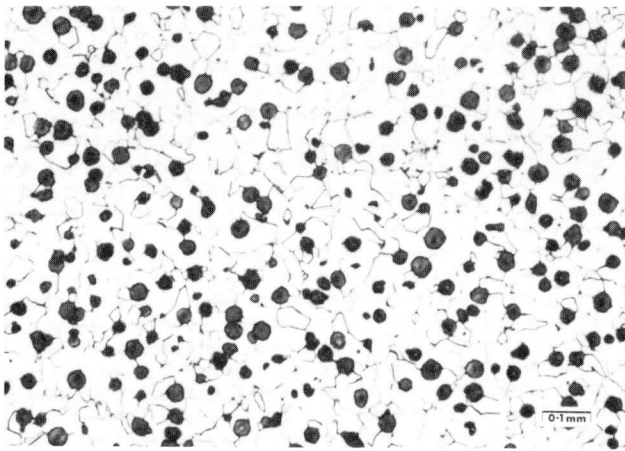

Figure 5.42 A typical microstructure of a permanent mould casting of ferritic spheroidal iron

parameters is crucial to avoid variations in wall thickness and fractures.

The mode of operation depends on the type of mould, method of mould preparation and the availability of molten iron. The basic mould consists of a metal shell or die with a lining and removable end plates to retain the metal in the mould during casting. Different die lining methods have been developed. Commonly, propriety coatings consisting of a silicon base with a binder such as bentonite are sprayed on to a preheated die or spun in place on a cold die and then oven dried.

The most suitable machine design and equipment depends on the cycle time of the various operations. The more important basic elements are:

1. mould preparation time, i.e. the elapsed time between extracting one pipe and pouring the next casting in the same die;
2. pour cycle, i.e. the time to get the mould up to operating speed, pour, solidify and stop the rotation,
3. extraction time, i.e. the time to remove the casting from the mould;
4. time to clean the mould;
5. the number of dies available;
6. availability of iron;
7. the number of pouring stations or machines.

In many respects the art of centrifugal casting is a new and rapidly developing process. It has matured in some areas, for example in the mass production of iron pipe. Generally, the type of product that lends itself to centrifugal production has one or more of the following characteristics:

1. it will be essentially round with a hole in the centre;
2. the shape and size of the part will not be readily available by other manufacturing methods;

3. the quality and reliability of the part will be superior to that of a competitive product and
4. the cost of production will be competitive with other processes.

Continuous casting

This method of casting finds application in the production of bar stock in the shape of rounds, squares, rectangles, hexagonals and other simple shapes. The continuous casting machine comprises a metal receiver into which freshly inoculated iron is continuously fed, which, adjacent to the mould, usually a water cooled graphite mould provides not only a continuous supply of iron but acts as a feeder head eliminating liquid shrinkage defects. The solidified bar is continuously drawn from the mould and cropped into any standard convenient length, usually 6 or 8 feet.

A wide range of grey, alloy and spheroidal irons are available commercially. This basic stock is often attractive to a wide range of markets from the designer who requires speedy testing of new designs in a variety of materials and avoidance of pattern time considerations, to the producer of high volume simple shapes which demands high integrity. The high quality product is characterized by the finer surface graphite structures and superior machineability when compared to more conventionally cast irons.

Fettling

The finishing of castings after removal from the mould is probably the aspect of foundry processing which has received least engineering development or attention. Fettling has been reduced, largely through improved moulding methods. However, there is a limit and it cannot be eliminated entirely from the casting process. Much has been done for casting cleaning by developments of shot blasting machines. These machines now have sufficient capacity so that airless turbines can project shot with such velocity and volume that decoring can take place inside the cabinet. Separation systems remove the sand from the shot so that only clean shot is used for blasting. The continuous barrel type is proving very attractive for blast cleaning of small castings. The main requirement is for high quantity production to be fed into the machine continuously, thereby pushing the castings forward into the helix governing the discharge.

The term fettling constitutes the major part of finishing processing. It may be regarded as including:

1. cleaning, involving the removal of adherent sand and scale;
2. dressing, involving the removal of excess metal, heads and feeders, ingates and runners etc.;
3. annealing operations and heat treatment.

Because fettling arises out of the need to provide runner channels to enable molten iron to flow freely and enter the mould cavity, the ease with which fettling is carried out is decided by the fundamental properties of the iron being fettled. In the case of grey, mottled or white irons, the runners, feeders etc. may be simply knocked off due to the low ductility of the irons involved. Spheroidal irons are more difficult when it comes to ingate detachment. Gating systems often take account of fettling difficulties by incorporating a wedge-shaped ingate to ease the problem and avoid casting rejection due to the metal 'breaking-in' during removal from the casting by mechanical means. Grinding and finishing of castings has developed only insofar as wheel speeds have been increased to enable stock removal to be carried out at greater speed where pads remain from feeder junctions.

In recent years when discussing fettling, the discussion has been directed to the greatest problem, that of the environmental control. It is not only the dust created by the grinding operation which produces health hazards, but also the level of noise created by the operations. Noise may be generated by mechanical and pseudo-mechanical equipment. It is often in association with other hazards, e.g. cranes, blowers, exhaust fans, compressors etc. In order to reduce or obviate noise it may be necessary to:

1. replace old machines by quieter machines;
2. modify the design perhaps by incorporating material of high damping capacity or low ring such as nylon;
3. fit noise suppression mounts so that structure-borne noise is reduced;
4. make machines automatic so that they can be isolated in booths.

References

1. BERRY, J.T. and PEHLKE, R.D., Computer aided design for castings, *A.F.S. Trans.*, **92,** 101 (1984)
2. CLEGG, A.J., Foundry CAD/CAM developments; an overview, *Brit. Foundryman*, **79,** 397 (1986)
3. DE L. DAVIES, V., Computer aided design, *Proceedings of Perspectives in Metallurgical Development*, Metals Society, London, p. 101 (1984)
4. CLYNE, T.W., Modelling of heat flow in solidification, *Mat. Sci. Eng.*, **65,** 111 (1984)
5. PEHLKE, R.D., MARRONE, R.E. and WILKS, J.O., Computer simulation of solidification, *Monograph*, A.F.S., Des Plaines, Il., p. 232 (1976)
6. ERICKSON, W.C., Computer simulation of solidification, *A.F.S. International Cast Metals Journal*, **5,** 30 (1980)
7. SULLY, L.J.D., The thermal interface between casting and chill moulds, *A.F.S. Trans.*, **84,** 735 (1976)
8. HO, K. and PEHLKE, R.D., Mechanisms of heat transfer at a metal mould interface, *A.F.S. Trans.*, **92,** 587 (1984)
9. WINTER, B.P., OSTROOM, T.R., HARTMAN, D.J., TROJAN, P.K. and PEHLKE, R.D., Mould dilation and volumetric shrinkage of white, grey and ductile cast iron, *A.F.S. Trans.*, **92,** 551 (1984)
10. ZENG, X.C and PEHLKE, R.D., Analysis of heat transfer at metal sand mould boundaries and computer simulation of the solidification of a grey iron casting, *A.F.S. Trans.*, **93,** 275 (1985)
11. PEHLKE, R.D., The interface in computer simulation of heat transfer in metallurgical processes, *Metals Eng. Quarterly*, **11,** 9 (1971)
12. GARCIA, A. and CLYNE, T.W., A versatile technique for characterisation of metal/mould heat transfer and correlation with thermal and structural effects, *Proc. Solidification Technology in the Foundry and Casthouse*, Inst. of Metals, Warwick (1980)
13. HANSEN, P.N., Solidification and related structure as a function of the metal/mould boundary temperature, *Proc. Solidification Technology in the Foundry and Casthouse*, Inst. of Metals, Warwick (1980)
14. CRUI, J.S. and CHEVRIES, J.C., Finite element and thermal contact resistance modelling, *Proc. Solidification Processing*, Sheffield, (1987)
15. HARTLEY, J.G. and PATTERSON, J.A.I., The influence of temperature, moisture content and binder content on the thermal conductivity of dried bentonite bonded zircon and silica sands, paper presented at A.F.S. Casting Congress (1983)
16. DESAI, P.V., BERRY, J.T. and KIM, C., Computer simulation of forced and natural convection during filling of a casting, *A.F.S. Trans.*, **92,** 519 (1984)
17. PEHLKE, R.D., BERRY, J.T., ERICKSON, W.C. and JACOBS, C.H., The simulation of shaped casting solidification, *Proceedings of Solidification of Metals*, Sheffield (1977)
18. DE L. DAVIES, V. and MOE, R., Computed feeding range for gravity die castings, *Proceedings of Solidification of Metals*, Sheffield (1977)
19. HEINE, R.W. and UICKER, J.I., Risering by computer

assisted geometric modelling, *A.F.S. Trans.*, **91**, 127 (1983)

20. BERRY, J.T., PEHLKE, R.D. and DESAI, P.V., Recent developments in shaped casting solidification simulation, *Proc. Solidification Processing*, Sheffield, (1987)

21. HONG, C.P. and UMEDA, T., Numerical simulation of solidification processes by boundary element method, *Proc. Solidification Processing*, Sheffield (1987)

22. HANSEN, F.B., The Hubert Project, *Proc. Solidification Processing*, Sheffield (1987)

23. RENKONEN, A., Casting design using solidification simulation based on the F.E.M. heat transfer analysis, *Proc. Solidification Processing*, Sheffield (1987)

24. OHNAKA, I., Computer simulation of solidification of castings, *Proc. Solidification Processing*, Sheffield (1987)

25. HAMAR, R. and HERNANDEZ-OCANDO, F., Feeding effect and casting optimisation, *Proc. Solidification Processing*, Sheffield (1987)

26. HUANG, H.C., LEWIS, R.W. and MORGAN, K., Multigrad approach to finite element simulation of casting problems, *Proc. Solidification Processing*, Sheffield (1987)

27. HAMILTON, E., Tooling for lost wax investment castings, *A.F.S. Trans.*, **93**, 903 (1985)

28. SHEEHAN, J. and RICHARDSON, D., The H.M.P. process – a new method for producing plastic patterns, *A.F.S. Trans.*, **92**, 203 (1984)

29. ANDERSON, G. and CROWLEY, T.J., Precision foundry tooling utilising CAD/CAM and the H.M.P. process, *A.F.S. Trans.*, **93**, 895 (1985)

30. PATZ, M., Lost foam process update, *A.F.S. Trans.*, **93**, 901 (1985)

31. HOULT, F.H., The H process of repetition casting manufacture, *A.F.S. Trans.*, **87**, 237 (1979)

32. GERHARDT, P.C., Jr., Computer applications in gating and risering system design for ductile iron castings, *A.F.S. Trans.*, **91**, 475 (1983)

33. FLINN, R.A., *Fundamentals of metal casting*, Addison Wesley, London, p. 41 (1962)

34. AUBREY, L.A., BROCKMAYER, J.W. and WIESER, P.F., Dross removal from ductile iron with ceramic foam filters, *A.F.S. Trans.*, **93**, 171 (1985)

35. WEBSTER, P.D. and YOUNG, J.M., Computer aided gating systems design, *Brit. Foundryman*, **79**, 276 (1986)

36. KARSAY, S.I., *Ductile iron III: Gating and risering*, QIT. Fer. et Titane Inc. Publication, p. 44 (1981)

37. WHITE, R.W., Risering of ductile iron castings, *Foundry*, **88**, 96 (1960)

38. BISHOP, H.F., MYSKOWSKI, E.T. and PELLINI, W.S., A simplified method for determining riser dimensions, *A.F.S. Trans.*, **63**, 271 (1955)

39. BISHOP, H.F. and ACKERLIND, C.G., Dimensioning of risers for nodular iron castings, *N.R.L. report 4737* (1956)

40. CORBETT, C.F., Methoding of castings using a microcomputer, *Brit. Foundryman*, **76**, 117 (1983)

41. GOUGH, M.J. and MORGAN, J.J., Feeding ductile iron castings – some recent experiments, *A.F.S. Trans.*, **84**, 358 (1976)

42. RUDDLE, R.W., A computer program for steel risering, *A.F.S. Trans.*, **90**, 227 (1982)

43. GOUGH, M.J. and CLIFFORD, M.J., Feeding nodular iron castings, *Proc. Soldification Technology in the Foundry and Casthouse*, Inst. of Metals, Warwick (1980)

44. HEINE, R.W., Riser design for mould dilation, *A.F.S. Trans.*, **73**, 34 (1965)

45. AMRHEIN, R.F. and HEINE, R.W., Experiences with riser design, *A.F.S. Trans.*, **75**, 659 (1967)

46. HEINE, R.W., Feeding paths for risering castings, *A.F.S. Trans.*, **76**, 134 (1968)

47. ROBERTS, R., LOPER, C.R. Jr. and HEINE, R.W., Riser design, *A.F.S. Trans.*, **77**, 373 (1969)

48. HEINE, R.W., Riser base and connection design for white iron castings, *A.F.S. Trans.*, **77**, 559 (1969)

49. LOPER, C.R. Jr., BANNERTEE, P. and HEINE, R.W., Risering requirements of ductile iron castings, *Grey Iron News*, May, 1964

50. HEINE, R.W., Risering principles applied to ductile iron castings made in green sand, *A.F.S. Trans.*, **87**, 65 (1979)

51. HEINE, R.W., Design method for tapered riser feeding of ductile iron castings in green sand, *A.F.S. Trans.*, **90**, 147 (1982)

52. HEINE, R.W., UICKER, J.J. and GATTENBEIN, D., Geometric modelling of mould aggregates, superheat, edge effects, feeding distances, chills and solidification microstructure, *A.F.S. Trans.*, **92**, 135 (1984)

53. HEINE, R.W. and LEEDOM, C.P., Comparing the functioning of risers to their behaviour predicted by computer programs, *A.F.S. Trans.*, **93**, 481 (1985)

54. BROWN, D. and HEINE, R.W., Riser conversion from white iron to ductile iron casting, *A.F.S. Trans.*, **90**, 73 (1982)

55. KARSAY, S.I., Risering methods for gray and ductile iron castings, *A.F.S. International Cast Metals Journal*, **5**, 45 (1980)

56. CORLETT, G.A. and ANDERSON, J.V., Experiences with an applied risering technique for the production of ductile iron castings, *A.F.S. Trans.*, **91**, 173 (1983)

57. LOPER, C.R. Jr., Sequential filling of mould cavities, *A.F.S. Trans.*, **89**, 405 (1981)

58. RASSENFOSS, J.A., Mould materials for ferrous castings, *A.F.S. Trans.*, **85**, 583 (1977)

59. ATTERTON, D.V. and ROBERTS, W.R., Moulding materials, *Brit. Foundryman*, **64**, 421 (1971)

60. NICHOLAS, K.E.L. and ROBERTS, W.R., Mould compaction, moulding sand composition and their influences on the quality of iron castings, *Brit. Foundryman*, **72**, 69 (1979)

61. ADAM, W.L., Pressure moulding with standard synthetic sand – 25 years of progress, *A.F.S. Trans.*, **88**, 231 (1980)

62. DISYLVESTRO, G., MUNN, H.A. and DIMMER, M., Stability – ability – profitability in green sand moulding – an industrial comparison, *A.F.S. Trans.*, **90**, 699 (1982)

63. SCOTT – WEBSTER, W., High pressure moulding, *Brit. Foundryman*, **75**, 167 (1982)

64. SINDERMANN, H., Green sand compaction by air impulse

moulding: theory and practical experience, *A.F.S. Trans.*, **93**, 241 (1985)

65. MOORE, W.H., Ramming characteristics of moulding sand, *A.F.S. Trans.*, **86**, 1 (1978)

66. BROOME, A.J., Mould and core coatings and their application, *Brit. Foundryman*, **73**, 96 (1980)

67. LOWER, G.W., MOORE, D.J., PASTERNAK, L., RUNDMAN, K.B. and WEIRES, D.J., Properties of moulding sands: Effects of sand size distribution and mixing energy input, *A.F.S. Trans.*, **91**, 559 (1983)

68. SMIERNOW, G.A., DOHENY, E.L. and KAY, J.G., Bonding mechanisms in sand aggregates, *A.F.S. Trans.*, **88**, 659 (1980)

69. SHIH, T.S., GREEN, R.A. and HEINE, R.W., Evaluation of green sand properties and clay behaviour at 7 to 15% bentonite levels, *A.F.S. Trans.*, **92**, 467 (1984).

70. SHIH, T.S., HEINE, R.W. and GREEN, R.A., Evaluation of green sand properties and clay behaviour at 8 to 15% bentonite levels: Part II, *A.F.S. Trans.*, **93**, 689 (1985)

71. BEELEY, P., *Foundry Technology*, Butterworths, London (1976)

72. KUPCRUNS, S.K. and SZRENIAWSKI, J., Hardness as an indirect measure of the physical properties of densified moulding sands, *A.F.S. Trans.*, **91**, 191 (1983)

73. SCHUMACHER, J.S. and HEINE, R.W., The problem of hot moulding sands – 1958 revisited, *A.F.S. Trans.*, **91**, 879 (1983)

74. BETHKE, L.R., Trend analysis in green sand control, *A.F.S. Trans.*, **91**, 423 (1983)

75. SODERLING, L. and NEITNER, S., Applied spectrophotometry in methylene blue clay determinations, *A.F.S. Trans.*, **92**, 77 (1984)

76. BETHKE, L. and MASARIK, J., Trend analysis in green sand control – a computer program, *A.F.S. Trans.*, **92**, 77 (1984)

77. ATTERTON, D.V., The carbon dioxide process, *A.F.S. Trans.*, **64**, 14 (1956)

78. NICHOLAS, K.E.L., The CO_2 – silicate process in foundries, paper presented at BCIRA Conference (1972)

79. ATTERTON, D.V. and STEVENSON, J.V., The carbon dioxide process: a look back – a look forward, *A.F.S. Trans.*, **89**, 55 (1981)

80. STANBRIDGE, R.P. and WILLIAMS, R., Recent advances in the technology and application of silicate bonded sand, *A.F.S. Trans.*, **87**, 77 (1979)

81. OWUSA, Y.A. and DRAPER, A.B., Inorganic additives improve the humidity resistance and shakeout properties of sodium silicate bonded sand, *A.F.S. Trans.*, **89**, 47 (1981)

82. MacDONALD, R.M., Sodium silicate is back again, *A.F.S. Trans.*, **88**, 451 (1980)

83. GOTHERIDGE, J., A look into the future wider applications of the sodium silicate/carbon dioxide process through a better understanding of the basic principles and the new technology, *A.F.S. Trans.*, **87**, 765 (1979)

84. GEORGE, R.D. and SMITH, G., Advances in the silicate CO_2 process, paper presented at BCIRA Conference on Advances in Technology and Binders for gas hardened moulds and cores (1986)

85. CARLSON, G. and THYBERG, B., Core production by the sodium silicate warm air process, paper presented at BCIRA Conference on Advances in Technology and Binders for gas hardened moulds and cores (1986)

86. HUUSMANN, O. and LEMKOV, J., Hot air curing of sodium silicate powder bonded sand for cores, paper presented at BCIRA Conference on Advances in Technology and Binders for gas hardened moulds and cores (1986)

87. SVENSSON, I. and LEMKOV, J., Further developments of silicate binders for mould and core production, paper presented at BCIRA Conference on Advances in Technology and Binders for gas hardened moulds and cores (1986)

88. SVENSSON, I., Kinetics and properties of dry curing of sodium silicate binders, *Brit. Foundryman*, **79**, 223 (1986)

89. PIOTROWSKI, K.J. and SVENNSON, I., On the ester curing of sodium silicate based binders, *Brit. Foundryman*, **78**, 117 (1985)

90. MARTIN, G.J. and ENNIS, C.S., CO_2 silicate process – the state of the art, *Brit. Foundryman*, **78**, 376 (1985)

91. SKUBON, M.J., Microwave curing of core binders and coatings, *A.F.S. Trans.*, **86**, 183 (1978)

92. LIEDEL, D.S., Reclamation of sodium silicate bonded sands, *A.F.S. Trans.*, **93**, 429 (1985)

93. STEVENSON, J., Reclamation of sodium silicate bonded sand, paper presented at BCIRA Conference on Advances in Technology and Binders for gas hardened moulds and cores (1986)

94. BOLTON, G.J., DOLES, R.S., McNALLY, F.M. and SHAKELFORD, D.A., An introduction to the SAH binder system, *A.F.S. Trans.*, **91**, 675 (1983)

95. BOLTON, G.J., Reclamation of a new inorganic, no bake binder, *A.F.S. Trans.*, **91**, 95 (1983)

96. HIGHFIELD, J.W. and LAFAY, V.S., The mechanism, control and application of self setting sodium silicate binder systems, *A.F.S. Trans.*, **90**, 201 (1982)

97. ANON., Crossfield soluble silicates for the foundry industry, Booklet, Crossfield Chemicals Warrington, England (1986)

98. KIM, Y.D., Development of zero nitrogen furan no bake binders, *A.F.S. Trans.*, **89**, 113 (1981)

99. LUKACEK, G.S., FONTAINE, R.E. and RUFFER, N.J., Humidity – its effect on no bake binders, *A.F.S. Trans.*, **91**, 455 (1983)

100. MONROE, R.W., Use of iron oxide in mould and core mixes for ferrous castings, *A.F.S. Trans.*, **93**, 355 (1985)

101. ROBINS, J. and TORIELLO, L.I., Cold box process performance of improved binder system, *A.F.S. Trans.*, **78**, 157 (1970)

102. SCHOEN, J.R., BERANT, D.J., ROWLAND, R.A., LUTHER, N., SCHULZE, W.E. and CHEEK, W. Cold box systems engineering, *A.F.S. Trans.*, **85**, 545 (1977)

103. WARD, C.A. and COX, G.M.A., Todays tooling requirements for automatic production of precision cold box cores, paper presented at BCIRA Conference on Advances in Technology and Binders for gas hardened moulds and cores (1986)

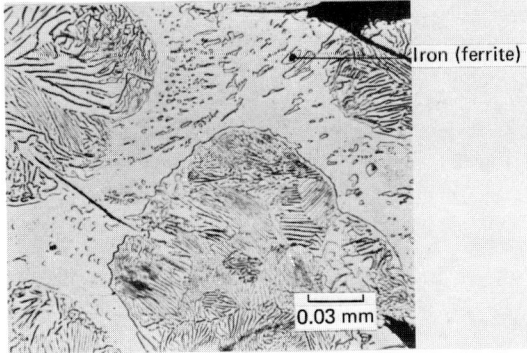

Figure 6.6 shows dendrites of pearlite and interdendritic ternary eutectic which appears to have two constituents when etched in 4% picral. (Courtesy BCIRA).

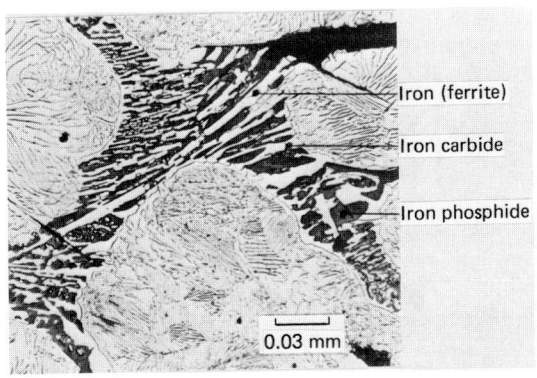

Figure 6.7 etched in Murakami's reagent, shows that some, but not all, of one constituent is Fe_3P. The use of a third etchant, hot alkaline sodium picrate, which darkens Fe_3C but leaves ferrite and Fe_3P unetched, shows that the interdendritic ternary eutectic consists of ferrite, Fe_3P and Fe_3C. (Courtesy BCIRA).

Grey flake irons

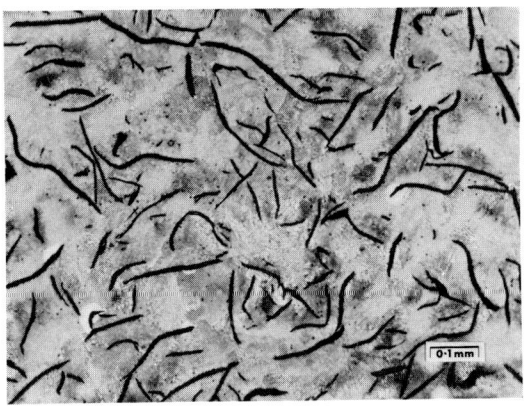

Figure 6.8 is taken from a high grade flake iron of composition

2.9% C, 1.5% Si, 0.70% Mn, 0.12% S, 0.05% P

It shows a microstructure of type A graphite flakes in a pearlitic matrix. Etchant 4% picral. (Courtesy BCIRA).

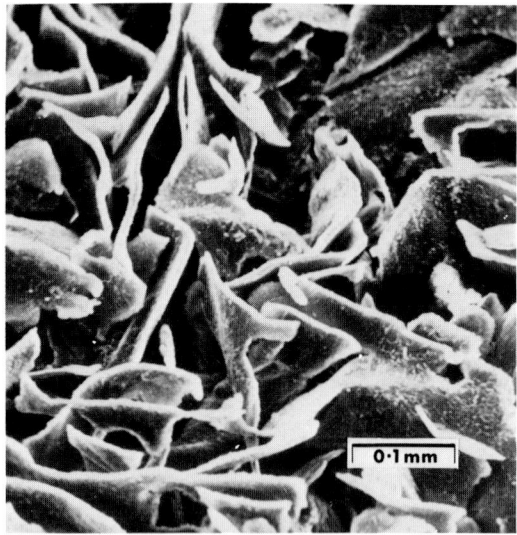

Figure 6.9 is a S.E.M. micrograph of the structure of the iron shown in *Figure 6.8*. It shows the highly anisotropic nature of the graphite flakes, which are interconnected and branch frequently. (Courtesy BCIRA).

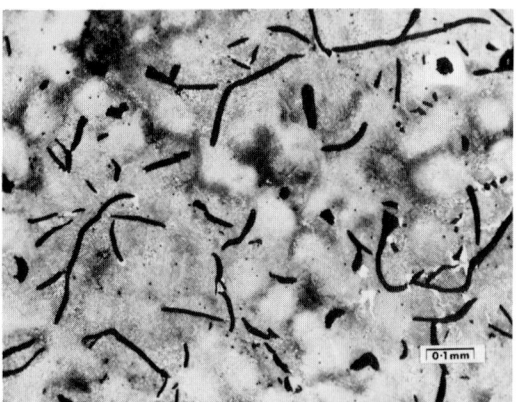

Figure 6.10 shows the microstructure of a compacted graphite iron. It displays short and thick graphite flakes in a pearlitic matrix. Etched in 4% picral. (Courtesy BCIRA).

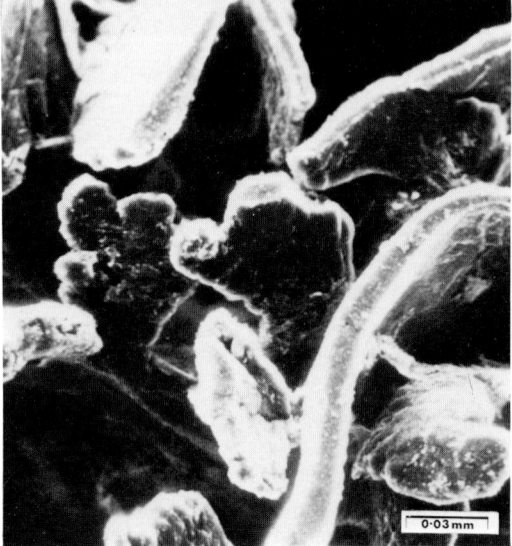

Figure 6.11 shows an S.E.M. micrograph of a compacted iron made by using a Mg–Ce–Ti addition to control the graphite morphology. Compared to *Figure 6.9* the flakes are thicker and the edges are blunt. In common with *Figure 6.9* the flakes are interconnected. (Courtesy BCIRA).

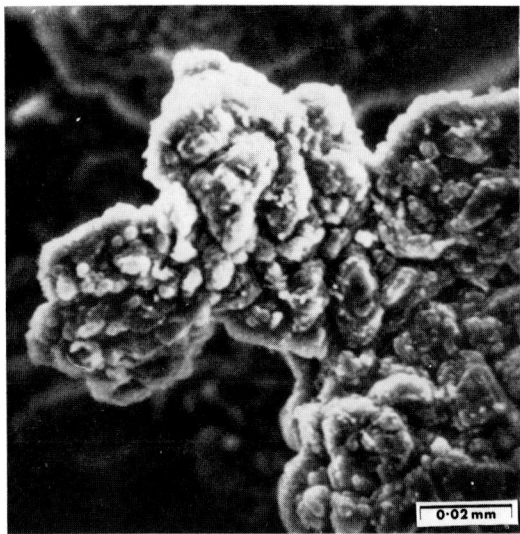

Figure 6.12 shows a S.E.M. micrograph of compacted graphite in a thick section casting of composition

3.7% C, 1.6% Si, 0.66% Mn,
0.01% S, 0.05% P, 65 p.p.m. N

The N segregates to eutectic cell boundaries during solidification where there is sufficient in solution to influence the growth of graphite flakes. Flake thickening is promoted by the growth of hexagonal hillocks on the surface of the flakes. (Courtesy BCIRA).

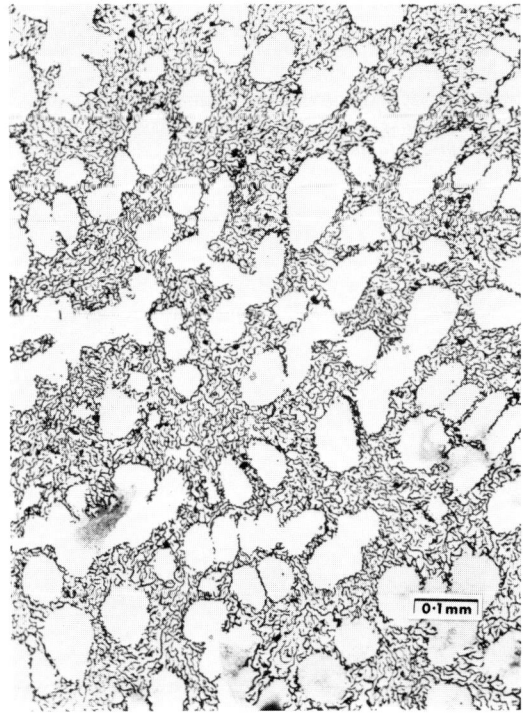

Figure 6.13 shows a microstructure of proeutectic dendrites and fine undercooled interdendritic graphite in the unetched condition. This type D graphite is found in rapidly solidified irons and irons containing Ti. (Courtesy BCIRA).

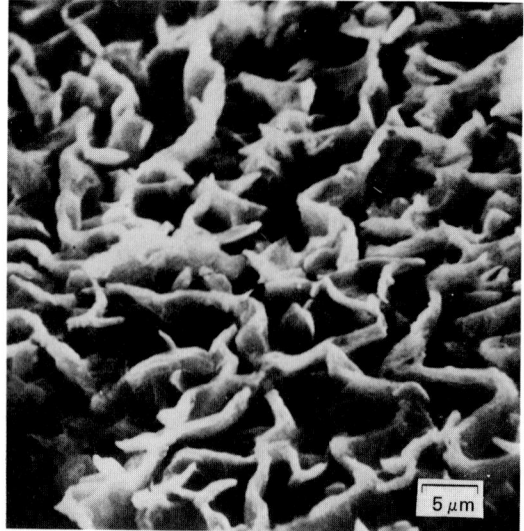

Figure 6.14 is a S.E.M. micrograph taken from the iron shown in *Figure 6.13*. It shows that undercooled graphite is of flake form and interconnected, but is much finer than that shown in *Figure 6.9*. This is due to the presence of a greater undercooling during growth which results in more frequent branching of the graphite. (Courtesy BCIRA).

Abnormal graphite flake forms

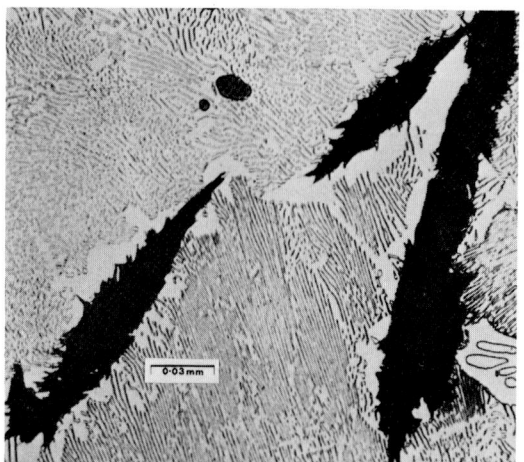

Figure 6.15 is taken from a heavy balance weight made of grey flake iron contaminated with 0.003% Pb. It shows an abnormal flake morphology known as 'sooty' or 'spiky' graphite. It occurs in irons containing Pb and H. Pb may be introduced from contaminated scrap. It may be introduced from moisture in ladies or green sand moulds. This degenerate form of graphite is found in heavy section castings or in areas of light castings that have been cooled slowly. 'Spiky' graphite causes a significant reduction in strength. Etched in 4% picral. (Courtesy BCIRA).

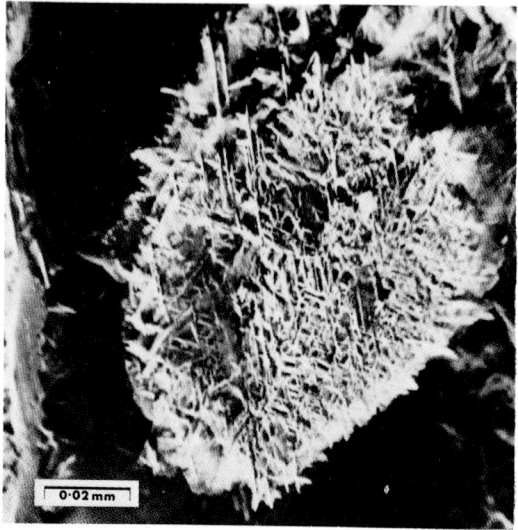

Figure 6.16 is a S.E.M. micrograph which illustrates that 'spiky' graphite shows a Widmanstätten mode of growth at the flake surface. (Courtesy BCIRA).

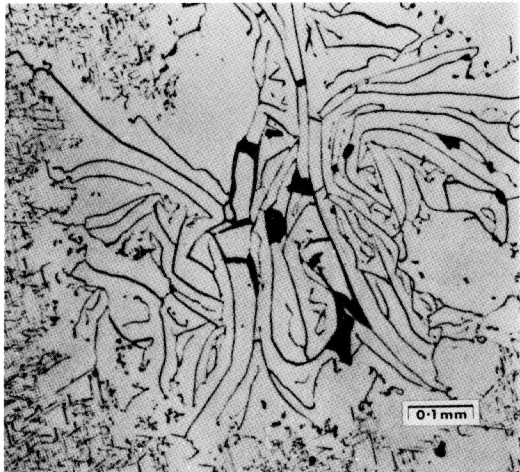

Figure 6.17 Austenitic irons can dissolve more Pb than unalloyed irons and display more extensive Widmanstätten graphite patterns. The photograph shows Widmanstätten graphite in an austenitic iron which contains

16.6% Ni, 6.0% Cu and 0.03% Pb

It was cast from a damp ladle. Etched in 5% nitric acid in acetone. (Courtesy BCIRA).

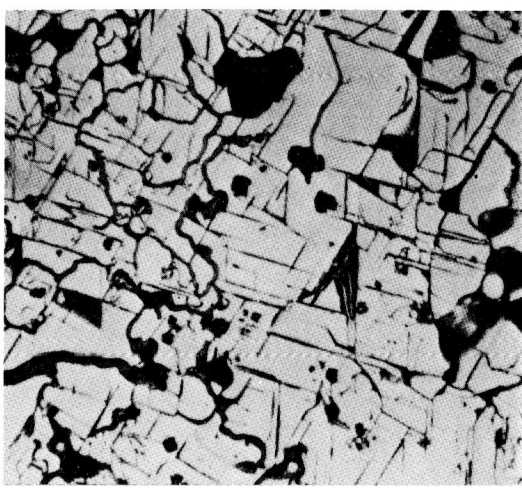

Figure 6.18 shows extensive Widmanstätten graphite formation in high Si (15%) iron cast from a damp ladle. Etched in etchant of 1 part HF, 2 parts HNO_3 and 3 parts glycerine. (Courtesy BCIRA).

Coral graphite

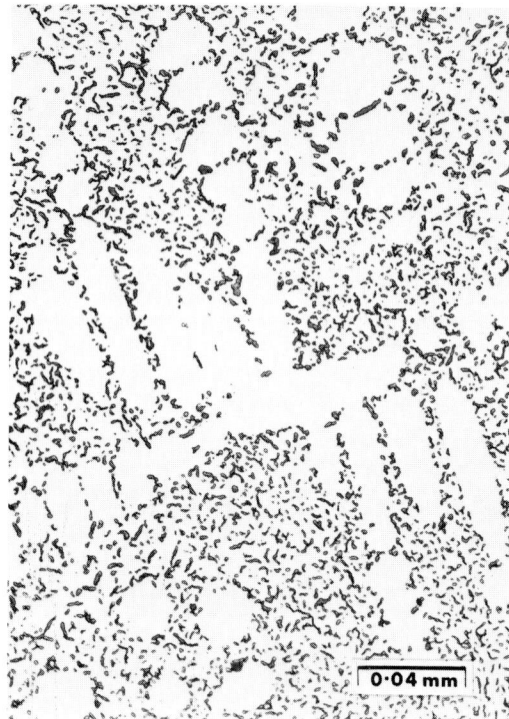

Figure 6.19 is a section taken from a rapidly solidified pure Fe–C–Si alloy. It shows proeutectic dendrites and interdendritic graphite eutectic in the unetched condition. (Courtesy BCIRA).

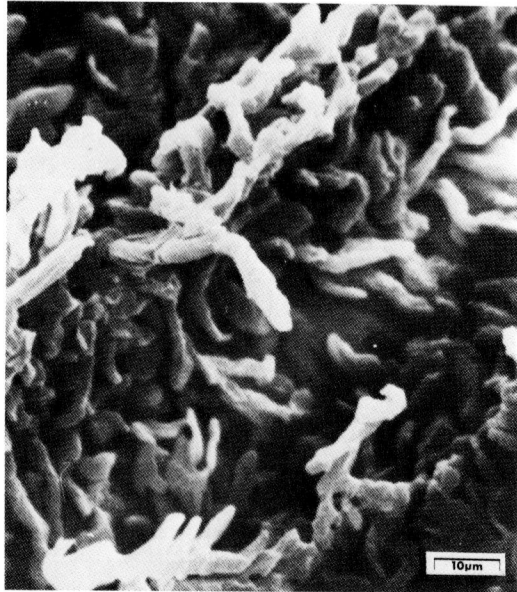

Figure 6.20 is a S.E.M. micrograph taken from the same alloy as that in *Figure 6.19* and shows that the eutectic is of a fibrous nature. (Courtesy BCIRA).

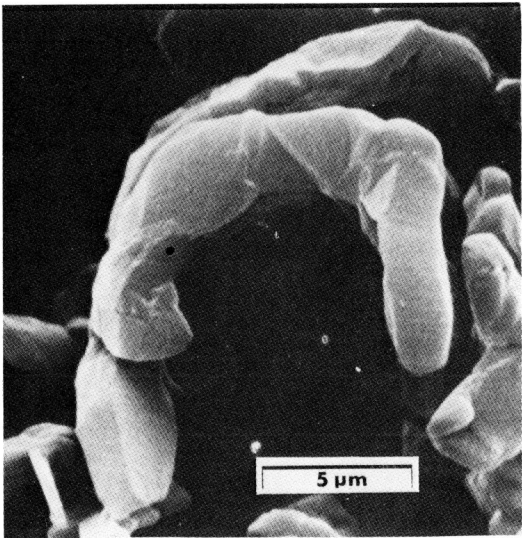

Figure 6.21 is a S.E.M. micrograph from a similar specimen and shows a step instability on the graphite surface which has travelled to the edge of the crystal and then continued over the edge to the opposite face creating a graphite rod or scroll. The driving force for this mode of growth is constitutional undercooling, resulting from the build up of Si in the liquid ahead of the graphite interface. (Courtesy Georgi Publishing Company).

Spheroidal graphite

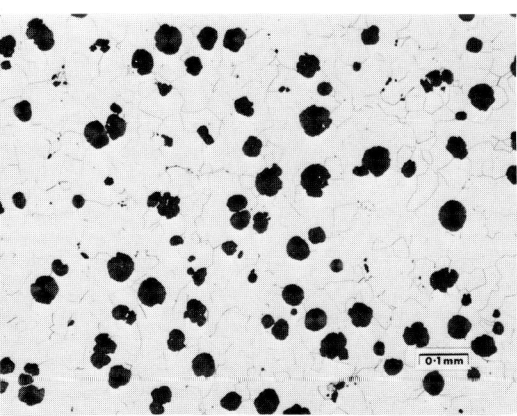

Figure 6.22 is taken from a ferritic spheroidal iron of composition

3.2% C, 2.3% Si, 0.3% Mn, 0.01% S, 0.07% P, 0.05% Mg

C.E.V. = 4.2

This irons shows a divorced eutectic of well-shaped spheroids of graphite in austenite which has transformed to ferrite during cooling. Etched in 2% nital. (Courtesy BCIRA).

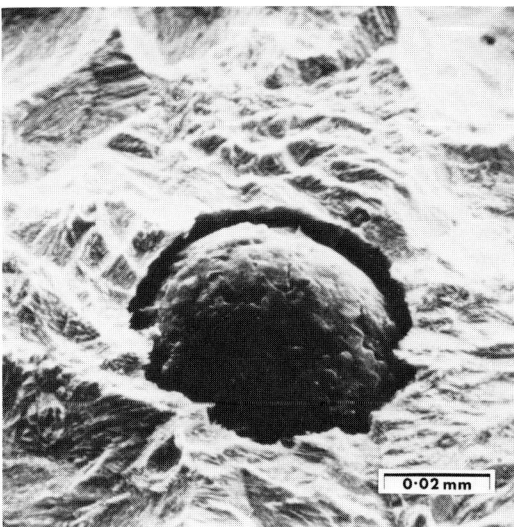

Figure 6.23 is a S.E.M. micrograph showing a spheroid on the fracture surface of a ferritic spheroidal iron. (Courtesy BCIRA).

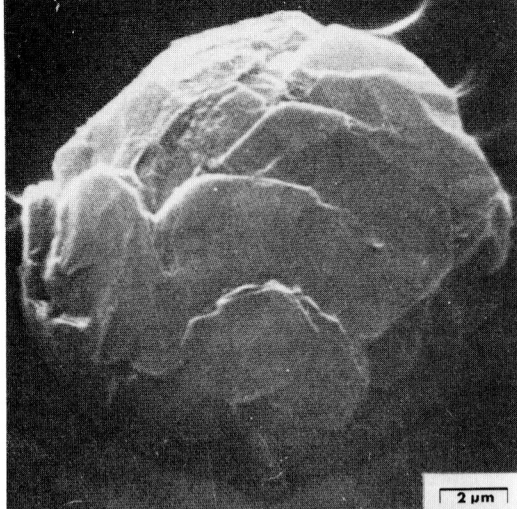

Figure 6.24 is a S.E.M. micrograph of a spheroid in a Ni–C alloy. It shows a cabbage-like form. This suggests that growth has occurred by extension of graphite in the 'a' direction. (Courtesy Georgi Publishing Company).

Figure 6.25 is a S.E.M. micrograph of a spheroid showing a star-like form. The growth conditions have been such as to promote radial growth by spiral growth in the 'c' direction. (Courtesy Georgi Publishing Company).

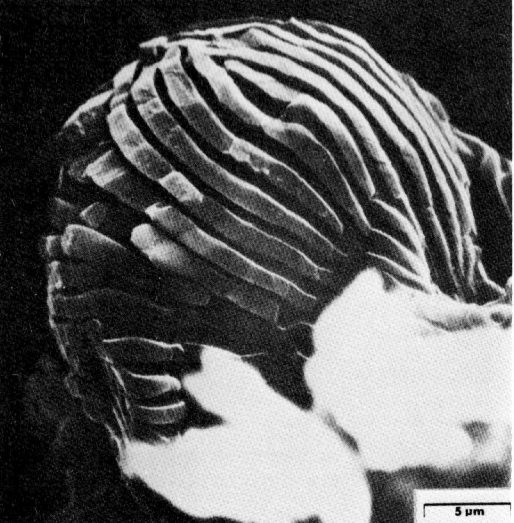

Figure 6.26 is a S.E.M. micrograph showing spheroidal growth occuring by the circumferential propagation of elongated steps on a graphite surface. (Courtesy Georgi Publishing Company).

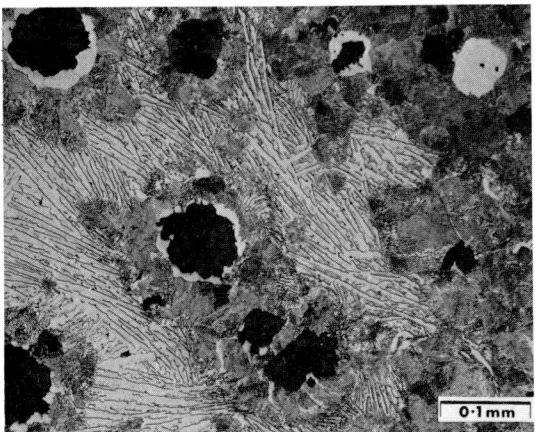

Intercellular carbides in spheroidal irons

The mechanical properties of spheroidal irons are reduced if the graphite is degenerate and if carbides are present in the intercellular boundaries. Undesirable graphite shapes resulting from incorrect compositional balance have been shown in Chapter 1 *Figure 1.12*. Intercellular carbide formation has been discussed in Chapters 1 and 3. There are several reasons for carbide formation.

Figure 6.27 shows long carbide plates typical of intercellular carbides. These are caused by excessive spheroidizing agent, as in this case, or form in areas of inverse chill. Etchant 2% nital.

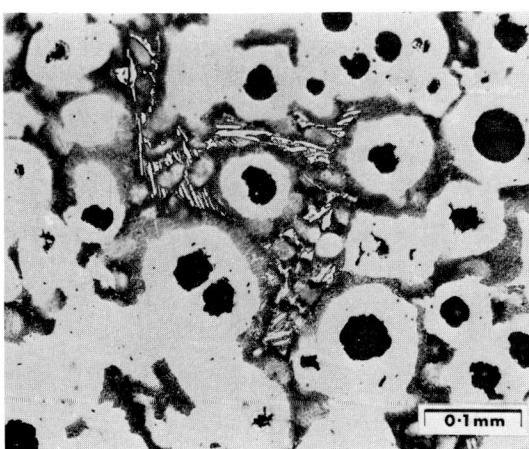

Figure 6.28 shows carbide plates in a heavy section casting caused by the segregation of alloying elements such as Mn, Cr and Mo to the intercellular areas during solidification. Etched in 2% nital.

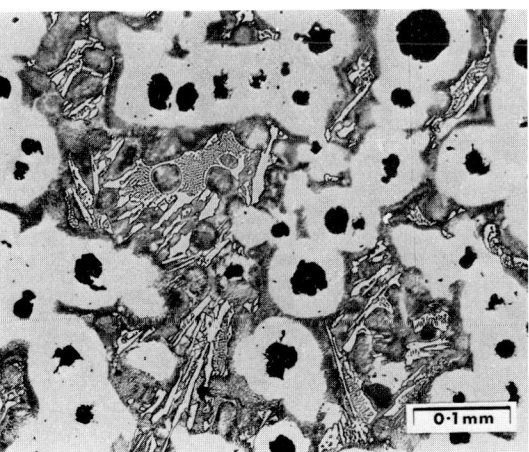

Figure 6.29 illustrates the most common cause of intercellular carbide formation, the incorrect metallurgical condition for the cooling rate. This section was taken from a thin section of gear cast in a shell mould. The carbide is continuous with islands of iron dispersed in it. Etchant 2% nital.

(a)

Malleable cast irons

Figures 6.30a, b and *c* show the variation in structure from the surface to the centre of a Whiteheart malleable iron of composition

3.0% C, 0.75% Si, 0.24% Mn, 0.22% S, 0.06% P

It was produced in the traditional way by heating from room temperature to 1060 °C, holding at 1060 °C for 80 hours and cooling slowly to room temperature.

Figure 6.30a shows the structure of the surface. A wide decarburized layer exists with oxide penetration and some scattered carbides. Etched in 2% nital.

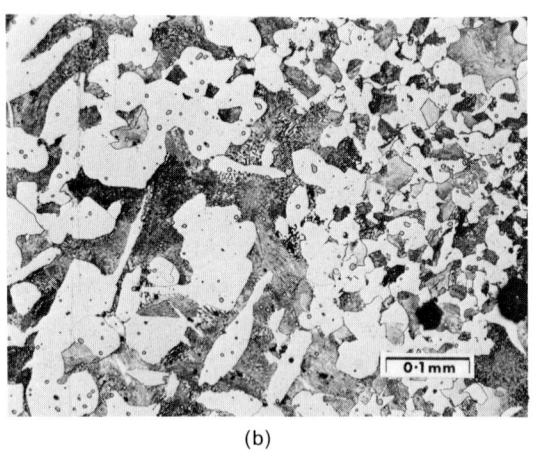

(b)

Figure 6.30b shows the carbon content increasing from surface to core as evidenced by the increased pearlite content. Etchant 2% nital.

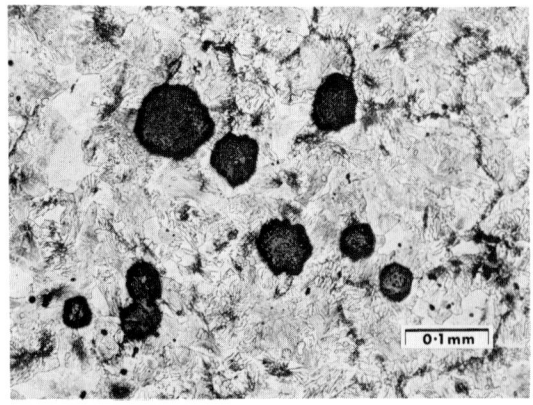

(c)

Figure 6.30c shows the core structure which is temper C spheroids in a matrix of pearlite with some ferrite. For a given section size and malleabilizing treatment, the temper C structure depends on the Mn:S ratio. If sufficient Mn is present to combine with all the S the temper clusters are of an aggregate type. However, S promotes spheroidal growth. As the Mn:S ratio decreases the clusters become more spheriodal. Etchant 2% nital.

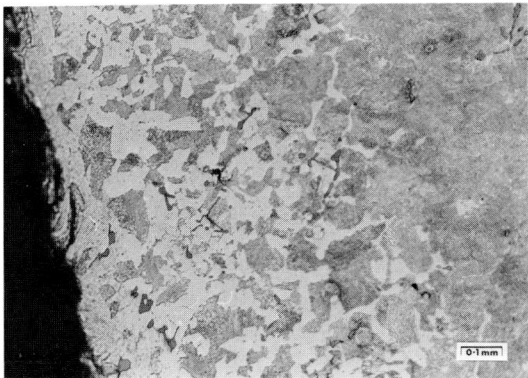

Figures *6.31, 6.32* and *6.33* show the variation of structure from surface to core of a Blackheart pearlitic malleable iron composition

2.6% C, 1.1% Si, 0.12% S, 0.03% P, 1.1% Mn

It was produced by heating from room temperature to 980 °C, holding at 980 °C for 8 hours, cooling to 760 °C in 3 hours, from 760 °C to 700 °C in 17 hours followed by relatively quick cooling to room temperature.

Figure 6.31 shows that some decarburization has occured at the surface, but it is much less than in the Whiteheart iron. Etched in 2% nital.

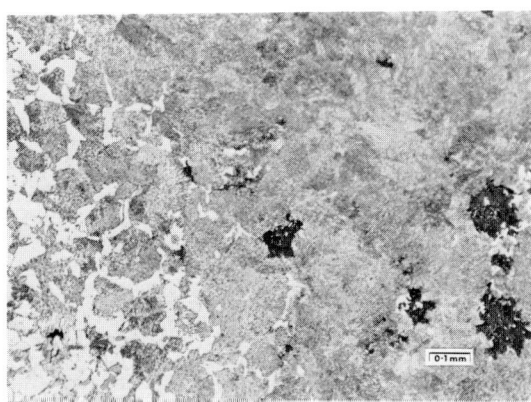

Figure 6.32 overlaps the area of *Figure 6.31*. It shows that the C content increases rapidly just below the surface. Etchant 2% nital.

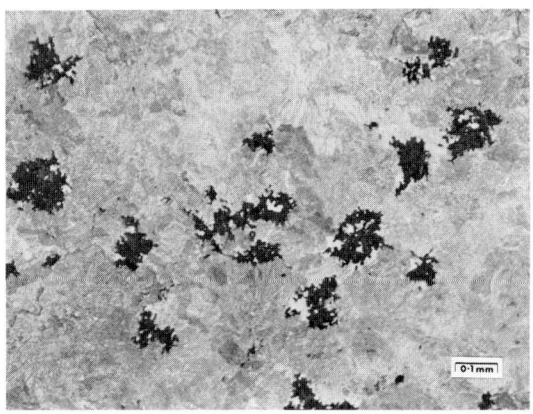

Figure 6.33 shows the structure of the core. This consists of very irregular C clusters (little free S) in a pearlitic matrix with some ferrite areas and MnS particles. Etched in 2% nital.

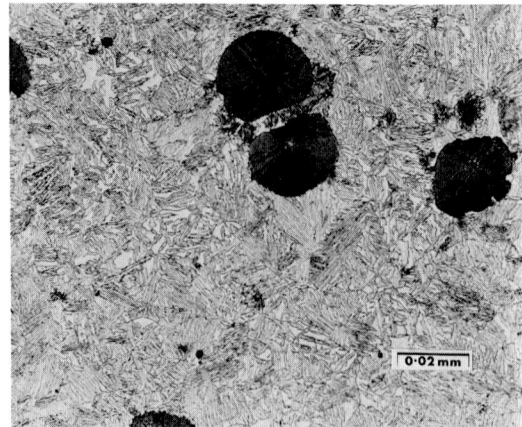

Austempered spheroidal irons

Figures 6.34 and *6.35* show the difference between upper and lower bainitic spheroidal irons.

Figure 6.34 shows an upper bainitic spheroidal iron austempered above 330 °C to give a matrix of carbide-free ferrite plates with retained austenite. This structure is characteristic of an austempered iron satisfying the proposed BCIRA grade ADI 950/6. See *Table 4.8* in Chapter 4. Etched in 4% nital. (Courtesy BCIRA).

Figure 6.35 shows a lower bainitic spheroidal iron austempered below 330 °C to give a matrix of ferrite needles containing carbide particles. This iron has a higher tensile strength and hardness, but lower ductility and toughness. It would satisfy the proposed BCIRA grade ADI 1200/1. Etched in 4% nital. (Courtesy BCIRA).

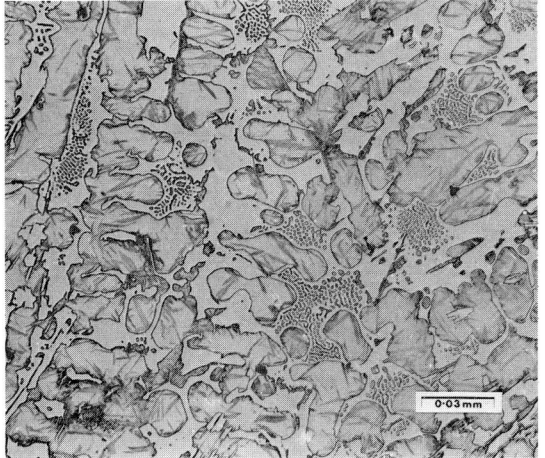

Alloy cast irons

This series of micrographs illustrates typical as-cast structures of alloy irons described in Chapter 1.

Figure 6.36 shows a section from a Ni-hard martensitic white iron of composition

2.95% C, 0.68% Si, 0.78% Mn, 0.06% S, 0.05% P, 4.21% Ni, 1.86% Cr

The structure consists of proeutectic austenite containing martensite needles with interdendritic, austenite-martensite/carbide eutectic of ledeburite form. Etched in 4% picral.

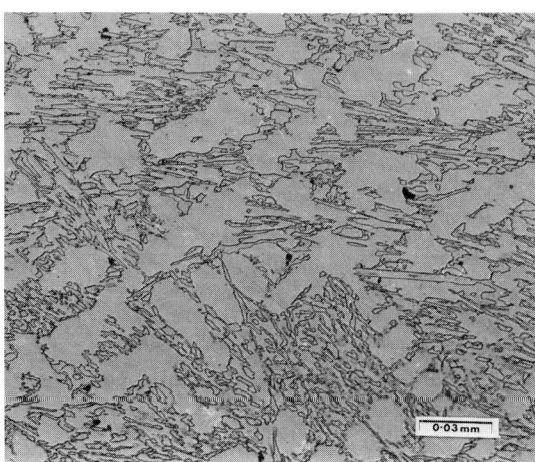

Figure 6.37 shows the structure of a high chromium iron of composition

2.75% C, 0.29% Si, 1.0% Mn, 0.04% S, 0.04% P, 27.5% Cr

The structure consists of massive carbides in slab or needle form in a ferritic matrix. Etched in alcoholic ferric chloride. Murakami's reagent will darken the carbides.

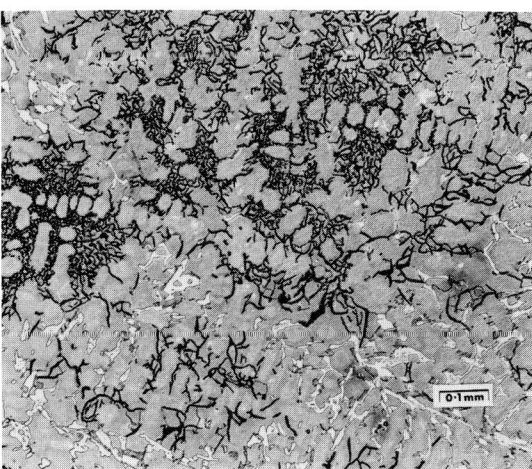

Figure 6.38 shows the structure of a non-magnetic austenitic iron of composition

3.31% C, 2.20% Si, 6.0% Mn, 0.03% S, 0.05% P, 11.60% Ni

This structure shows cored austenite dendrites with fine interdendritic flake graphite and interdendritic carbide. This alloy is used for castings in electrical equipment. Etched in 2% nital.

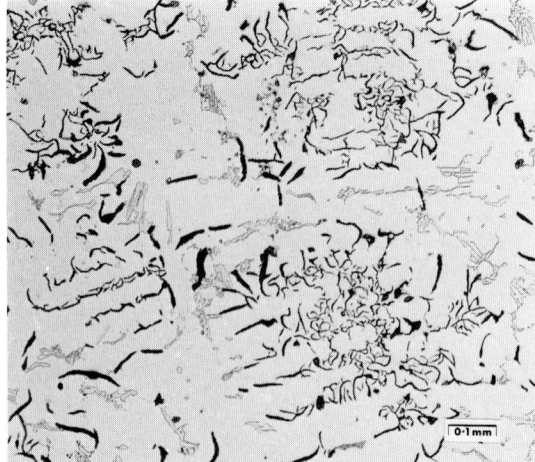

Figure 6.39 shows the structure of a Ni-resist austenitic cast iron of composition

2.54% C, 1.51% Si, 0.93% Mn, 0.5% Cu, 0.03% S, 0.03% P, 32.6% Ni, 3.24% Cr

This structure shows primary austenite in dendritic form with interdendritic carbide and flake graphite. Etchant alcoholic ferric chloride.

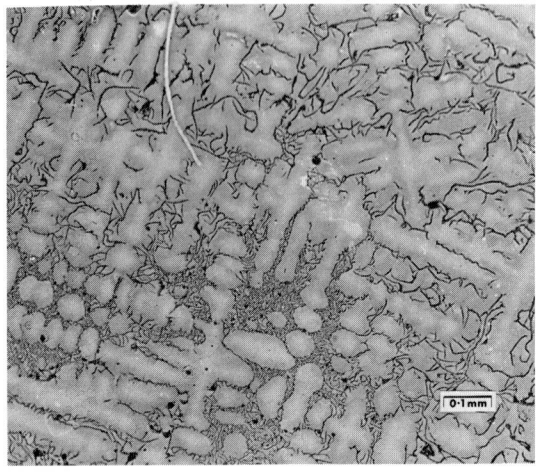

Figure 6.40 shows the structure of the high-Si heat resisting iron Silal of composition

2.15% C, 5.0% Si, 0.35% Mn, 0.03% S, 0.02% P

The structure consists of cored dendrites of ferrite with undercooled interdendritic flake graphite. A few white areas of high Si carbide are present. Etched in hot alkaline sodium picrate.

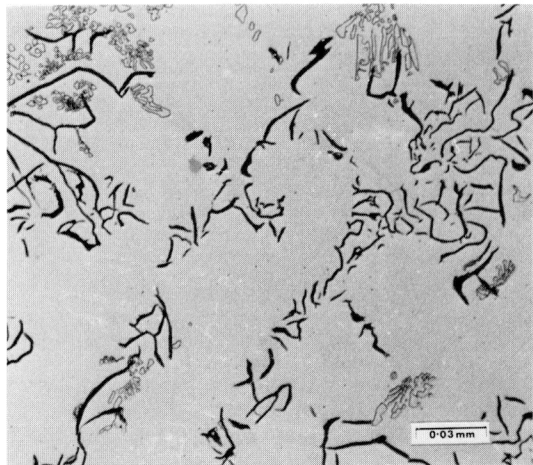

Figure 6.41 shows the structure of Nicrosilal austenitic cast iron of composition

1.92% C, 5.6% Si, 1.1% Mn, 0.03% S, 0.05% P, 1.32% Ni, 0.12% Cr, 1.02% Mo

This structure shows an austenitic matrix with fine medium graphite flakes and some eutectic carbide. Etched in alcoholic ferric chloride.

Low temperature brazing of cast irons

Figures 6.42 to *6.45* show microstructural features of a brazed joint between ferritic spheroidal irons using a Ag-based brazing alloy of composition

49% Ag, 16% Cu, 23% Zn, 4.5% Ni, 7.5 Mn

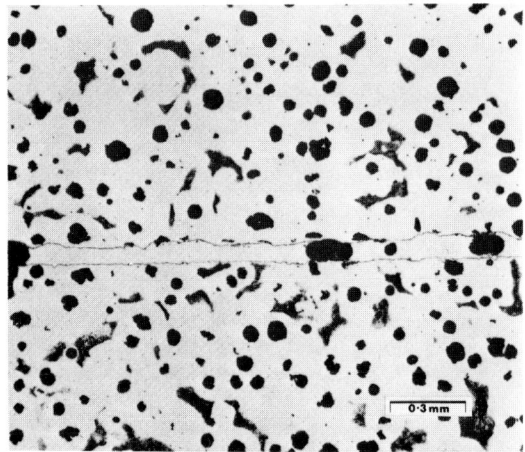

Figure 6.42 shows a brazed shear joint revealing small voids in the joint line. Etchant 4% picral. (Courtesy BCIRA).

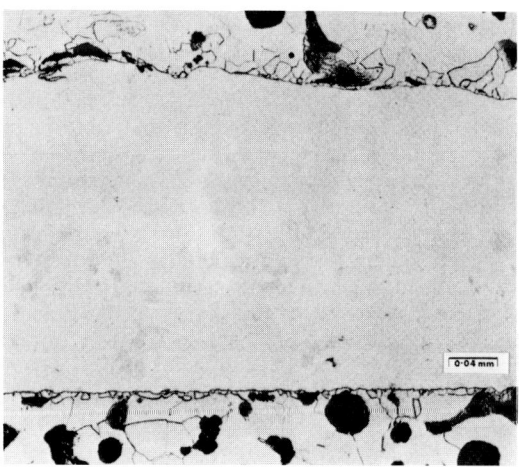

Figure 6.43 shows the interface between the brazing alloy and the ferritic spheroidal iron in a sound joint. Good wetting is evident. A surface layer of fine ferrite grains is evident in the iron. This is attributed to recrystallization of a joint surface prepared by machining. Etchant 4% picral. (Courtesy BCIRA).

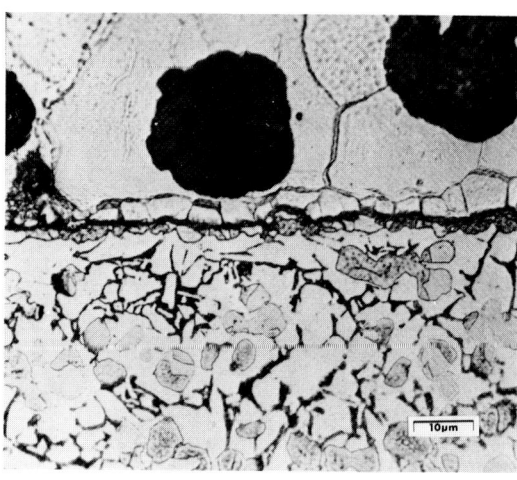

Figure 6.44 shows that the structure of the brazing alloy is modified close to the interface. This is attributed to solute diffusion in the interface area during brazing. Etched in alcoholic ferric chloride. (Courtesy BCIRA).

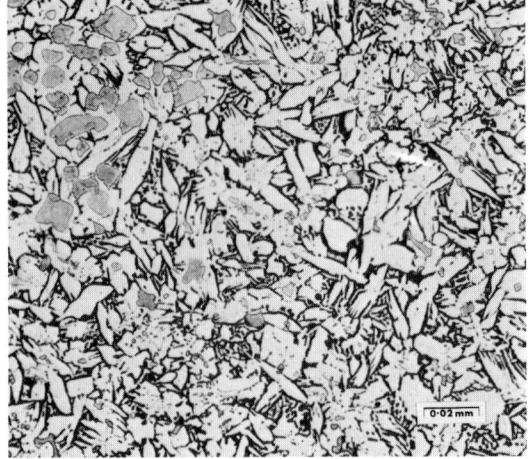

Figure 6.45 shows the structure of the brazing alloy. Etchant alcoholic ferric chloride. (Courtesy BCIRA).

Welding of cast irons

Figures 6.46a and *b* show typical structures observed close to the interface between weld pool and parent iron when ferritic spheroidal iron parts are tungsten inert gas-welded with a Ni–Fe electrode. Both micrographs show a weld deposit that consists of small graphite spheroids in an austenitic matrix. The parent metal fusion zone shows graphite spheroids and primary carbides. The heat affected zone (HAZ) shows martensite and other transformation products depending on the welding conditions and postweld heat treatment. The parent metal structure of graphite spheroids in a ferritic matrix is below the HAZ zone.

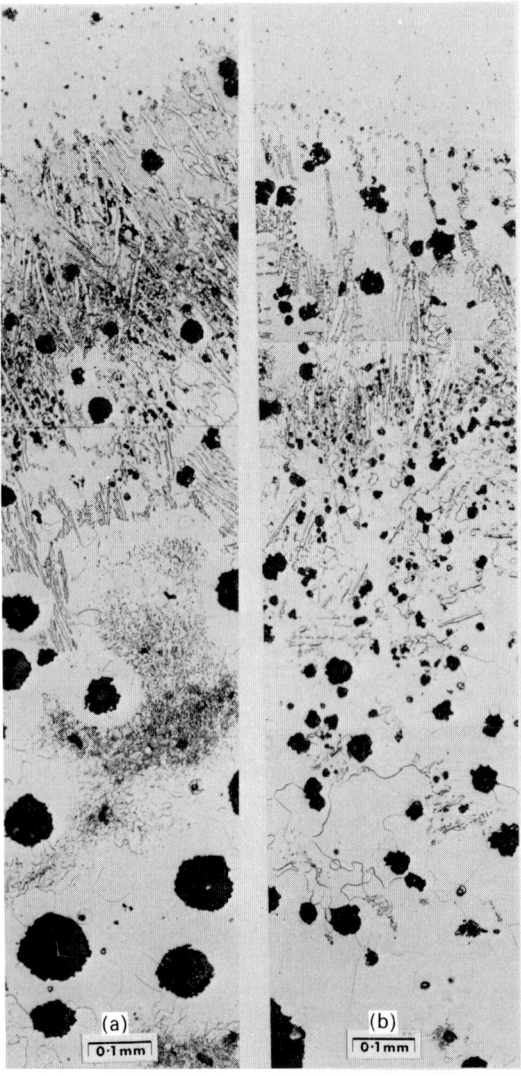

Figure 6.46a shows the influence of tempering in the HAZ in the root pass of a multipass weld. Secondary graphite formation is evident but the primary carbides in the fusion zone remain. Etched in 2% nital.

Figure 6.46b shows the features of the last pass weld in the same joint. Etched in 2% nital.

Index